THE SILVER SEEKERS:

They Tamed California's Last Frontier

BOOKS BY REMI NADEAU

City-Makers

The Water Seekers

Ghost Towns & Mining Camps of California

Los Angeles

California: The New Society

Fort Laramie and the Sioux

The Real Joaquin Murieta

Stalin, Churchill and Roosevelt Divide Europe

The Silver Seekers

THE
SILVER SEEKERS
They Tamed California's Last Frontier

by
REMI NADEAU

CREST PUBLISHERS

COPYRIGHT © 1999 By REMI NADEAU

Library of Congress Catalog Number 98-73976

PRINTED IN THE UNITED STATES OF AMERICA

First Edition

ISBN 0-9627104-7-4

Cover Design by Margaret Nadeau

All rights reserved. No part of this publication maybe reproduced or used in any form or by any means—graphic, electronic or mechanical, including photocopying, recording, taping or information storage and retrieval—without written permission of the publisher, except in the case of brief quotations in reviews.

Crest Publishers
P.O. Box 22614
Santa Barbara
CA 93121-2614

To Remi Robert Nadeau,
who helped make this
book possible

TO THE READER:

A few explanations will be useful:

1. Chronology. Though the chapters each tell a complete story, they are arranged in the chronological order of their beginnings. Thus they generally parallel each other through to the 1880s. In chapters two, three and four, which all begin in the early 1860s, the order is purely arbitrary. The Bodie chapter is placed near the end because, although its origins are in the early 1860s, its real boom begins in 1877.

2. Prices. Dollar amounts given in this book are the actual figures at the time and are not corrected for subsequent inflation. To understand the true values or purchasing power in today's terms, the reader should multiply the dollars by roughly 20 times.

3. Ancestor. The Remi Nadeau who ran a mule team freighting business in the book's time period was the author's great, great grandfather.

4. Non-fiction. This is a history book, not fictionalized history. It is drawn almost entirely from primary sources, including legal documents, newspaper reports, diaries, letters, sworn testimony and other court records, and reminiscences of eye-witnesses. The conversations are as given in such records, and are not fictionalized.

5. Sources. To avoid the distraction of end-notes in a book for general readership, a new format for identifying sources is adopted. This is a multi-bibliography in which the sources are given for each subject, major character or episode.

CONTENTS

PREFACE   1

    In California... the Sierra Nevada forms the border line between the two metals, gold prevailing on the Western slope and silver on the east and beyond.

    — Hubert Howe Bancroft, Chronicles of the Builders

1. SILVER IS KING!   3

    Just as California's Gold Rush faded, the fabulous Comstock strike drew thousands of silver seekers back over the Sierra to the Great Basin—and to Eastern California.

2. BELSHAW'S MILLIONS   9

    Silver, like gold, is where you find it. As early as 1860, pioneer prospectors found it in the Coso Range, south of Owens Lake. The Coso excitement in turn led to one of California's largest silver discoveries and its wildest silver town—Cerro Gordo—high in the Inyo Range overlooking the lake. Victor Beaudry and M.W. Belshaw got there early, bought the key mines, built the furnaces, and then locked with challengers in the state's biggest silver battle. When the smoke cleared they were still the bullion kings of Inyo County, but the underground prize was playing out.

3. THE RISE AND FALL OF LEWIS CHALMERS   43

    South of Lake Tahoe, silver was struck on the East Carson River just before the Civil War. Thousands rushed in from California and Nevada Territory, creating a cluster of mining towns and a new Alpine County in the crown of the Sierra. In 1867, as the Alpine treasurehouse was declining, Lewis Chalmers arrived with British capital to stir new life. Near Silver Mountain he created the "Chalmers Mansion", married his young housekeeper, and with her reigned as the first couple of Alpine society. Until the mid-Eighties he divided his time between London, where he raised more money from investors, and Silver Mountain, where he spent it on new ventures. Then, just as suddenly as he appeared, he vanished, leaving his wife and Alpine County in bittersweet decline.

4. THE TOWN WITH NO LAW   65

While other silver camps flared and died, Benton lived high for fifteen years. But not without sharp ups and downs. Pap Kelty preempted the whole town, and almost got away with it. And the two biggest mining companies—the Comanche and the Diana— squared off in a dispute that began with an underground explosion and ended with a grand bilk of Comanche creditors.

5. MINING THE STOCKHOLDERS   87

It takes a gold mine to work a silver mine.
— Old Spanish proverb

6. THE JINGLE OF NADEAU'S TEAMS   95

The empire created by the Silver Seekers was strung together—made into a fluid, mobile community 400 miles long—by the men with wheels and teams. Different stage men ran their passengers into Eastern Sierra from north, west and south. But marketing the silver product and returning with life's necessities was, from the late 1860s to the early 1880s, a task dominated by one freight outfit with more than a thousand mules run by a Canuck named Nadeau.

7. THE FIGHTING McFARLANES   119

South of Death Valley and north of nowhere, mines had to produce rich ore to overcome the distance to market. At Ivanpah and later at Providence, the ore was indeed rich. The four McFarlane brothers were in the vanguard of those arriving to take it—braving waterless wastelands, sandstorms, Indians and those interminable distances. As isolation spawned the wildest and wickedest camps in California, survivors also had to get past the worst that other men could do to them—as John McFarlane discovered too late.

8. THE SILVER SENATORS   133

No camp had more illustrious fathers and a more soaring reputation than Panamint, and deserved them less. When Senators John P. Jones and Bill Stewart of Nevada bought into the silver discovery just west of Death Valley, the rush was on. When they invested a fortune in mines and mill, the boys were sure they were onto a good thing, and for a time Panamint rode high, wide and wicked. As mines and mill closed down, some claimed it had been a stock promotion. But the answer lay in something simpler—bad judgment.

9. PAT REDDY FOR THE DEFENSE   155

The silver deposits at Darwin had been sitting under everybody's noses as they rushed to Cerro Gordo and Panamint. But Darwin had rich ore that was easily worked, and a handy location near Owens Lake and the Bullion Trail. Suddenly it was a magnet for all the characters, good and bad, who had followed silver up and down Eastern California. Leading them was Pat Reddy, miner turned lawyer, who grabbed and exploited the biggest mine. Pat fought for law and order, but he could not prevent the violent climax to one of California's first mine labor disputes.

10. ENTER GEORGE HEARST   177

When George Hearst bought into the Modoc Mine in the Argus Range, he helped create Lookout, the unique "town on top of a mountain." Down below, the Minnietta Belle Mine added to the excitement. Across Panamint Valley, it took ten big kilns to supply the charcoal for the furnaces. With Hearst's furnace and Lookout's saloons roaring round the clock, times were wild and good, but late at night you had to be careful not to fall over the side.

11. THE CONSCIENCE OF GENERAL DODGE   185

Mammoth it is, and mammoth it was—at least in the minds of General George S. Dodge and his friends who bought the mine and fostered the timberline town near the crest of the High Sierra. For a time Mammoth City and Mammoth common stock were on everybody's lips from Bodie to San Francisco. Within four years of its discovery, Mammoth was dead and so was General Dodge.

12. BAD MEN OF BODIE   205

Though silver was king in Eastern California, the biggest mining district was more gold than silver. Bodie was a paradox in other ways, too. Named for Bill Bodey, it spelled his name wrong. It was discovered in 1860, but its first big rush came 17 years later. Pat Reddy, one of its leading characters, repeatedly championed the rule of law, but when the Miners Union ordered George Daly out of town, Pat advised him to leave. In fact, while Bodie had a squad of lawmnen, its name became synonymous with lawlessness.

13. THE LAST BONANZA 243

With Calico located reasonably close to San Bernardino, and with its ores rich and easily worked, it was bound to be a profitable camp despite the dropping price of silver in the early 1880s. What was also good was the extent of its riches, making it the most productive silver district in California, and the arrival of the railroad at its front door. What was bad was that the same riches fostered more greed than usual in the Western camps, and for a time some of the boys stood each other off with Winchesters.

14. SILVER WAS KING 269

The tumult and the shouting dies; The Captains and the Kings depart.

— Kipling

APPENDIX 273

1. **The Story of Silver, 273**
2. **Miner's Jargon, 276**
3. **Weapons, 279**
4. **How to get there, 281**

THOSE WHO HELPED 283

SOURCES 287

INDEX 318

PREFACE

> *In California... the Sierra Nevada forms the border line between the two metals, gold prevailing on the Western slope and silver on the east and beyond.*
>
> — Hubert Howe Bancroft, *Chronicles of the Builders*

A decade after the California Gold Rush of '49, thousands stampeded by foot and horseback over the Sierra to the call of Nevada silver. First to the Comstock Lode, then to every new strike in the Great Basin, the Silver Rush raged from the Civil War to the mid-1880s. In Eastern California, from Tahoe to the Mojave, men and sometimes women searched, fought and even died for the treasure. Before their mines faded, they had tamed California's last frontier.

Though it paralleled the Nevada excitements, California silver formed a well-defined empire, bound in by the Sierra on the west and the forbidding Great Basin on the east. In this geographic wedge full of natural wonders, from tallest mountains to deepest valleys, developed a special society full of equally wondrous people. Arriving with overflowing energy and hope, they were fired still higher by the unimagined beauty around them. Under any rock might be another Comstock! No scheme was too extravagant, no risk too great. And failure? Laugh and start over.

So this was a rollicking society fueled by endless optimism, childlike credulity, the gambling mentality, boundless good humor. It was a fluid society—many moved up and down this frontier with each new strike. They sold each other worthless mines, plunged into mining stocks on outlandish rumors, paid their stock assessments in the hopes of future rewards. Many of them jumped each other's claims and held them by right of shotgun.

Leading the pack were the strongest of will and the quickest of wit—men who found opportunity and seized it. Few were miners, and if they were they turned to other pursuits—buying mines that others discovered, controlling mills and furnaces, transporting bullion from, and supplies to, the mines. These reaped the millions, fought other titans for control, and remain the players on the stage of history.

The outward trappings of civilization came early to this last frontier. New counties were created, mining districts formed, courts established, judges and sheriffs elected. Roads were built, mostly north and south in the shadow of the Sierra. Stage drivers and mule team freighters rolled in to claim their share of the silver without digging for it.

Throughout this wilderness of magnificent distances, the law was over the horizon. Men took safety in their own brawn and their own guns—dangerous in a land where saloonkeepers and wild women were among the first to arrive. Even leading citizens—merchants, freightmasters, mine superintendents—got involved in shootings. And juries responded to the universal plea of self-defense.

But other forces were afoot to tame this frontier. Hand presses hauled in by mule teams gave voice to irate editors calling for order. From north, west and east the Iron Horse burst in with 19th Century technology. Good women, arriving by stagecoach, were devastating to the old disorder. First children, then schools and churches, doomed the reckless hour.

And suddenly, with the mines faltering and the price of silver plummeting, the scene itself faded. The characters made their exits. Even the stage sets vanished. Every silver camp that had howled with revelry soon became—and still is—a ghost town. Only the farm and ranch communities that had sprung up to serve the mines now remain. It is civilization, not frontier. Also remaining is the splendid stage itself—the shadowy canyons of the desert, the sparkling crags of the Sierra—that once echoed to the tumult and the shouting.

1. SILVER IS KING!

Just as California's Gold Rush faded, the fabulous Comstock strike drew thousands of silver seekers back over the Sierra to the Great Basin—and to Eastern California.

Into the mountain-bound mining camp of Grass Valley, California, rode a weary traveler late in June 1859. He had jogged more than 150 miles over the massive Sierra Nevada from the Washoe country, in western Utah Territory. With him, mostly as a curiosity, he carried some odd-looking chunks of gold-bearing ore.

Next day Melville Atwood, the local assayer, tested the rock. What he discovered made him doubt his calculations. For besides the gold content, which ran about $1,000 to the ton, the specimens contained a much higher value in silver—over $3,000 per ton!

What was more, the stranger confided, over in Washoe the discoverers were extracting the gold and throwing the rest away! Since California's big strike more than a decade earlier, prospectors had not even thought of looking for anything but gold. An earlier test of Washoe ore by assayer James J. Ott in Nevada City had failed to reveal the silver.

Within hours of Atwood's assay, the neighboring towns of Grass Valley and Nevada City were boiling with excitement. First to learn the news was Judge James Walsh, an old hand in California mining and a friend of the ore-bearing stranger from Washoe. Near midnight he banged frantically on the door of another friend. Quickly they piled provisions on a mule, mounted their horses, and spurred out of Grass Valley. Ever the opportunist, Walsh hoped to buy one of the mines before the owner learned about the silver.

Not far behind them clambered a desperate party in pursuit, some traveling on borrowed money, others on borrowed horses. Within two days a clattering column was surging through the pine-forested Sierra, some on horseback, some afoot, all bent forward like hounds on the scent. Riding in the van was the tall, muscular figure of George Hearst, then a rising young mining man of Nevada City. With him was Atwood the assayer, who had confided the news and joined the rush.

When this vanguard arrived in the barren hills of Washoe, the original miners still knew nothing of their ore's silver content. The two discoverers, Peter O'Riley and Pat McLaughlin, were washing out the gold with their rocker, letting the rest of the ore roll down the side of Sun Mountain.

One of them sold his share for $3,500 to Hearst, who was so anxious to buy that he rode back over the mountains to Nevada City to raise the money.

Judge Walsh paid $11,000 for the interests of one Henry Comstock, a local prospector who had fast-talked the two discoverers into giving him a share. To seal his bargain with Walsh, Comstock took ten dollars as a down payment for what would later be worth millions. Then he bragged to his fellow miners that he had fooled "the California rock shark!"

He thought enough of the discovery, however, to call it Comstock's Lode wherever he went. And so talkative was he that his name became attached to the greatest single deposit of precious metal ever discovered in the United States.

By the summer of 1859 Walsh and Hearst were shipping ore over the Sierra—ore so rich that it could be carried 160 miles by muleback and another 80 by steamboat, and could still be smelted in San Francisco at a good profit. By October the growing shipments were attracting attention as they passed through the Mother Lode. Early arrivals in Washoe were writing back that the mines were the richest in the country. California newspapers were quoting assay figures of thousands of dollars per ton. Before long, bars of Washoe silver were hauled through the streets of San Francisco and displayed in bank windows before the eyes of gathering crowds. With silver valued at $1.35 an ounce, the attraction was irresistible. All at once California rang with a new cry:

"Silver in Washoe!"

When this cry burst upon the boys in the fall of 1859 they were especially vulnerable. It had been a long summer, and in Sierra canyons the placer and hydraulic mines were idle for lack of water. To this restless crew, disillusioned with gold as the placers failed, the silver call was a trumpet blast. It conjured the wealth of ancient kings in the Levant, the riches shipped for centuries from Peru and Mexico to the Spanish crown. All at once mules, horses, flour, picks and shovels were in fevered demand.

"From the crack of day to the shades of night," exclaimed one San Franciscan, "nothing is heard but Washoe."

It made no difference that the new strike was located on the very desert through which most of them had suffered on their way west to California. They only knew it was "Forty-nine all over again!"

With "Washoe!" thundering like a battle cry, the rabble army converged on Sierra passes. From San Francisco they swarmed onto the decks of river steamboats, sprawling wherever they could find room between bales and boxes, jabbering about Washoe in a dozen tongues. At Sacramento those who could pay the fare took the puffing Iron Horse—first on the West Coast—a few miles further, then staged onward in six-horse wagons.

At Placerville, snugly tucked in the Sierra foothills, stagecoaches and mule trains were booked days in advance. Streets and hotels, saloons and restaurants, were thronged with a noisy crowd of expectant millionaires. As the line of glory-hunters moved through Placerville's streets each morning, clattering along with shovels, picks and washpans, there rose from the throats of bystanders the inevitable shout: "Go it, Washoe!"

Up into the pines the adventurers thronged, making an unbroken line of men, mules and wagons from Sacramento Valley up the South Fork of the American River and over the mountains at Echo Summit to Carson Valley. Those who hiked or rode muleback slogged in the ruts of freight wagons, jumping out of the trail to avoid being knocked down when a pack train brushed relentlessly past.

Among the worst hazards on the trail were the wayside taverns, where the travelers piled in on one another in frantic quest for board and bed. At Strawberry Flat, hundreds of travelers congregated each night, flooding the barroom and jostling each other for a place near the dining-room door. Next morning, after a night's sleep in a room with 250 companions and a bracing wash at the horse trough, the silver-hunters were on their way. At Genoa, first settlement reached on the east side of the Sierra, accommodations were even more formidable. Lodgers were packed like stowaways—two and three in a bunk, the unfortunate ones curling up on saloon floors, behind store counters, between packing boxes, and even on the tops of nail kegs.

Early in November a storm struck the Sierra, covering it with the snow and ice that had already brought tragedy to many California-bound emigrants in previous years. But this deterred only the faint-hearted and the sane. One storm after another raked the Sierra in one of the fiercest winters on record. Numberless animals and a few men met death in blizzards and avalanches, but still the most daring pressed on, driven by visions of Washoe silver.

As the spring thaw approached, all of California seemed to rally at the foot of the Sierra. The winter's isolation had left the Comstock so short of supplies that prices were soaring. Rival freight-packers raced to be first across with whiskey and other "necessities". As early as February they were laying blankets in the snow for their animals to walk on, taking them up behind the train as it advanced and spreading them on the path ahead.

By early March even the stages were running again, but passengers had to walk much of the way, holding the coach to keep it from rolling down the mountainside. Above Strawberry Flat they trudged on by foot, braving fierce winds and shoveling a path before them across Echo Summit.

Worse hazards stalked the other Sierra routes opened to accommodate the tide. From California's Northern Mines the adventurers stormed up the tempestuous Yuba River, joined a mule train at Downieville, and bent onward along narrow trails that hung hundreds of feet above the foaming river. From the Southern Mines they ventured through giant redwood groves, over Ebbetts Pass to the East Fork of the Carson River.

Near the summit of this remote passage the stampeders encountered more than rough trail. Two of them stayed up all night waving firebrands to protect a load of bacon from three grizzly bears that, as one man recalled, "were grumbling and gnashing their teeth."

By April 1860 some 150 Californians were arriving in the Washoe country every day. Estimates of its population reached 10,000 that spring, with thousands more on the way. Those remaining in California were investing every spare cent in Washoe mining stocks.

From other directions they came—a few Mormons from Salt Lake, many more who were "busted" in the Pike's Peak gold rush. But most were Californians—not only its drifters and schemers, but the very flower of its people.

First to arrive were the mining men, the Walshes and the Hearsts, who knew ore and hoped to buy promising leads with their own or someone else's money. Marching after them was the whole lusty crew that made up frontier society—promoters and speculators, traders and gamblers.

Among them were followers of the oldest profession, for not even Sierra snowstorms could bar Washoe to the fair but frail. Others had even less visible means of support—the thieves and cutthroats who customarily joined new stampedes.

But the backbone of the throng was the common miner—the same who for ten years had been pouring into California from the East, and had been rushing to every new strike on the coast. Honest, hard-working, hail-fellow, he was the sinew of frontier society. Shovel on shoulder, he walked or rode over Sierra passes, eager to possess a new country. At least he would gain a better wage as a miner there than in the declining gold fields. At most he would strike his own vein of silver and return to the States a millionaire.

As the main body of stampeders arrived in the spring of 1860 they found Virginia City at once the most miserable and the most exciting place on earth. Night and day the saloons and gambling houses filled the air with a constant din of oaths and laughter, rattling dice and clinking coins. Through the spring of 1860, Washoe was teeming with would-be tycoons, all talking of "lodes", "dips", "angles", "indications", and trying to sell one another shares in a claim.

By April the Comstock mania reached its height. One visitor estimated that only one inhabitant in fifty was actually mining the earth. A few more were out prospecting for new leads, while the largest proportion was engaged in buying and selling shares. When two friends met on the street the customary salutation, rather than a handshake, was to thrust ore samples at each other. Instead of asking after one's health, they would inquire about assays, claims and outcroppings.

But the Comstock Lode was basically sound to the extent of a third of a billion dollars. Its riches helped to finance the Union side in the Civil War. Through the Sixties and Seventies its wealth was the first fact of economics on the Pacific Coast.

And was the Comstock the only silver bonanza in the Great Basin? Many of the thousands, arriving too late to share in the big strike, rushed over the desert like ants on a hill to be the first at a new lode. Sure enough, across the new territory and state of Nevada, they struck silver at Austin, Eureka, White Pine, Pioche—each with its own stampede and its frenzied hour.

Nor were the silver excitements confined to Nevada. Another silver empire lay next door in the shadow of California's High Sierra.

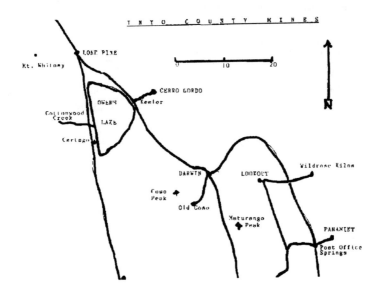

2. BELSHAW'S MILLIONS

Silver, like gold, is where you find it. As early as 1860, pioneer prospectors found it in the Coso Range, south of Owens Lake. The Coso excitement in turn led to one of California's largest silver discoveries and its wildest silver town—Cerro Gordo—high in the Inyo Range overlooking the lake. Victor Beaudry and M.W. Belshaw got there early, bought the key mines, built the furnaces, and then locked with challengers in the state's biggest silver battle. When the smoke cleared they were still the bullion kings of Inyo County, but the underground prize was playing out.

Early in 1860, when Californians heard the cry of Washoe silver, it rang a familiar bell—the lost Gunsight mine! Most of the boys had heard of the Forty-niners who said they had found, near Death Valley, a mountain of rock that glistened with silver. When they had reached the coastal settlements one of them had asked a blacksmith to make a sight for his rifle with a piece of the pure silver. Some of the party, followed by others, had returned to the region but they never found the "Lost Gunsight".

Now, fired anew by the Gunsight legend, nine mining men left Oroville, on the Feather River in California's Northern Mines, in mid-March, 1860. Leading them was Dr. Erasmus Darwin French—physician, soldier and miner. Born in New York state in 1822, he had come to California with Gen. Stephen W. Kearny's 1846 expedition in the Mexican War and had fought in the battle of San Pasqual.

When the war ended in 1848, French had adopted California and joined the rush to the gold fields. In September 1850 he had ventured into the uncharted Mojave Desert with three companions and an Indian guide. Then, through what would later be called Darwin Wash, they explored Panamint Valley but failed to find the Gunsight. French settled into the pastoral life, first in San Jose, then in Chico.

But miner's fever was in his blood. Now in 1860, he resolved to revisit the desert, and since he had been there once, the rest of the party looked to him as the leader.

"The guide of the silver seekers," declared Oroville's *Butte Democrat*, "thinks he knows almost the very spot where it can be found."

Instead of riding over the Sierra toward Washoe, where everyone else was heading, they boarded a steamboat down the Feather, the Sacramento and the San Joaquin to Stockton. From there, well mounted and outfitted, they rode south to Visalia, then the principal settlement in the lower San Joaquin Valley. They found it a "beautiful and thriving town" composed mostly of one-story brick buildings. There they recruited three more men, bought more provisions, and rode eastward into one of the least explored corners of California.

Skirting the lower end of the Sierra through Walker Pass, French's party reached the gold mining camp of Keysville in the second week of April. Here, earthen breastworks gave evidence of an Indian scare four years before, and the locals filled the prospectors with warnings. But they moved on, "prepared to give them a round turn, if necessary," as one of the stalwarts wrote home.

"We have twelve men in our company, and can discharge one hundred and seventeen shots in a minute."

North of the Mojave in the Coso Range east of Little Lake (then called Little Owens Lake), they found neither Indians nor the Gunsight, but silver ore-bearing ledges that to them justified the name, Silver Mountain. In early May they sent a messenger to San Francisco with ore samples to be assayed. At month's end he returned with promising results. By this time the boys had staked some ninety claims and, meeting at Haiwee Meadows on May 28, 1860, formed the Coso Mining District. And they named some natural sites—Darwin Falls and Darwin Canyon—after their leader's middle name.

By late June, Visalia was seething with news of the Coso strike. A visitor from Red Bluff wrote home with such a glowing account that more men, postponing the summer harvest, left for Coso. In the silver camp the number of claims had jumped to 150, and a party of men rode westward with the usual selected ore samples. In Visalia they incorporated the Coso Mining Company, sent more specimens to assayers in San Francisco, and started selling stock certificates. When assay results came back, they showed silver at exciting values of from $287 to $1,117 per ton.

At this news the lower San Joaquin came alive. Prospecting parties filled the streets of Visalia, mercantile stores were doing a brisk trade in provisions, and new buildings were going up to the joyful tune of saw and hammer.

By late July, when a returning miner was interviewed by the San Francisco *Alta California*, the story was getting bigger. The party had panned some earth and found it half gold. Five hundred men were on the spot digging for silver and gold: "there is room for thousands." And inevitably, the whole silver region was "not excelled by Washoe itself."

Spurred by such tales, men were forming mining parties and riding over the Sierra nearly every day by the summer of 1860. The frenzy danced still higher when some of them brought back more specimens from the Coso Range assaying over $2,000 per ton.

People unable to travel to Coso were eager to invest. They were quickly accommodated by new stock corporations—the Great Western, Pioneer, Rough and Ready—some of which sprang from actual shafts and tunnels in the Coso Range. A small quartz mill was built at what is now called Old Coso. Through 1861 and '62, prospectors were combing the Argus, Slate and Panamint ranges eastward to Death Valley, forming new mining districts and launching new stock companies.

Northward in Owens Valley, gold-bearing ledges also beckoned. But pioneer miners were fighting the native Paiutes and as one early arrival said, "We cannot prospect and watch Indians at the same time."

When the U.S. Army arrived and founded Camp Independence in July 1862, Owens Valley was safer, at least for a time. A soldier from the camp discovered gold ore eastward in the Inyo Range, giving rise to an embryo camp, San Carlos. Other gold discoveries created new towns—Bend City, Chrysopolis, Owensville—on the east side of Owens Valley. Southward, between them and the Coso diggings, lay the biggest bonanza of all, waiting to be discovered.

By the mid-Sixties the richest of the Coso findings had played out, and in 1866 hostile Indians burned the quartz mill. As the Yankee discoverers trudged out, their place was taken by Mexican miners, experienced in working the silver ores of Sonora. At the same time they struck gold ore, and Coso became more a gold camp than a silver camp

With either metal they crushed the rock with their *arrastras*, in which burros dragged grinding stones around a circular enclosure. Then they roasted the gravel in *vasos*, which were small stone-and-adobe ovens that caught the molten silver and other metal which, being heavier than the rock, sank to the bottom. The metal was cooled in molds, and the product was sold and shipped as bars of bullion.

In 1864 three of these Sonoran miners headed by Pablo Flores were prospecting northward in the Inyo Range, east of Owens Lake. Near the top of Buena Vista Peak, at more than 8,400 feet elevation, they struck a silver-bearing ledge. The place would soon be called *Cerro Gordo*, literally, "Fat Hill", a natural designation by miners from the lower Coso Range.

But before they could reap their reward, the whites and the Paiutes renewed fighting in Owens Valley. In the wide-ranging hostilities, the Indians killed two of the Mexicans and captured the third, Pablo Flores, whom they took prisoner.

In January 1865 the whites massacred forty-one Paiutes in their camp at the mouth of Owens River as it emptied into Owens Lake. The panic among the Indians gave Flores an opportunity to escape. Returning to Cerro Gordo, he and some new Mexican companions struck more silver leads, built their arrastras and vasos, and began shipping bullion.

In the next two years most of the claims that would later become famous—the San Felipe, the Ygnacio, the San Lucas—were staked by Mexicans. José Ochoa, who owned the San Lucas Mine over the divide from Cerro Gordo in the canyon leading to Saline Valley, went further than his companions Hiring some of them as employees, he drove the mine deeper into the mountain and shipped ore to the new ten-stamp mill across Owens Valley at the gold camp of Kearsarge, east of Independence in the Sierra.

Ore rich enough to be hauled nearly sixty miles and still yield a profit did not escape the attention of Yankee miners returning to Owens Valley as the Indian war ended. Up the eight miles from Owens Lake to Buena Vista Peak they rode or walked. On April 15, 1866, the men at Cerro Gordo met and formed the Lone Pine Mining District, named for the town already alive north of Owens Lake. But Cerro Gordo itself was fast becoming a town, destined for fame in its day as California's biggest silver producer.

One of the earliest to scent the opportunity at Cerro Gordo was Victor Beaudry, the sutler at the army post, Camp Independence. From Montreal, the French Canadian had joined the Gold Rush in 1849. A merchant in San Francisco and then in Nicaragua, he next joined his brother, Prudent Beaudry, in the sleepy adobe town of Los Angeles. Early in 1861 he rushed with other miners to the gold placers in San Gabriel Canyon, and was promptly elected president of the mining district.

At the outbreak of the Civil War, U.S. troops were stationed in Los Angeles to counter secessionist sympathizers, and Beaudry applied his business experience to become the sutler for the garrison. By the fall of 1861 he had opened another sutler's store among the troops stationed in San Bernardino. And when the Indian war in Owens Valley brought a garrison of bluecoats to found Camp Independence in 1862, Beaudry followed his army friends and opened a sutler's store there in October 1866. Before the year was out he also launched the pioneer merchandise store in Cerro Gordo.

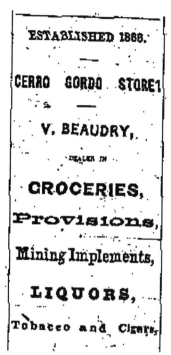

Short of stature and long of moustache, Beaudry cut a natty figure with his well-tailored suits among the uncouth miners. Always the likeable Frenchman to Owens Valley people, he was called "a jolly good fellow" by a friend in his home town of Independence. Later, in Cerro Gordo, he sponsored a brass band, which the boys named in his honor.

But Beaudry followed austere principles in his private life. One spring he was visiting friends in Independence when the girls of the house rushed outside to pick flowers from the cactus. With no warning from Beaudry, they proceeded to prick their fingers on the spines. When they returned in tears, Beaudry sternly observed:

"The only way to teach children is through experience."

Beaudry was equally exacting in business practice. Late in 1861 the first troops stationed in Los Angeles had been transferred to the East; since they had not been paid in California and had run up considerable credit with Beaudry, he left on the ship with them to collect when they were paid in New York.

In Cerro Gordo, the wily Frenchman was soon sharing in the treasure by a process that was enriching many a storekeeper in the Western mines. He extended liberal credit to the miners, mostly Mexicans, until in some cases their only recourse was to mortgage part or all of their mines to him. Then, when they were unable to meet the payments, he foreclosed.

In November 1867, Beaudry won a judgment of some $1700 against two of the pioneer Cerro Gordo miners, Joaquin Almada and José Ochoa, and was awarded in lieu of cash a piece of the Union Mine, a galena or lead-bearing lode. Then in February 1868 he foreclosed on Blas Mendez for parts of sixteen lodes, including the Union galena mine and the San Felipe quartz ledge. Within two years he filed foreclosures on footage in fifteen more lodes, among them the Merced and the Ygnacio, against five defendants including Pablo Flores, the original Cerro Gordo discoverer.

As early as May 1868 Beaudry owned a half-interest in the Union, which was destined to be the bonanza of the hill. Assured of plentiful ore, he built a smelter of his own, its rock chimney rearing assertively at the lower edge of town. Firing it up in August, with the blast from bellows operated by hand, he began turning out gleaming bars of silver-lead bullion. Soon he installed a mechanical bellows driven by horsepower, and later a larger, steam-driven furnace. Victor Beaudry, the little Canuck storekeeper, was becoming the silver king of Buena Vista Peak.

By this time Cerro Gordo was no longer Beaudry's secret. In May 1867 a Mexican prospector from Owens Valley rode into Virginia City with tales of Cerro Gordo's riches. Known as a man of reticent nature, he astounded the bustling mining capital with his shouts of laughter and his eager display of silver quartz samples. After a careful assaying their value became known on the streets, and Virginia City echoed with the news. Soon the strike was known in every mining center on the Coast.

Among the first to arrive was Judge James Walsh, the same who had led the rush from Grass Valley to Washoe in 1859, and had bought the interests of Henry Comstock. At Cerro Gordo he took an option on the San Ygnacio from Pablo Flores and others in February 1868.

Another early arrival was Mortimer William Belshaw, a practical engineer with a wealth of experience in silver mining. His great grandparents had emigrated in 1770 from County Antrim, Ireland, to New Haven, Connecticut. Later the family took up farming in central New York, where M.W. Belshaw was born in 1830. He worked his way through school as a teacher and then in a watch and jewelry shop, receiving a bachelor's degree from Geneva College.

Belshaw was clerking in a toll collector's office on the Erie Canal when the gold fever caught him early in 1852. Joining the rush by steamer via the Isthmus of Panama, he arrived in San Francisco not quite twenty-two years old. Within three years he opened a watch and jewelry shop in Fiddletown, bought gold dust from the miners and became the local agent for Wells Fargo Express. By 1858 he could return to New York and bring his hometown sweetheart, Jenny Oxner, back to California as his bride.

Of medium build and husky frame, Belshaw cut a formidable figure, adding weight through the years in the custom of successful Victorian entrepreneurs, and eventually topping 250 pounds. Foregoing the moustache customary in his day, he made up for it with abundant "muttonchop" sidewhiskers. He was one of the few Argonauts with a college education, and his tremendous capacity for learning helped to make him one of the few to succeed at mining. Although a man of quick and positive movements, he could work patiently and long toward a single goal.

More than anything else, Belshaw was an individualist, a man who made bold steps while those around him faltered. In 1862 he left California, hurried to the mines of Sinaloa, Mexico, and returned two years later a master in the working of silver ores.

Eastern California Museum and Stephen Ginsburg

Between 1865 and 1879, California's biggest silver and lead producer was Cerro Gordo, high in the Inyo Range overlooking Owens Lake. In top picture, taken from the road entering town, the slag dump from M.W. Belshaw's furnace is at left center. In bottom picture, taken above town and looking down the canyon, smoke is from stack of Victor Beaudry's furnace.

California State Library
M.W. Belshaw became the "Bullion King" of Cerro Gordo and defended his control against all comers.

Eastern California Museum
This is a day's run of silver-lead bullion bars at Belshaw's furnace in Cerro Gordo in the mid-1870's.

The Belshaws were living in a handsome home on Jackson Street in San Francisco when word reached him of the silver strike in Cerro Gordo. Leaving his wife and two sons, he took the stage for Owens River to test his silver knowledge. With him was a partner, Abner B. Elder, a native of western Ohio, a veteran of the Civil War, and another graduate of the silver mines of Mexico.

Reaching Cerro Gordo in April 1868, Belshaw and Elder found a swarm of miners already digging into the side of Buena Vista Peak. The quartz mines were so rich, they found, that the impatient silver seekers were throwing away as useless any ore worth less than $200 per ton. Each miner was happily working his own piece of "the Hill", insisting that it was one of many separate ledges, and not connected to a master lode owned by someone else. They had, wrote one arrival, "a big rope fastened to a convenient tree in the camp with which they will hang the first man who talks 'one ledge' to them."

But Belshaw was attracted by the galena ledges, for their forty to eighty percent lead content was essential in smelting the silver quartz ores. He knew that the man who built a smelting works would control Cerro Gordo. While others were buying the famed silver mines, Belshaw announced that he sought only galena deposits, as "the lead veins could control the working of the silver ores."

It was not long before Belshaw realized that the Union Mine, located on the slope above town, tapped the biggest lode of galena on the Hill. Already Victor Beaudry was acquiring one-half of this lode. But one of its remaining owners, Joaquin Almada, proved an easy target for Belshaw's glib tongue, and on May 6 sold him a third interest. The price was a one-fifth share in a smelting furnace that as yet existed only in Belshaw's fancy.

Having closed this masterful transaction, Belshaw extracted several tons of galena ore, had it smelted into silver-lead pigs in the nearby Mexican vasos, loaded them into several light wagons, and with A.B. Elder headed south on the dusty road across the Mojave Desert for Los Angeles. At San Pedro harbor they transferred them to the sidewheel coastal steamer, *Orizaba*. Arriving in San Francisco, they made their way to 402 Front Street where, if Belshaw knew his man, they could get the financing to build a smelter at Cerro Gordo.

Egbert Judson was one of the few to make a huge fortune in California mining. He was born in 1812 in Syracuse, New York, where he may have known young Belshaw before joining the Gold Rush in 1850. Two years later he opened the first assay office in San Francisco. Next he became a pioneer in hydraulic mining in the Northern Mines, and the principal owner of the famed Kennedy Mine in Jackson.

Starting in 1855 Judson devoted most of his energy to the mine explosives business, and in 1867 his chemical company tested dynamite under license from its inventor, Alfred Nobel. Out of this came the Giant Powder Company, with Judson the founder and chief owner. At the height of his career he owned mining, chemical and real estate enterprises from New Jersey to Alaska.

Like Victor Beaudry, Judson was a bachelor and popular among a wide circle of friends—certainly one of the prize catches in San Francisco. Mild-mannered and affable, he could be a bulldog when aroused. In 1856 more than 100 armed squatters forcibly settled on a rancho he partly owned in the redwoods north of the Russian River. They announced they would "kill anyone who should attempt to stop their trespassing."

Judson knew how to handle the challenge. Riding north to Healdsburg, he visited the office of Col. L.A. Norton, a lawyer, real estate agent and general troubleshooter. Clearly the man for the job, Norton confronted the squatters and got them out without a shot fired.

Now, in the spring of 1868, Judson listened to Belshaw's tale and agreed to finance his enterprise. Together they formed the Union Mining Company, with A.B. Elder as a lesser partner. Without delay Belshaw and Elder returned to Owens Valley to start operations.

Belshaw's first problem was to bring in heavy machinery for his furnace. Throughout July he and Elder were grading a wagon road up the rugged eight-mile ascent from Owens Lake. The route was so winding that Cerro Gordans joked of having to be drunk to drive it. In the narrowest section there was no room for wagons to pass. It became customary, when two wagons faced each other, for the smaller one to be disassembled and the animals tied to rocks while the larger wagon passed.

In this constricted passage of what was called the "Yellow Grade", Belshaw set up his toll gate and began collecting a dollar for every two-horse wagon, and a quarter for every horse and rider, entering Cerro Gordo. His control of the silver treasure had begun.

Strange-looking machinery was soon swaying up the tortuous grade by jerk-line mule team for Belshaw's furnace, whose red brick foundation was taking shape above camp near the Union Mine. Belshaw had designed a smelter to produce four tons of bullion per day, an unheard-of output by the lead-smelting process.

Such an unwieldy ore charge could not be uniformly heated by existing means, and Belshaw found he must break new trail in furnace technique. He invented the "Belshaw water jacket"—a great boiler-like cylinder, double-walled with scalding water surging between, and lined on the inside with a new fire-resistant clay found only a few hundred yards below camp. Within this metal cauldron, perhaps five feet in diameter and fifteen feet long, ore mixed with charcoal reached the melting point. An improvement over previous water jackets, it was adopted in other silver towns of California and Nevada.

In September 1868 Belshaw fired up his steam-driven blast furnace, charged his mammoth water jacket with ore and charcoal, and began producing silver-lead "base" bullion at a faster rate than the United States had ever known. Working under A.B. Elder as chief smelterman, the furnacemen kept the fire roaring round the clock for months at a time.

By day the smoke darkened the sky and filled the street with cinders; by night the vermilion flame cast a glow over the town and lighted the miners' way as they changed shifts. Under the withering heat the ores melted and the heavier lead and silver elements settled to the bottom to be tapped and run off into molds. The capacity soon rose to 120 bars per day, mostly lead with some silver content, each one eighteen inches long, shaped like a loaf of bread, and marked with the name "Union". Their weight averaged eighty-five pounds, their value from $20 to $35 each.

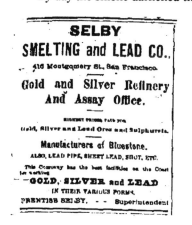

Belshaw began regular shipments on December 1, 1868, sending his bullion across the Mojave Desert and the Coast Range in a three-to-four week wagon trip to San Pedro harbor. There it was transferred to the steamer *Orizaba*, which carried fifteen tons per trip and unloaded three days later at the San Francisco wharves. It was then delivered to the smelting works of Thomas H. Selby, one of San Francisco's earliest hardware merchants.

Since 1865, Selby had operated a shot tower for making shotgun and other ammunition, using lead brought from Europe; and in the fall of 1867 had built a refinery at the foot of Hyde Street, assuring San Franciscans that "all gases or vapors will be blown...in the direction of Goat Island."

The Selby Works now became the principal market for Cerro Gordo's base bullion. For Selby could do what Belshaw's furnace could not—separate the silver from the lead. The silver he would deliver to the local U.S. mint. With the lead he would feed his shot tower. What surplus metal his works could not handle he bought and shipped via Cape Horn to the smelters of Swansea, Wales.

Under Belshaw's energy, Cerro Gordo was roaring with activity by the end of 1868. By increasing miners' wages to attract new labor, he made Cerro Gordo a "four dollar camp", traditional mark of a full-fledged mining town.

Such excitement was not lost on the West's mining fraternity. By the fall of 1868 Cerro Gordo's population had jumped from 200 to 750. More were on the way over two new stage routes—a weekly service through Aurora from Reno on the Central Pacific Railroad, and another weekly from San Francisco via Visalia and Walker Pass (fare: $60). In November the town had several stores, two restaurants, a billiard saloon, but as one resident wrote, "no lawyers, no preachers, no gamblers." Inevitably, he added:

"This is probably the richest silver mining district in the world."

By the spring of 1869, another local booster reported:

"Roads are being constructed, town lots staked off, buildings going up, shafts sunk," and Belshaw's furnace "turning out the bullion faster than it can be carried away."

The camp's main street was filled with grizzled sourdoughs and fast-talking promoters, merchants and vagabonds, men who had turned up at every new strike from Coloma to Aurora.

Since Cerro Gordo's rise coincided with the end of the Indian wars, farmers were resettling in Owens Valley to supply produce to this and other mining camps in the Eastern Sierra. By the summer of 1867 Lone Pine was sprouting anew as the entrepôt for Cerro Gordo. Farmhouses were going up in the valley, and a flour mill was erected at Bishop Creek.

By the end of 1869 farmers around Lone Pine were marketing hay in Cerro Gordo at from $65 to $75 per ton. Throughout the valley there were at least 1,500 settlers, diverting side streams, irrigating crops, and running more than 7,000 head of cattle. In Cerro Gordo they marketed not only hay, but barley, corn and potatoes.

Thus the mining camps brought a well-rounded system of commerce that was extending up the length of the Eastern Sierra. It included not only miners, but merchants, freighters, farmers and ranchers. And in the valley, at least, it brought pioneer women and children who, in turn, would bring schools, churches, libraries, and the social amenities.

As early as 1866 the new county of Inyo was born, carved from parts of Mono, Fresno and Tulare Counties. This in turn brought law, land titles, taxes, public roads, a courthouse and public officials. Civilization—perhaps, but Inyo was still frontier.

Beginning in the 1870s two competing stagecoaches lurched into Cerro Gordo every day from Lone Pine, where they connected with the tri-weekly line from San Francisco, the semi-weekly from Reno and a new weekly from Los Angeles.

But the greatest traffic into camp was the brigade of pack burros parading down the slopes along switchback trails, bringing ore and charcoal to the smelters, timbers for the mines, and water from nearby springs for the restaurants, stables and the steam engines of Belshaw's and Beaudry's furnaces.

By 1870 sturdy frame buildings, with canopy-covered porches and high false fronts, were taking the place of rock-and-canvas shacks along the main street. Among them were several general stores, saloons, restaurants, blacksmith and other shops, doctors', lawyers' and assayers' offices. More were rising as fast as lumber and shingles could arrive by mule team from the sawmills at Big Pine and Bishop Creek.

Other cabins and shops were springing up in the nearby "suburbs" of Belmont, on the east side of Buena Vista Peak, and Lower Town, on the Yellow Grade a mile below Cerro Gordo. On either side, the slopes were dotted with the rock or board shanties of the miners, some of them "half house and half gopher holes," roofed with canvas and stuck to the slopes with rough adobe. Scattered among them were the Mexican vasos and the mines that pocked the hillside till it looked like a prairie-dog town.

Cerro Gordo's social life centered about its several saloons and its two "hurdy-gurdy" houses, or dance halls. The Cosmopolitan Saloon, with its two billiard tables hauled up the Yellow Grade by straining mule teams, was the acknowledged vortex of society. Behind its swinging doors political caucuses and miners' meetings were held, prospecting parties made up, and the latest silver strike celebrated.

At opposite ends of town were the dance houses of Lola Travis and Maggie Moore, where the floors vibrated every night to the stomp of hob-nailed boots and the jig time of the harp and fiddle. Square dances predominated and they were short in duration to enable the bartender to dispense more frequent drinks.

Cerro Gordo had none of the civilizing graces—no church, no newspaper, and only a few wives of mine superintendents or merchants. The fancy women of the dance halls were the only female companions. As in other Western towns the men knew them by nicknames—Mud Hen, Horned Toad, the Waterfall, Featherlegs, and the Fenian.

The latter, whose real name was May Merritt, was so popular that she divided her time between Cerro Gordo and Lone Pine, where the furnacemen had to sojourn frequently to clear the lead poison from their systems. The Fenian had a ferocious temper, a weakness for whiskey, and a large black dog that always accompanied her. On a visit to Los Angeles she beat up a man for calling her an Irish bitch, broke an Italian boy's harp for refusing to play an Irish song, was put in jail, escaped from jail, and was allowed to leave town on the promise that she would never return.

One night while staggering among the cabins on the mountainside above Cerro Gordo she lost her footing and fell through a roof, causing the Chinese card players within to flee the shack in terror. During the election campaign of 1872, at a time when the editor of the *Inyo Independent* was absent, the printer's devil followed the custom of headlining his own political candidates at the top of the editorial page; sure enough, it was:

"FOR VICE PRESIDENT, THE FENIAN OF LONE PINE."

It was the combination of these painted ladies and "Forty-rod" whiskey that most often brought guns into play. For the *Inyo Independent* the frequent affrays on the Hill were an inexaustible source of news. Shootings became so commonplace that a fledgling doctor, arriving in March 1871, left town the same night without his baggage. As he rode down the Yellow Grade he poured out his story to the first man he met:

"My friend, I came here to buy a stock of drugs and practice medicine, but damn me if I want an interest in a shooting gallery!"

In 1872 another doctor settled for a time in Cerro Gordo and reported that shootings were so frequent that most of the houses displayed bullet holes. One patient surrounded his bed with sand bags four feet high for protection; the good doctor had to use a step ladder to attend him. Whiile seldom visiting the dance halls, the doctor nonetheless was called to them frequently to attend the victims of shootings and knifings.

Once the doctor took a young newcomer to one, "to note the effect on him, especially as a shooting might take place at any time." When asked the name of one of the girls, the doctor replied, "the Horned Toad". Overhearing this, she drew a stiletto and was about to stab the doctor when The Fenian grabbed her wrist. Then one of the Horned Toad's friends came after the doctor with a knife. A prominent local mining man, George Snow, drew his revolver and shot the man dead. The shooting became general, the room filled with smoke, the lights went out, and the doctor's young friend left the dance hall and Cerro Gordo.

When George Snow was brought before the Justice of the Peace for a hearing, His Honor concluded, "The court fines you thirty dollars, and if the fine is not paid in thirty days the court will go your security."

Such lawless law was not unusual in Cerro Gordo's justice court. One night a pile of wooden poles was stolen—an unforgivable crime in this lumber-shy camp. One inoffensive citizen was dragged before the Justice of the Peace as a suspect. Not a speck of evidence could be found against him, but the worthy magistrate had a duty to perform.

"You may not have stolen the poles," he admitted, "but there has been poles stolen, and I must make an example of somebody, so I fine you twenty-five dollars."

In the fall of 1871 the more civic-minded citizens of Cerro Gordo, deploring its lawlessness, sought a means of attracting the boys away from the dance halls. The honest miner, thought these reformers, must be induced to give up those revelrous nights at Lola Travis's and Maggie Moore's. The quick-time of the harp and fiddle must be abandoned for more cultural, high-minded pursuits.

And so the Cerro Gordo Social Union, a literary and debating club, was founded in November 1871. The largest, indeed the only, hall in town was taken over for the meetings, and the novelty of the pastime captured the camp's exuberant heart. Week after week the boys flocked to the debates, the girls found their floors deserted, and the club's founders congratulated themselves on their uplifting efforts.

Enthusiasm swelled when a full house listened one night to a heated political wrangle. The champion orator of the camp was holding the boys in silence with his eloquent phrases—"vile machinations," "secret intrigues," "crawling reptiles"—when he abruptly stopped to take a breath. In that moment the seductive strains of the harp and violin drifted in the windows from Lola's place.

Without ceremony, as though under a spell, the boys rose simultaneously and made for the door. Up to Lola's hall they tramped, while the breathless speaker stared at empty benches. Only the club's faithful secretary, suffering from a lame foot, remained in his place.

A cruel blow, one which it could not survive, had been dealt the Cerro Gordo Social Union. Hilarity returned to the dance halls, and the camp grew more turbulent than ever.

As joint owners of the controlling Union Mine, Belshaw and Beaudry were the silver kings of Cerro Gordo. But since 1869 their domain had been invaded by the Owens Lake Silver-Lead Company. Building a rival furnace on the lake shore at Swansea, the Eastern-owned concern had started buying silver ores from independent miners, and by ending the monopoly of the silver kings, had boosted prices. Beginning in 1871, it made heavy purchases in Cerro Gordo mines and offered a direct threat to Belshaw and Beaudry.

At the same time Belshaw was strengthening his position by acquiring other galena mines and buying the remaining shares in the Union. Believing that Cerro Gordo was, in fact, a "one ledge" camp, he meant to show eventually that all other mines were adjuncts of his pioneer Union. The advance of the Owens Lake company of Swansea simply made Belshaw redouble his drive to control the whole "Hill".

Belshaw's first move was to obstruct the Swansea company's ability to supply itself with ore. He left his toll road in utter neglect, while each rain carved ruts and piled boulders until the Owens Lake ore wagons could travel down the Yellow Grade only half loaded, and at caterpillar speed. This not only doubled tolls and hauling costs but threatened to cause a shortage of ore at the lower furnace.

True, the rutted roadway also hindered the wagons hauling down Belshaw's and Beaudry's products, but since these were base-metal bullion, not untreated ore, their traffic was only a fraction of that shipped down by the Owens Lake company. And since their weight was on the wagon-bed floor, they were not as top-heavy as the ore wagons

Exasperation at this outrage led to a challenge by John Simpson, who owned half of the Omega Tunnel through which the Swansea company extracted its ore. Simpson was a short, pipe-smoking Britisher, blue-eyed, sandy-haired, and scornful of anything American. He was schooled in the mines of the Mother Lode, and though irritable and demanding, was recognized as an able mine superintendent. Having arrived in Cerro Gordo a month before Belshaw, he had crossed swords more than once with the bullion kings.

On the night of August 3, 1871, Simpson opened his attack. Riding down the tortuous canyon, he drew to a halt at Belshaw's tollhouse below camp. According to custom, the keeper demanded the 25-cent toll. Simpson refused to pay. Spurring his horse through the gateway, he rode on down the Yellow Grade. The keeper then rushed up to Cerro Gordo for a deputy to hunt down the offender. Simpson was arrested at Lone Pine and taken back to camp.

His purpose, of course, was to bring the situation before the courts, and he now demanded trial by jury. As this was a hot issue in Inyo County, ninety men were examined before an impartial panel was selected. After a stormy two-day session in the Independence courthouse, the jury handed down a verdict of "not guilty".

The factious Briton had spent considerably more in lawyers' fees than the two bits in question, but he had brought the contest to the attention of the county supervisors. At their next meeting they reduced Belshaw's toll rates to correspond to the poor condition of his road.

At the same time the citizens around Owens Lake began subscribing funds for a new road to be built by the Swansea company along a route surveyed by Simpson. Early in September the road builders tackled the slopes of the Inyo Range with picks and shovels, carving the road that, except for its present starting point at Keeler, is still used today. Near Cerro Gordo they left the rain-washed Belshaw road and cut around the north side of the canyon. But within a mile the rugged ground forced the roadbed back into the narrow ravine, back over Belshaw's route, through two shadowy defiles around which a burro path could not be graded. When wagons began passing over the new road, surveyed and graded four-fifths of the way to Cerro Gordo by the Owens Lake contractors, Belshaw continued to collect his tolls.

"Nobody can cut around him," the *Inyo Independent* claimed in describing the situation, "nor can opposition ore buyers afford to haul over his road."

With this the *Independent's* editor, Pleasant A. Chalfant, joined the Swansea company in fighting Belshaw.

"When an exclusive franchise falls into the hands of a man whose soul is not above a five cent piece, then the public may expect to be choked until that five cent piece is forthcoming."

Belshaw had refused $8,000 for the road, Chalfant pointed out, but for tax purposes declared its value not over $1,000.

"We earnestly recommend that the county take it at that price for the public benefit." Belshaw would never improve it, concluded the editor, "as long as he thinks he can thereby prevent any other individual or company from getting a foothold in what he conceives to be his especial domain—at the mines of Cerro Gordo."

When the county supervisors met at Independence in August 1872 they were presented with petitions signed by almost every taxpayer in Owens Valley. They asked the board "to buy this toll road and make it a free one, or if that cannot be done, to reduce the tolls to the minimum allowed by law." Their lawyer maintained that Belshaw had "worked too hard for several years in collecting tolls on his mountain trail and not enough in trying to make it good enough for wagons, and consequently he ought to be made to take a rest."

Opposing him in defending Belshaw was young Pat Reddy, quick of wit and quick of tongue, already a power in Inyo County and destined for fame as a criminal lawyer throughout the West.

"If it had not been for the owner of that trail," Reddy retorted, "practically there would have been no Cerro Gordo, or need of any road at all."

With belligerent John Simpson on the board as the member from Cerro Gordo, it was not surprising that the three supervisors favored the Owens Valley petitioners. Within a few days they reduced Belshaw's toll rates again to half the original charge. But before they could bring themselves to appropriate the road entirely, John Simpson resigned his post and left Inyo for a mining assignment in the Mother Lode. The boys on the Hill, showing where their own sympathies lay, thereupon almost unanimously elected a new supervisor: Mortimer W. Belshaw.

That quite definitely ended the fight for the gateway to Cerro Gordo.

But the Owens Lake company was not alone in its fight against Belshaw. Since 1870 the owners of the San Felipe mine had disputed his claim to what he called the Union lode, thus challenging his drive to dominate the whole Hill. Their original claim was a silver-quartz lode, situated near the Union shaft on the slope above town. In February 1870 they also purchased the Omega Tunnel from John Simpson and began extending it through to their silver vein.

Within a few weeks Belshaw noticed, on the dump of the Omega Tunnel, galena ore that could only come from his own Union lode. Ignoring the "No Admission" sign over the entrance, he stormed into the black depths and ordered the San Felipe miners away from his vein.

Belshaw's outburst was heeded for a few days, but soon galena once more appeared on the dump. He now waited for a chance to accost one of the owners.

In April he saw Gustave Wiss, a German doctor who was the principal owner of the mine, at the mouth of the tunnel, showing the San Felipe holdings to a prospective buyer. Belshaw charged down from his furnace and met the two on the ore dump by the tunnel entrance. Standing on the disputed galena, he asked Wiss why he did not work his own mine rather than "forcibly going into another man's mine."

"If that is the Union mine it could be proved legally," snapped the doctor. He claimed it as the San Felipe.

To Belshaw this was preposterous. The Union was a galena deposit, containing both silver and lead. The San Felipe was a pure silver-quartz vein.

"This is your San Felipe mine, and this is my Union mine," he roared, pointing up the Hill. "You have forty or fifty feet to run into your tunnel before you strike the San Felipe mine; you are now taking ore out of this, the Union mine."

Time would tell, retorted Wiss, whether or not the vein belonged to him. Meanwhile, if Belshaw meant business, he could sue.

"I could sue a beggar and catch a louse," quipped Belshaw.

He stalked away with the warning that they could never haul that ore from the dump, "nor have any of the profits or benefits of it."

Wiss knew well that Belshaw made no empty threat. Let a San Felipe crew load a single wagon with that ore and a body of Belshaw's Union miners and furnacemen would descend on his dump in force.

But soon Belshaw was taking steps to seize the San Felipe and remove all doubt of his title to the biggest silver-lead deposit yet discovered in California. The San Felipe company had almost ruined itself running the tunnel through many feet of barren rock to reach the silver lode—a common mistake in undercapitalized mines.

Pat Reddy's younger brother, Edward A. "Ned" Reddy, had worked for the San Felipe and was owed back pay. In October 1870 he brought a judgment suit against the mine, causing a public sale of part of the mine. Belshaw was the buyer. In May 1871, young Reddy filed another suit for the rest of the mine and assigned it to Belshaw. Behind this, one could see the fine hand of Ned's older brother Pat, who was Belshaw's lawyer.

Victor Beaudry also took a hand. When John Simpson had sold his Omega Tunnel to the San Felipe he took back a mortgage. With the payments delinquent, Beaudry acquired the mortgage and started foreclosure in February 1871. He and Belshaw hauled the galena ore dump to their smelters.

Now, if the San Felipe company did not redeem its property by paying off the judgments in six months, the mine and tunnel would be turned over to the bullion kings and the San Felipe would lose all claim to the great bonanza of the Hill.

Not until early November 1871, only a month before the December 5 deadline, did the San Felipe moved to reclaim its property. Meeting in San Francisco, the directors sent M. Allison Wheeler to Inyo with the funds to pay off the judgments. A Forty-niner from New York, Wheeler had turned up at every big strike from the Comstock to Mexico, earning a wide reputation as an expert mining engineer. He had been the interested party who had accompanied Wiss when Belshaw upraided him at the tunnel mouth—an incident that had not deterred Wheeler from buying a share in the San Felipe.

Armed with a bag of more than $4,000 in gold coins, he left San Francisco in frantic haste. After a 400-mile train-and-stage ride, he stepped into the sheriff's office at Independence on November 29, with six days to spare. Sheriff Abner B. Elder, Belshaw's former partner, was out of town; his undersheriff sat behind the desk absorbed in local matters.

"I came up on behalf of the San Felipe Mining Company of San Francisco," began Wheeler, "to redeem the property that has been sold under execution."

The officer looked up in surprise, slowly absorbing the situation. Ownership of the San Felipe, and perhaps of the Union and the whole Hill, was about to be secured. He replied that he was ready to take a trip to Lone Pine and asked if the transaction could be deferred a couple of days. No, answered Wheeler, acutely aware of that December 5 deadline—the business must be concluded here and now. Taking the bag of gold from his pocket, he counted out the required amount and demanded a receipt.

"While you are writing your receipt," agreed the undersheriff, "I will go and see to my horse."

He stepped across the street, and Wheeler knew he was scurrying for the office of Pat Reddy, Belshaw's attorney. When he returned Wheeler presented the receipt.

"I will not sign that now," said the officer.

Wheeler was dumfounded. Apparently they meant to deny his redemption.

"You have got the money," he cried, "and I must have a receipt."

"Counsel ordered me not to sign it now."

In desperation Wheeler stepped to the door, called in a local citizen, and, taking his name, bade him bear witness that he had offered the money to the undersheriff. But Wheeler knew it was a doubtful assurance of redemption, one that would probably wilt before Pat Reddy's glib tongue in an Inyo courtroom.

In a moment the undersheriff had swung into the saddle and resolutely ridden out of town toward Lone Pine, leaving the San Felipe money in his office. M. Allison Wheeler watched him go in silent wrath, knew he would not return before December 5, and realized now that, in Inyo County, Belshaw's word was law.

Once the deed to the San Felipe had been delivered into their hands, Belshaw and Beaudry believed themselves the undisputed kings of Cerro Gordo. But at this time the Swansea company had finished perfecting another title. It had bought the Santa Maria mine, last claim to the only body of galena left outside the Belshaw-Beaudry interests. While the Omega Tunnel had not yet been turned over to the Bullion Kings, the Swansea company was extracting Santa Maria ore through it to supply the smelter at Owens Lake with the needed lead flux.

In July 1872 Belshaw and Beaudry notified the Owens Lake Company at Swansea: No more galena ores could be taken from the Omega Tunnel, as they were the property of the Union Mining Company. So when Belshaw's edict reached the San Francisco office of the Owens Lake company, its manager promptly took train and stage for Inyo County.

Galen M. Fisher was not the man to bow before Belshaw's mandates. His father was a New England sea captain, and as his mother accompanied her husband on his voyages, Galen happened to be born in Buenos Aires. Reaching California in the late 1850s, he amassed a fortune in banking, insurance and real estate. From 1872 he was a leading citizen of Oakland, where he built and sold more than 100 houses and would later become city clerk and treasurer.

This was the man who now stormed up the Yellow Grade to Cerro Gordo. First he stalked into the office of Victor Beaudry and demanded the meaning of the notice. Beaudry replied that the Santa Maria lode was part of the Union claim. And he gave Fisher to understand that he and Belshaw "owned the whole Hill." The Union, he declared, is "an older title than the Santa Maria by ten days."

When Fisher accosted Belshaw he received the same answer: The Swansea workings were "a portion of the Union mine."

The infuriated Owens Lake manager had no intention of seeing the Swansea furnace shut down for want of lead. If the Union title was older than the Santa Maria, he would secure a title older than either—the San Felipe.

Ordering his furnace superintendent to keep using Santa Maria ores, Fisher returned to San Francisco. From Gustave Wiss and M. Allison Wheeler he bought a half-interest in the San Felipe, even though their ownership was doubtful. But he promised to finance legal steps to recover the mine from Belshaw and Beaudry. He would challenge that farcical refusal of Wheeler's redemption and retrieve the San Felipe from the bullion kings. That done, he meant to prove that the San Felipe, an older claim than the Union, actually made *him* the ruler of Cerro Gordo. The high-stake game over Inyo silver was approaching a showdown.

Fisher promptly sent a counter-notice to Belshaw and Beaudry, demanding surrender of the Union mine. When this was ignored, he brought up his legal guns. In January 1873, Fisher filed suit against the bullion kings for more than a million dollars in rents, profits and damages, and recovery of the San Felipe tunnel and mine, which he claimed included the great Union bonanza.

Owens Valley people waited through the spring while the contending factions squared off for a vicious battle in the Independence courthouse. When John Simpson, who bore scars of an earlier fight with the rulers of the Hill, heard of the suit he gave bitter warning to the San Felipe group:

"They will beat you at the trial and then kick you out of town."

On July 10, 1873, the "Big Suit" opened in Independence with fast-talking Pat Reddy heading Belshaw's defense counsel. The first issue, a ruling to be decided by the judge, was whether or not Wheeler had made a valid redemption of the San Felipe mortgage. Pat Reddy claimed that Wheeler had shown no authority, and had therefore been rebuffed by the undersheriff. But Fisher's lawyers called in local witnesses to the transaction and secured a ruling that Wheeler had actually redeemed the property.

Ownership of the San Felipe was now restored to its previous owners, and the question remained whether or not it included the Union lode. This decision was up to the jury, and with a million dollars and the Union mine hanging in the balance, the issue was the talk of Inyo County. To the Owens Lake company and manager Galen Fisher, it meant a final battle in the struggle for Cerro Gordo.

Belshaw and Beaudry, of course, claimed that since the San Felipe was a silver-quartz vein, it had no connection with the Union galena deposit. Just before adjournment on July 15, Fisher's lawyers obtained a court order allowing them to enter the ground between the San Felipe and Union shafts and strip the surface debris from the outcrop. They intended to show that the galena vein extended between the two mines.

With time running short before the trial closed, the order was dispatched that night by pony rider to the San Felipe agent in Cerro Gordo, forty miles southward. Next morning a crew of San Felipe men marched with picks and shovels to the slope north of camp. They found an armed body of Union miners holding the ground. Obviously, Belshaw had also sent a message from Independence. The San Felipe crowd was told it could not trespass on Union ground without a fight, order or no order.

Down to Independence rode the San Felipe agent. Drawing up that night at the courthouse, he outlined the situation to the trial judge. The worthy magistrate then called in Belshaw for contempt of court in resisting his order and admonished him against further interference.

Back to Cerro Gordo rode the agent. Next morning, July 17, the San Felipe men began stripping the surface between the two shafts while the whole camp came up the Hill to watch.

Meanwhile, Pat Reddy and the Union lawyers were making headway in the courtroom. Bentura Beltran, a well-known and respected Mexican prospector, testified that he had originally located the San Felipe as a silver-quartz vein, and had later discovered and claimed the nearby Union galena vein. He was supported by the first mining recorder for the district, who swore that, when recorded, the San Felipe was not a galena vein but a silver-quartz lode. Unless those San Felipe miners uncovered a lead vein connection between the two mines, it appeared that Belshaw and Beaudry would win the battle.

Final examinations and defense plea occupied the day of the 18th. Except for the closing plea of the plaintiffs scheduled for next morning, the trial was over. Still the San Felipe miners had not appeared from Cerro Gordo to testify, though word passed through Independence that they had proved a galena connection between the two shafts.

That night, while the jury was confined in the courthouse, a group of Belshaw's men stationed themselves before the door to prevent the San Felipe faction from entering and presenting the latest evidence.

The San Felipe miners had, in fact, uncovered a continuous lead belt from their mine northward to within sixteen feet of the Union shaft, beyond which they dared not dig for fear of tumbling rocks into the mine to the danger of the men below.

When court opened on the 19th, M. Allison Wheeler and several miners were present and ready to testify to the lead connection. Belshaw's lawyers were quick to remind the judge that all testimony had been closed the previous day. He was forced to refuse the evidence. But one of Fisher's attorneys, in his final plea, referred to the miners' testimony that had been denied. This vein connection, he pointed out, fixed the ownership of the galena bonanza with the San Felipe claim, an older title than the Union by six months.

Mort Belshaw, suddenly rising from his chair, bellowed his defiance. He declared it untrue that the miners had established the connection, and he had witnesses to deny it. When the courtroom had quieted down, the judge reproved the San Felipe lawyer for alluding to the subject.

At 11:30 a.m. the jury retired for a three-hour deliberation. In midafternoon they filed back to their places and announced the verdict—"for plaintiffs."

Belshaw's lawyers objected that it was not a proper verdict. The judge patiently read the charge, by which the Union mine and a million dollars would be forfeited to the Owens Lake company, and the foreman confirmed that this was the jury's decision. Belshaw's counsel then demanded that the panel be polled. Each juror, rising in his place, declared the verdict to be his own. There was no question that the Owens Lake-San Felipe faction had just won the biggest silver-lead mine in California.

The people of Inyo, their sympathies with the bullion kings who had built Cerro Gordo, thought the decision an appalling injustice. Chalfant of the *Independent,* siding now with Belshaw, declared his pioneer endeavors had "done more to benefit this county than all other primary enterprises combined."

But Belshaw and Beaudry had not abandoned the fight. In a few weeks Pat Reddy got a stay of proceedings, which suspended the execution of the verdict pending his efforts to secure a new trial in the state Supreme Court. The San Felipe and the Union mines remained in the hands of the bullion kings while the issue dragged on. The fight for Cerro Gordo was not finished.

Meanwhile, Belshaw and Beaudry were stepping up production to drain the mountain's wealth before they could be forced to give up the Union mine. They had already installed a steam hoisting works at the shaft head and attracted more miners with increased wages of $5.00 per day. Now they rebuilt their furnaces to double the bullion capacity, and Cerro Gordo fairly rocked and rumbled to the heavy beat of industry.

But the bullion kings were beset with other limitations over which they had less control. Mule-team freighters hauling their silver-lead bullion to market could not keep up with the furnace output. Wood needed for fuel in the furnaces and timbering in the mines was running short as the piñon pines were cut down on every slope of Buena Vista Peak. And a water shortage in this semi-desert hampered growth. Without snow water, the steam boilers for the furnaces could operate only intermittently through the summer months.

If any one of these obstacles could not be overcome, Cerro Gordo's output would not grow, but shrink.

Freighting the bullion involved a seventeen-day trip by mule teams to Los Angeles, then a three-day voyage by steamer up the coast to San Francisco. In 1872 the superintendent of the Owens Lake company's smelter at Swansea, James Brady, built a little steamboat and named it the *Bessie Brady*, after his daughter. First commercial lake steamer west of Salt Lake, it took on the bars of bullion at the foot of the Yellow Grade and, chuffing across the waters, dropped them at Cartago Landing and saved three days out of the wagon trip.

Then freightman Remi Nadeau, who had hauled Cerro Gordo bullion for years, had a proposition for Belshaw and Beaudry. If they would put up $150,000 for constructing way stations and better roads, as well as buying more mules and wagons, he could move the accumulated bullion and catch up with the furnaces. Forming the Cerro Gordo Freighting Company in June 1873, Nadeau cleared the piles of silver-lead bars at Cartago and matched the redoubled output of the furnaces within a year's time.

Another enterpriser stepped in to solve the wood shortage. Colonel Sherman Stevens had left his New York home at the age of sixteen and entered the fur trade in the Great Lakes. Joining the Gold Rush in 1851, he settled in Owens Valley before the Indian war was over, and was one of the first to invest in a mine at Cerro Gordo.

Now this tall and stately pioneer saw in the Sierra Nevadas an almost unlimited source of timber. In January 1873 he got a $25,000 loan from the Owens Lake company, with the agreement that he would always deliver its fuel at 25 cents a cord less than he charged Belshaw and Beaudry. Then he built a sawmill at the head of Cottonwood Creek and a wooden flume to shoot the timbers down to the bullion trail. From there, wagons hauled them to Cottonwood Landing, the *Bessie Brady* towed barges filled with the logs across the lake to Cerro Gordo Landing, and mule teams pulled them up the Yellow Grade. It was an elaborate and costly way to get fuel to the furnaces, but with the trees gone in the Inyo Range, it was the only way.

Water—or lack of it—was the third obstacle to Cerro Gordo's growth. From springs several miles away, pack burros were carrying in water at from 10 to 15 cents a gallon retail, and in winter the supply was enhanced with snow water. In 1870, Belshaw had tapped a local spring with a pipeline, but in winter it frequently froze and broke. Cerro Gordo was still stunted from lack of water.

However, further north into Saline Valley, Miller Springs gushed from the rocks with some 120,000 gallons per day—more than enough to slake Cerro Gordo's thirst.

Tackling this challenge was a local mine owner, Stephen Boushey, with the help of Los Angeles bankers. He laid 11½ miles of pipes, buried them to prevent freezing, and installed a pump to raise the water 1800 feet over the top to Cerro Gordo. Starting in May 1874 water poured into town at five cents a gallon retail. Now the boys could send two shirts a week to the laundry, and could take a bath without feeling guilty. Most important, the furnaces could maintain full production throughout the year.

Thus with the help of three enterprisers, the bullion kings could extract and market Cerro Gordo's riches at capacity speed—as much as possible before they could be lost in the courts. With mines, furnaces and mule teams all operating to the utmost, Cerro Gordo in 1874 sent 5,290 tons of ingots—worth some $2,000,000—to Los Angeles. This almost equalled the total shipment for the two previous years.

Though his opponents apparently held the high cards in the grim game they were playing, Belshaw was in fact reaching into the earth and making off with the stakes. Always sociable and hearty despite his callous business methods, Belshaw now found himself entertaining more often at his hillside house, and making more frequent trips "below" to see his wife, Jenny, in San Francisco. With his younger half-brother, John T.C.S. Belshaw, he opened a general store in Cerro Gordo. And despite the San Felipe lawsuit hanging over their heads, he and Beaudry felt confident enough to join other entrepreneurs in projecting a railroad across the desert to haul Cerro Gordo's rising output.

Through these halcyon years in the middle Seventies, Belshaw got himself into further adventures. In July 1873 he and a companion, W.A. Goodyear, rode mules to the top of what was then thought to be Mt. Whitney, the highest mountain in the States. But it was obvious to them that what appeared to be the highest from down in the valley at Lone Pine was actually shorter than several other peaks. Filling a tin cup with water to its brim, they leveled it and sighted across the top. By this and other means—performing triangulation sightings from Cerro Gordo and financing another expedition to the top of the Sierra—Belshaw helped to establish the real Mt. Whitney.

Then, returning from one of his frequent trips to San Francisco, Belshaw was sitting on the box beside the driver as the stagecoach pulled into the station at Coyote Holes, at the eastern mouth of Walker Pass. There they were stopped by the notorious bandits, Tiburcio Vasquez and Cleovaro Chavez, one on each side of the road with leveled Henry rifles. The driver was about to lay whip to the team and dash on for Indian Wells.

"Stop!" shouted Vasquez. "Hold up your hands!"

Belshaw, with little stomach for an adventure, advised the reinsman to halt. The stagecoach swayed to a stop, the passengers were forced to line up in the road, and while Chavez covered them, Vasquez went through their pockets. Belshaw gave up a silver watch, $20 in gold, and a new pair of boots. This meager haul, especially from the silver magnate of Inyo County, was a clear disappointment to the road agents.

"Belshaw," warned Vasquez, "if I ever catch you on this road and you haven't a thousand dollars for me, your travels will be ended."

A few moments later, the jingling of bells announced the arrival of two northbound teams of the Cerro Gordo Freighting Company, of which Belshaw was part owner. Surprised by bandit rifles, they were robbed of their coin and marched to a hillside to join the other victims and the station attendants, who had been previously disarmed by the bandits. Then Vasquez and Chavez mounted up and whirled southward on the Bullion Trail.

Three months later, after cutting a swath of robberies in Southern California, Vasquez was captured by a sheriff's posse in the brush-covered hills of what later became Hollywood. He was taken by steamer up the coast to San Francisco, where he waited in jail a few days before being sent to San Jose for trial.

At his cell door, Vasquez recognized a familiar face. Mortimer W. Belshaw, always the businessman, came directly to the point. Speaking in Spanish, he asked Vasquez what he had done with the watch he took from him at Coyote Holes, adding that he would willingly pay for its return. Pleased at the recollection of that bold exploit, Vasquez replied that he thought the watch was in the hands of Cleovaro Chavez, who was still at large. At that Mort Belshaw decided he had no urgent need for the timepiece; he would forego a meeting with that esteemed *hombre*.

With the solutions to its problems and the surge of bullion production, Cerro Gordo gained new business buildings on its main street and a population of a thousand people. The Union Hotel was erected to compete with the two-story American House, which John Simpson had built in 1871.

Cerro Gordo's greater prosperity naturally brought an influx of the lawless fraternity. Scarcely a month passed without another shooting affray. The enterprise of Belshaw and Beaudry had spawned a wild town that lived by the revolver—one of the worst in the American frontier. As the *Inyo Independent* commented:

"A good calaboose or a little judicious hanging is much needed upon Cerro Gordo hill."

One of the bloodiest outbursts occurred in early February 1873, when two men were desperately wounded in Maggie Moore's dance hall at the lower end of town. A few minutes later shots rang from the camp's upper end, and a Mexican was carried out of Lola Travis's house with a ball in his stomach. Guns barked again the next night, and two antagonists fell wounded in an exchange of shots.

"Cerro Gordo is a prolific source of the 'man for breakfast' order of items," observed the *Independent*.

When the grand jury convened in March 1873, County Judge John A. Hannah, dean of the Inyo bench, reviewed the unpunished reign of crime in Cerro Gordo and launched a bitter tirade against "these lawless ruffians, who with murder in their hearts and the implements of death strapped upon their persons, congregate in public places, ever ready to discharge their death-dealing weapons upon the unoffending and unarmed citizens."

Following this outburst, a truce-like quiet settled over the front for a few months. But by mid-October the *Independent* detailed another fatal affray with the laconic introduction:

"Our local shooting item for the current week reaches us from Cerro Gordo."

On November 6, at Maggie Moore's house, two men fell dead before another burst of gunfire.

"This makes five men shot, four killed outright, in this county in as many weeks."

When word came of another affray in Cerro Gordo in the middle of the month, editor P.A. Chalfant had to change the score to seven men in seven weeks.

"Pistols continue to crack and good men go down before them as though neither law nor society valued men's lives any more than those of so many wild animals."

Still the revolver ruled "Bullion Hill", and releases on self-defense remained the prevailing custom. Chalfant conceded that nobody mourned when bad men killed each other.

"But frequently innocent parties are the victims; and if there is law to do it these affrays should be stopped on that account, if no other."

By this time, Belshaw and Beaudry were benefiting from the troubles of their arch-rival, the Owens Lake Silver-Lead Company of Swansea. For months in 1872 and '73 its income had been cut drastically by the failure of freighters in hauling its bullion to market. Now it was so heavily involved in mine purchases and lawsuits in its race with the bullion kings that solution of the wood and water shortages gave it small comfort.

At the end of 1873 the company was in financial straits, and some of its Eastern investors foreclosed a $98,000 mortgage. Early in March 1874 the last ingots were run out and the Swansea furnace shut down. Late in July a summer cloudburst struck the Inyo Range and buried Swansea, furnace and all, with several feet of sand and debris. The very elements themselves had written the finale to the weary contest between the Owens Lake company and the bullion kings.

But Belshaw's adversaries still had a formidable weapon in the San Felipe mine and their lawsuit against his Union mine. Even here lawyer Pat Reddy was making headway in seeking a new trial before the state Supreme Court in Sacramento. In May 1875 he wired the news to Inyo:

"We have won in the San Felipe."

While this simply meant another round in the fight, Cerro Gordo erupted in celebration. Belshaw ordered a general holiday, and while the boys gathered around a great bonfire blazing in front of the American Hotel, the bullion kings made speeches from the balcony.

At a high point in the excitement someone threw his hat into the flames. Belshaw and the others followed the example, and somebody suggested they raid Beaudry's store for new hats. The hilarious miners, Beaudry among them, trouped across the street and descended on his shelves. When they burst out of his store they were wearing hats too big or too small, and several rough miners were topped with high stovepipes. The last to ransack the counters came out wearing gingham sunbonnets and paraded around the fire. Then the bullion kings threw a free dance and supper, and Cerro Gordo did not rest till daybreak.

Yet the celebrants were already suffering some doubts. For three months mining had been suspended on the Union. In April, Beaudry's furnace had shut down, although Belshaw's continued to run on ore from the Union dump. Remi Nadeau's Cerro Gordo Freighting Company reduced its schedule to one team every other day, and the Cerro Gordo stage abandoned its daily trips for a tri-weekly service. Even Lola Travis took her tittering flock down the Yellow Grade for the booming new town of Darwin.

Throughout Inyo, the rumor spread that the great body of galena that had upheld the camp for years was exhausted. But some people suspected that Belshaw and Beaudry were simply trying to make the San Felipe crowd, already limited in funds, decide that the prize was not worth further struggle. A Supreme Court battle would be costly, and the bullion kings had the treasure to outlast their foes.

If such suspicions were true, the stratagem worked. Word reached Inyo in December 1875 that the suit was being settled by negotiations in San Francisco. The Union Consolidated Mining Company of Cerro Gordo was created on January 13, 1876, holding title to both the Belshaw and Beaudry smelters and the Union, San Felipe, Santa Maria and the rest of Buena Vista's strategic mines. Galen M. Fisher and his associates in the San Felipe company were given a third of the 100,000 shares of stock. Two-thirds were taken by Belshaw, Beaudry, and their San Francisco partner, Egbert Judson. They held three of the five seats on the board of directors, with Judson as president, Beaudry as treasurer, and Belshaw as superintendent of operations. And they would receive one percent of the profits every month as salaries, dividends or not.

The end of the savage struggle was celebrated on January 1, 1876, with a grand ball in Lola Travis's abandoned dance hall. Sure enough, the occasion was marked by an outbreak of violence in the customary style of Cerro Gordo's boom years.

During the festivities an American shot a Mexican with whom he had quarreled, and was pursued by a mob of the victim's *compadres.* Wounded by a volley of shots, he barricaded himself in a nearby saloon, which the crowd quickly surrounded. At length another American, Loyal Merritt, gained admittance and arrested him despite entreaties by other citizens who feared for the man's life. While Merritt was escorting him across the street to the office of Justice John R. Hughes, the excited Mexicans closed in and killed their quarry with a shot through the head.

At the inquest next day someone testified that Merritt had boasted of receiving money from the Mexicans for his part in the affair. Justice Hughes was obliged to level his well-known shotgun to prevent the Americans from shooting the offender on the spot. Merritt was banished from town and arrested in Darwin as an accessory to the murder by the mob. When another mob in Darwin threatened to lynch Merritt, Pat Reddy and a few friends faced down the angry men with a plea for law and order. In the trial at Independence, Reddy prosecuted the case and sent Merritt to prison.

Beginning in January 1876, Belshaw and Beaudry made hasty preparations for a return to full-scale output. A force of miners was returned to the Union mine, and Beaudry's furnace was fired up early in February for the first time in ten months. The Cerro Gordo Freighting Company revived its old schedule of two mule teams a day, and the Lone Pine stage resumed its daily trips up the Yellow Grade.

Throughout Owens Valley, these developments were received with a wink of the eye. Sure enough, people said, Belshaw and Beaudry had shut down mine and furnace only long enough to win a favorable settlement of the suit.

For the next ten months both furnaces were roaring, filling the air again with the din and smoke that spelled prosperity, and 1876 rivaled 1874 in peak bullion production. But the Union mine, its interior excavated in great vacant chambers at every level, was in fact yielding the last of its treasure.

In December 1876, Belshaw shut down his furnace for the last time, though Beaudry's continued to produce an average of 120 pigs a day for three more years. A new Union shaft was sunk in quest of blind lodes, but by the middle of 1877 it had gone down 900 feet without striking ore.

Still more troubles were in store. On the night of August 14, 1877, the town was aroused to find the new Union building and the entire hoisting works wrapped in flames. The miners on the night shift, unable to pass through the fire at the shafthead, escaped into San Lucas Canyon through a connection with the Omega Tunnel. A platform of timbers and earth hastily constructed near the 200-foot level saved the lower timbering, but more than $40,000 in mining equipment alone was destroyed.

By October the damage had been repaired, but when the directors met in San Francisco on February 9, 1878, Superintendent Belshaw presented a grim picture. The fire had cost the company a total of $75,000. The price of lead had fallen to the point where the lead content in the bullion no longer covered the cost of transportation to San Francisco, as in previous years. And the company debt had reached $110,000.

To carry on the exploration deeper in the Union mine, the board voted an assessment of $1.25 per share—a very stiff levy. Beaudry—after joining Belshaw and Judson in supporting the assessment—resigned from the board.

The Union Con assessment was promptly branded in Owens Valley as proof that the bullion kings were trying to "freeze out" Galen Fisher and the San Felipe interest. When such sentiments reached Belshaw's ears he was enraged. He would make, growled the old warrior, "the grass to grow in the streets of Lone Pine and Independence."

Chalfant of the *Inyo Independent* was quick to respond. Cerro Gordo was in hard times because "M.W. Belshaw has discouraged other capital in order to perpetuate his own monopoly." In another issue he wrote that the manager of the Union Con "affords no great amount of friendly feeling to either this paper or the community in general." When it was announced in January 1879 that Belshaw and Beaudry would be revisiting Cerro Gordo, the *Inyo Independent* observed pointedly:

"Many old friends there and hereabout would be glad to shake paws again with the genial Beaudry."

Within a month after the assessment was announced, it appeared that Belshaw was making good his threat. Miners' wages were reduced to $3.00 per day, and fully half the boys departed for more virile camps. The last stagecoach swayed down the Yellow Grade in April 1878, and the town was served thereafter only by a pony mail.

After the visit of Belshaw and Beaudry in February 1879, exploration was renewed in the Union. But the mine was scraping the barrel for pay dirt. Galena ore had grown so scarce that a refining works was erected to melt the bullion bars and extract the lead content to be used over and over in the reduction process in Beaudry's smelter. Late in October 1879 the Union mine was finally abandoned. On the evening of November 20 the boys gathered around Beaudry's works for what they called the "funeral". The blast was shut off, and the furnace exhaled its last steam.

Next day one of Remi Nadeau's mule teams hauled its last load down the Yellow Grade. Some 280 bars of base bullion from the smelter and a 420-pound mass of pure silver from the refinery rumbled downward to the long-familiar jingle of team bells.

Cerro Gordo itself was not quite dead. Other mines were still producing, their ores converted to bullion in vasos or other small furnaces. On Independence Day, 1879, the town mustered a rousing celebration, complete with music from the Beaudry brass band, an oration on a decorated grandstand, sack races and a burro race. And in September a grade school was opened by Miss Frazier, whom the *Inyo Independent* hailed as "an attractive young lady recently from the East." Her influence went beyond education. Miners and furnacemen washed their faces more frequently, attended taffy pulls, and generally competed for attention. As one resident reported:

"Since her arrival our morals have assumed a higher tone."

But in March 1880 Cerro Gordo suffered the fate most feared in Western mining camps. Around midnight on the 15th, the people were aroused by the cry, "Fire!" Maggie Moore's abandoned dance hall was wreathed in flames. Before the conflagration died, eight buildings on the south side of the street burned to the ground. Of those saved by frantic firefighters, two more were taken out in another fire three months later.

With nearly half the town gone, most of it without insurance, the spirit left Cerro Gordo. No attempt was made to rebuild. By August 1880 there were no more than ten Americans in town, though a number of industrious Mexicans, Indians and Chinese were still extracting ore out of Buena Vista Peak.

"Cerro Gordo," as one correspondent wrote, ". . .presents a picture of almost absolute quietude."

This time the old camp had finally died after fifteen lusty years. Estimates of the day placed her total output between $13,000,000 and

$20,000,000. But without counting the production of the Swansea works at Owens Lake, Cerro Gordo's actual yield between 1865 and 1879 was probably closer to $7,500,000—still much ahead of any silver camp in the Eastern Sierra.

Cerro Gordo's demise, capping the decline of other nearby camps, staggered Inyo's economy. True, the farmers of upper Owens Valley, their pioneer canals spreading green fields across the landscape, were by now self-supporting. Theirs was a life, if not idyllic, at least pleasantly civilized. In this farming enclave, 41 percent of the people were women.

By contrast, in Inyo County's mining region, 16 percent were women—typical of mining camps in general. And by 1880 the life had gone out of the Inyo camps. From a population of several thousand in the mid-Seventies, they were down to 400 in 1880. The loss was dramatized by the deserted stage stations from Mojave to Owens Lake, with a two-horse wagon bringing in an average of one passenger per day. In October 1880 one Eastern Californian wrote that Inyo County "is about the dullest in the State."

By this time the actors on the stage of Buena Vista Peak had already taken their exits. In 1875 Victor Beaudry had returned a millionaire to his home town of Montreal, where one of his brothers would soon be elected mayor. There in May 1876 he married Angelina LeBlanc, whom the *Los Angeles Times* described as "a lady of high social standing," while the *Los Angeles Express* reported:

"Miss LeBlanc belongs to a distinguished French-Canadian family, and is a most lovable lady."

Except for frequent trips to California, Beaudry tarried in Montreal until 1882, when he built a sumptuous two-story Victorian mansion on Bunker Hill in Los Angeles. It had verandas and balconies on two sides, bedrooms with dressing rooms and hot and cold running water, and a bathroom with indoor plumbing. Above the front door, the transom was painted with a cornucopia over the emblems of the United States and Canada, with "In God We Trust" lettered below.

Bringing his wife and children to this showplace, he joined his brother Prudent in real estate and other enterprises, including the Temple Street Cable Railway, which ran south on Spring Street through the growing residential district to Union Avenue.

Five years later the Beaudrys returned to Montreal. When they left town their friends gathered to see them off in the Southern Pacific's special palace car, "Carmello". Beaudry died in Montreal in 1888.

Mortimer W. Belshaw left Cerro Gordo in 1877 and, though he kept his position in the Union Con, settled with his wife and son in Antioch. There, with his usual energy, he founded the local water works, a department store, and the Bank of Antioch. As the free coinage of silver became a national issue, he contributed several magazine articles to the controversy. He headed a gold mine in Calaveras County, and with his old friend and partner, Egbert Judson, bought into the famous Kennedy mine in Amador County.

Then in 1881 Belshaw and Judson bought at sheriff's sale the Empire coal mine at Mount Diablo. Belshaw revitalized the company and superintended construction of a narrow-gauge railroad to Antioch to market the coal.

In April 1897 Belshaw was thrown out of his buggy in an accident on the streets of San Francisco. The incident was jarring but not believed to be serious. In reporting it, the *Antioch Ledger* jokingly commented:

"A man weighing 265 pounds can't afford to fall off any steeples."

Whether from this accident or, as one biographer indicated, from overwork in his many business duties, M. W. Belshaw died in Antioch in April 1898 at the age of sixty-eight. His wife, Jenny, returned to San Francisco, where she died in 1900.

Belshaw's son, Charles Mortimer Belshaw, graduated from Harvard, became president of the California Mining Association, a state assemblyman and then a state senator. But tragedy dogged his later life. His first wife, Miriam, committed suicide, and he and his second wife were both killed when their car rolled off an embankment at Half Moon Bay.

Egbert Judson, the explosives magnate who had financed Belshaw in Cerro Gordo, died a victim of his hazardous business. On July 9, 1892, a series of explosions wracked the Giant Powder Works at Fleming's Point, West Berkeley. The concussions rocked the entire Bay area, panicking people and horses and breaking hundreds of windows ten miles away in San Francisco. Judson was driving his buggy 500 feet from the powder magazine when it exploded with the largest concussion of all. He was thrown out of the buggy, striking his head. He lingered for six months before he died of heart failure at the age of eighty in his home at 15th and Valencia streets in San Francisco. The *Chronicle* called him the "last of the California millionaires."

Galen M. Fisher, who had challenged Belshaw's reign in Cerro Gordo, threw himself into commercial and political ventures on the west side of the Sierra. He owned ranches and various properties in five counties of San Joaquin Valley. In the mid-1880s he was city clerk and treasurer of Oakland. Before he died in 1889 he and his wife had eight children.

Darwin French, who had led the first expedition of silver-seekers to Inyo, took up farming as a homesteader in 1869 at Poway, San Diego County. Still the leader, he was elected a county supervisor and became chairman in 1870. At the age of 77 he wrote a short book filled with patriotic fervor entitled, *The Power of Destiny Revealed in our War with Spain and the Philippines.* French died at Ensenada, Mexico, in 1902.

Since 1880, more than one new company has tried to strike another fortune on Buena Vista Peak. Starting in 1909 an aerial tramway carried the ore down the side of the Inyo Range to the terminus of the Carson & Colorado Railroad at Keeler.

At the same time large deposits of high-grade zinc ore were discovered deep in the Union Mine. Beginning in 1911 the property was leased and then purchased by Louis D. Gordon, an experienced mining man, who built an improved tramway down the mountainside and made Cerro Gordo an important U.S. zinc producer.

When the author first visited the camp in 1939 the bucket tramway was still there, but Cerro Gordo was deserted. From time to time other enterprisers try their hand. Led by Jodie Stewart, a native Inyoite who owns the whole town, the Cerro Gordo Historical Society is rescuing the weathered buildings and preserving the artifacts from the days when the town was California's biggest silver excitement.

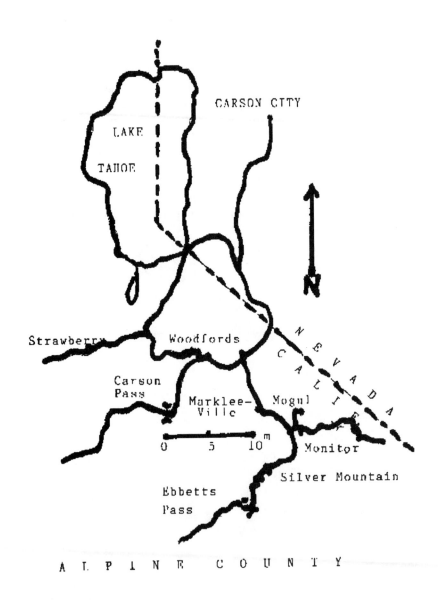

3. THE RISE AND FALL OF LEWIS CHALMERS

South of Lake Tahoe, silver was struck on the East Carson River just before the Civil War. Thousands rushed in from California and Nevada Territory, creating a cluster of mining towns and a new Alpine County in the crown of the Sierra. In 1867, as the Alpine treasurehouse was declining, Lewis Chalmers arrived with British capital to stir new life. Near Silver Mountain he created the "Chalmers Mansion", married his young housekeeper, and with her reigned as the first couple of Alpine society. Until the mid-Eighties he divided his time between London, where he raised more money from investors, and Silver Mountain, where he spent it on new ventures. Then, just as suddenly as he appeared, he vanished, leaving his wife and Alpine County in bittersweet decline.

The first rainbow hunters over the Ebbetts Pass route toward Washoe were blindly trampling silver under foot as they reached the canyon of the Carson River's East Fork. But in November 1860 three men—John Johnson, Wesley Poole and V. Harrison—chipped at some tortured outcroppings. What they found was silver ore.

Braving the Sierra winter, they formed the Silver Mountain Mining District on May 27, 1861. Others stopped to prospect—fourteen of them staying the next winter. Every week they floundered through deep snow for twenty miles back to Hermit Valley for provisions. Through the winter of 1862-3 some fifty men, many of them Norwegians, occupied what became Kongsberg, named for a silver mining town near Oslo. Located on Silver Creek a few miles from its junction with the East Carson, Kongsberg was soon changed to Silver Mountain.

Anticipating a rush when snows thawed in the spring of 1863, merchants ordered goods from San Francisco and Sacramento that were hauled by mule team up the Placerville road. Below Lake Tahoe the wagons branched southward over rough trails as far as a tavern operated by Jacob Marklee on the Middle Fork of the Carson. Since this was temporarily the head of "whoa navigation", with freight packed onward by muleback, the town of Markleeville sprang up as the depot for the region.

By the last of August 1863 it comprised nearly 100 buildings, including what was euphemistically called an opera house. Actually this was a hurdy-gurdy house where the boys could, for four bits, have a drink and choose a dancing partner from among half-a-dozen German girls. When Reverend Petit rode in over Carson Pass from Volcano and preached at three camps on the same Sunday, it was obvious that civilization had come to Carson Canyon.

The transformation was pictured by an observer who described a domestic scene: a husband was splitting a log to make shingles for the roof of a house; his wife, fashionably dressed in hoopskirt and fancy trimmings, was tossing flapjacks over an open fire; while a group of Indians silently watched the proceedings.

As expected in Silver Mountain, news of the silver strike reached the California gold fields and the rush began with the spring thaw in 1863. By the end of May the fifty men at Silver Mountain had swelled to more than 300, with 20 to 30 arriving daily. Nearly 100 claims were being worked by small companies of men tunneling to reach the ledges which struck downward from the outcroppings. Others were raising tents of canvas and brush, and houses of logs and shakes. In one of these an enterprising saloonkeeper was pouring drinks out of old pepper bottles into a few tin cups used by turns at two bits per drink.

At this point, supplies were packed in from three directions—over Ebbetts Pass from Murphy's and Angel's Camp, over Carson Pass from Volcano and Jackson, and from Genoa in Nevada Territory. On all three routes, hundreds of men were building wagon roads.

```
        GELATT & MOORE'S
   Pioneer Line of Stages
              FROM
      Genoa to Silver Mountain.
   VIA FREDERICKSBURG, WOOD-
      FORD'S, MARKLEEVILLE, MT.
      BULLION, MONITOR AND MOGUL.
   Leaves Genoa every MONDAY, WEDNESDAY, and
   FRIDAY, at 11 o'clock, A. M.  Returning, leaves the
   above places on alternate days, arriving at Genoa in
   time to connect with the Pioneer and Wilson's Line of
   Stages to all parts of California and Nevada.
              H. D. GELATT, } Proprietors.
              J. MOORE,     }
      J. MOORE, Superintendent, Genoa.
      September 30, 1866tfsf
```

First into the Carson Canyon was the road from Genoa, which halted for a time at Markleeville. Stagecoaches and freight wagons operated to the ends of the roads, but the mines were still reached by mule trains through the summer of '63.

Meanwhile, the boys were combing over the gullies and ridges in this wonderland of ledges turned on end. Some six miles southeast of Markleeville on Monitor Creek (named for the famed Union gunboat) they struck rich silver ore and the town of Monitor rose in a canyon so narrow there was room for only one street. To tap the wagon traffic the boys extended the road for a mile-and-a-half down to the East Carson, where it joined the road being graded from Markleeville to Silver Mountain.

Monitor thus became the first mining camp served by wagons, and promptly extended its buildings all the way down to the Carson. At that point the village of Mount Bullion appeared across the river almost overnight, while elsewhere in the watershed of the East Carson new strikes spawned still other camps and other mining districts—Raymond, Webster, Highland, and most important, Mogul, located two miles north of Monitor.

Suddenly the Sierra stillness that had been home to Indians and wild animals was shattered. The splashing of the Carson and the rustling of the pines had given way to the blast of the tunnel diggers and the whine of the sawmill.

By the first of August the earth was host to a thousand men, perhaps a dozen with wives and children. Those not digging their own mine were hired at $5.00 per day—a premium over the usual $4.00 in Virginia City. More were flocking in by express lines over all three routes. For the last few miles on each line, passengers stepped from stagecoaches and rode mules or horses on the trails.

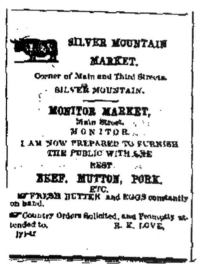

In Silver Mountain, town lots were selling briskly at $1,000 each. The boys were thronging into two hotels, several general stores, a livery stable, a dry goods store, and an ample supply of saloons among the forty log or board shanties in town.

"Tunnels and drifts are being sunk," wrote one newcomer that August, "and every few minutes the booming sound of a blast comes to the ear like a distant leisurely bombardment." Hundreds of miners were "scampering like a nest of disturbed ants."

"Nearly everyone," observed the visitor, "is, in his belief, in the incipient stage of immense wealth."

Though gambling and drinking were rife, this was still a moment of innocence. In Silver Mountain a Chinese prostitute had opened for business, but she departed when the people tore down her cabin and threw it in the creek. When the town was laid out a choice block was set aside for a public school. Claim jumpers drove posts into the ground to preempt the block. Outraged, the citizens held a public meeting, listened to fiery speeches, resolved to throw the jumpers out, and marched to the school site. There they uprooted the posts, built a rail fence around the block and cleared the ground of debris. Then they resolved to build a school immediately and to "hold it peaceably if we can, but forcibly if we must."

Meanwhile, California newspapers were fanning the frenzy. Late in August a correspondent of the *Sacramento Union* wrote glowingly from Silver Mountain. Men were selling shares in mines at feverish pace. "Thousands of 'feet' are sold daily at rates varying from one to one hundred dollars per foot. Every day brings its new excitement by new discoveries."

Early in October, at six in the morning, a woman correspondent named "Mountain Mary" climbed the stagecoach at the Mammoth Grove Hotel in the Calaveras Big Trees while the driver called, "All aboard for Silver Mountain!" After a gruelling but scenic ride over Ebbetts Pass (the last few miles by muleback), she reported to the San Francisco *Alta California:* Silver Mountain had 200 buildings, over 500 mining claims, and more than 100 tunnels being driven. Ore from the IXL mine was yielding $643 to the ton in gold and silver—a rich mine indeed.. Best of all, a dancing party in a log cabin included a supper that "would do credit to more antiquated places in the lower regions."

All this activity in the Carson watershed was pursued by several thousand people, but as yet there was no stamp mill or furnace, and no bullion being shipped. In their exuberance the boys demanded that a new county be formed. Campaigning in the September election, legislative candidates from Amador County promised one. In March 1864, Alpine County was created, mostly from Amador County, and Silver Mountain won the election for the county seat. Politics were fired with the Civil War frenzy, with Unionists overriding the Copperheads.

By this time the roadbuilders had reached the 15 miles from Markleeville to Silver Mountain. In early April a quartz mill and two sawmills were running near Markleeville, and another sawmill was being rushed to completion at the mouth of Silver Creek. An earlier sawmill near Silver Mountain, screeching round the clock, was cutting more than 7,000 feet of lumber per day, and the planks were hauled off before the sawdust could settle. Another quartz mill was under construction to reduce ore from the now-famous IXL mine.

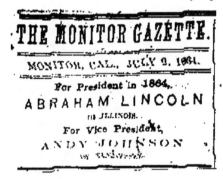

Not to be outdone, the town of Monitor was expanding as fast as lumber could arrive. By early June 150 buildings were up, and one of them housed the sprightly *Monitor Gazette*. Throughout Carson Canyon the boys talked of nothing but King Silver. No one doubted that the Alpine diggings would be "bigger than the Comstock!"

Among the earliest to arrive was Frank A.S. Jones, born in Wayne County, Ohio, in 1827. He had grown up in the Michigan woods, where he worked in lumbering and water traffic. In the Mexican War he served as a mounted rifleman. On his return he found Michigan becoming too civilized and joined the American Fur Company on the upper Missouri. He was running a lumber schooner on Lake Michigan when he got the California fever in 1860. Mining in California, Nevada and Utah, restless Frank Jones was in Alpine County as early as May 1862.

Socially, Jones was gregarious—active in both the Masons and Odd Fellows. His companions elected him Recorder of the mining district. He was ambitious, hard-working, opportunistic, proud, and a dogged adversary.

Together with his brother James, Frank Jones mined at Monitor and began buying feet in other mines. Starting in December 1863 they acquired positions in the Michigan Lode and 26 others, most of them parallelling each other in vertical position, with the outcroppings showing on the mountain behind the town of Monitor.

On January 20, 1864, the Jones brothers incorporated the Michigan Tunnel & Mining Company, with themselves as officers and Oscar F. Thornton as Secretary. Near the camp of Mt. Bullion on the East Carson they planned their tunnel, pointing toward all of these ledges, aiming to intercept them from below and stope out their claims with the help of gravity. To his brother, who was the outside man in search of funds, Frank wrote that there is "going to be a fortune for both of us if we can ever get money to have it opened."

But the summer of 1864 was a time of testing for the Alpine miners. Many hundreds of men were driving tunnels through barren rock to tap the ore-bearing ledges. But until they reached pay dirt, it was all expense and no return. When some tunnels did reach their ledge, most of the ore proved so stubborn that, except for the IXL ore near Silver Mountain, it could not be made to yield silver simply by crushing and the amalgam chemical process.

The same problems had been encountered in the Comstock, and the West was still learning what Europe already knew about working refractory ores. Furnaces were needed, and this required buying and hauling in more heavy machinery and boilers by mule team at five or six cents per pound. A hundred miles from the nearest foundries, Alpine was held captive by the Sierra's magnificent distances.

The boys toiled bravely on, despite a drop in miner's wages from $5.00 to $4.00 per day. Most of them, buoyed by their own gambler's optimism and the sanguine reports of the *Monitor Gazette*, believed firmly in Alpine. The IXL finally began shipping $2,500 in bullion per week from the quartz mill at Silver Mountain, its stock rising from $5.00 to $50 a foot in two months' time during the summer of 1865. But the more realistic began to think of selling whatever claims they owned.

"I have about made up my mind to abandon everything," Frank A.S. Jones wrote his brother at the end of September. The Michigan Tunnel was, he declared, "a complete failure."

"I think 2 Thousand Dollars would not buy this whole country now."

Yet Jones kept enough hope to stay on, driving the Michigan Tunnel whenever his brother sent funds. Realizing that the value of a mine increases with the distance from it, "Captain" James Jones sailed from San Francisco to London in the fall of 1865 to sell the Michigan Tunnel and Mining Company. It was a time when English investors, impressed by the riches of the Comstock, were anxious to put their money in American mines.

Calling himself "the first discoverer of all the prominent Mines" in the Monitor District, Captain Jones was not backward in presenting ore samples assaying at many hundreds of dollars per ton. He offered the Michigan property to a group of highly placed London mining investors. The mines were, he said, only 30 miles south of Virginia City, Nevada (actually, the closest were fifty miles), and he claimed a railroad would soon span the distance.

On February 23, 1866, the London investors formed the Imperial Silver Quarries Company, Ltd., with a capital stock of £500,000 ($2,500,000). On the same day they bought the entire property offered by the Jones brothers, who were to receive shares in the new company as partial payment.

In June, Captain Jones gave some ore specimens from two ledges at Monitor to the Imperial Company's assayer, Lewis Chalmers, who reported values in gold and silver of $6,440 from one and $2,175 from the other. Elated, the Imperial investors opened company offices in London's Palmerston Building, ordered the machinery for a 50-stamp mill, and sent it sailing round Cape Horn to San Francisco.

But in Alpine County nothing was happening. If the Jones brothers were retained to supervise the tunnel boring, little progress was made. And 140 tons of milling machinery was soon rusting on the North Point dock at San Francisco. By June 1867 the Imperial Company was, in the words of a principal investor, "almost at its last gasp."

Clearly the frustrated investors needed to take charge. Two of them, Henry Syme and Lord Earl Poulett, were put forward to head the company, with Poulett as chairman of the board. They resolved to send a man to California and set things right. For this they turned to their assayer, Lewis Chalmers. According to Poulett, he "has always commanded our esteem and confidence." With the title of Managing Director, Chalmers sailed from England on September 11, 1867.

Lewis Chalmers had not always been an assayer, or even a mining man. He was born in Fraserburgh, Aberdeenshire, Scotland, on March 9, 1825, the first son of the leading citizen. His father, Lewis Chalmers, Sr., held the position of Baron Baillie (chief magistrate) for nearly 40 years. An attorney by profession, the senior Chalmers was also the chairman of the town council and chairman of the Harbor Commissioners.

With an artificial harbor on Scotland's East Coast, Fraserburgh was a fishing town standing in the teeth of strong wintry gales; it developed people of resolute character. Though the elder Chalmers was criticized by government inspectors for his poor bookkeeping, he led the way in civic improvements, especially the breakwater that secured the all-important harbor. The *Aberdeen Journal* declared that "a more kind-hearted and upright man was not to be found in Aberdeenshire."

The young Chalmers studied law at Aberdeen University and practiced law in Aberdeen before returning to assist his father in Fraserburgh. When Lewis Chalmers, Sr., died in 1850, the son became Baron Baillie, with all its responsibilities, at the age of 25. In addition he became the local agent of the Bank of Scotland, the Justice of the Peace, and the factor (business manager) of Lord Saltoun, the principal owner of the town and the master of Carinbaly Castle nearby. Among Saltoun's investments managed by Chalmers were fourteen commercial and residential rentals, including Chalmers' own home in Fraserburgh.

Thus it was said of Chalmers that he "occupied the foremost position in Fraserburgh both publicly and socially." His office was "the chief business center of the town and district."

Chalmers' father had left a reputation difficult to match, but the son had additional qualities. As the *Aberdeen Journal* later described him:

"He was a man of most refined appearance, of great ability, and one of the most accomplished and polished public speakers who ever lived within the bounds of Fraserburgh."

Another observer called him "one of the best at any ceremonial or demonstration."

In 1850 Chalmers married a young woman of at least equal standing among Scottish families—Elizabeth Ann Gordon Cameron. She died in December 1851 after bearing a son, Lewis William. Within a year-and-a-half Chalmers married a second wife, Ellen Miller MacEwen, who bore four children before she died in 1863.

Despite these tragedies, Chalmers was able to maintain a brilliant social life befitting his position. It was later said of him that he was "for many years the great figure-head of social and public life—never since equalled in the eyes of the general community."

With his growing family and three servants, Chalmers decided the home in Fraserburgh was too small. In 1859 he built an imposing mansion as his factor's residence for Lord Saltoun's affairs, locating it in the woods out of town and calling it "Witch-hill". There his handsome carriages, his hunting horses and his regal lifestyle were the talk of Fraserburgh.

"He was," wrote one observer, "a man of exceptional liberality and goodness of heart."

In 1864 the Chalmers fortunes suddenly changed. As one chronicler put it, he "became involved in some financial speculations which ended badly and a crash followed." Lord Saltoun, who had been away for an extended period, came home. There he discovered that Chalmers had built Witch-hill as part of the Saltoun business, and was living a baronial life. Enraged, Saltoun fired Chalmers.

While Lewis still had other sources of income, the disgrace was too much to bear in a town the size of Fraserburgh. He left, as one reporter put it, "in a cloud of misfortunes." As he drove his carriage southward he stopped and turned for a final look at Witch-hill.

"I shall redeem my past," he is said to have declared, "and come back and claim my position again."

So, at the age of 39, armed with his business and legal experience and ample personal charm, Lewis Chalmers sought his fortune in London. In the firm of Johnson, Matthey & Company, "assayers and smelters to the Bank of England," he became a qualified practical assayer. In 1866 he joined the Imperial Silver Quarries Company, where his younger brother, John Chalmers, was already the chief engineer. By the end of the year Lewis became "Interim Secretary", and in September 1867 was appointed Managing Director at an annual salary of £1,000 (about $5,000). With that title and authority he sailed for America on September 11. Already, Lewis Chalmers was redeeming his past.

But from the beginning his mission struck trouble. For some reason, instead of sailing for the Isthmus of Panama, he landed in New York. There, for another unknown reason, he was delayed for days before taking steamer to the Isthmus and, after crossing on the railroad, sailing up the Pacific Coast.

It was early November before he reached San Francisco; then, by paddle-wheeler to Sacramento, railroad into the Sierra, and finally, stage to Alpine County, he stepped into the main street of Monitor on November 9. After ridding himself of the host of men trying to sell him mining claims, he contacted Oscar F. Thornton, secretary of the Michigan Tunnel Company, and showed him the papers transferring title to the Imperial Silver Quarries Company of London.

"I am here," he wrote in a formal letter on November 22, "to take possession of said property as agent."

Chalmers was astounded to learn that, after Imperial Quarries had agreed to purchase the Michigan Tunnel, Captain James Jones had turned over the controlling shares of stock to one B. F. Petiss. Unable to take possession, Chalmers tried to reach Thornton and Frank A.S. Jones, who was superintending some mine work in Silver Mountain. Unsuccessful, he finally contacted the attorney for Petiss and demanded the property. He would "make a formal application to the Court for possession, and also for damages against your client for every day during which the same is withheld."

As Chalmers understood the tangled story, the Jones brothers and Thornton had got a judgment against the Michigan Tunnel Company for work they had done, then had assigned this judgment to Petiss. The judgment could be redeemed and the controlling shares returned for $5,000.

To his brother John at Imperial headquarters in London, Chalmers sent an urgent wire on the transAtlantic cable that had been completed the year before:

"No possession till Jones [is] telegraph[ed] five thousand dollars. Meanwhile, withhold his share."

Finally, after more delay, Frank Jones announced he had received the $5,000, and paid it to Petiss. On February 18, 1868, Chalmers sent Petiss' quit-claim deed to the Alpine County recorder, and the Imperial Silver Quarries at last owned the property it had paid for twice. Lewis Chalmers had gone through a primer course in the ways of California mining.

As for the Michigan property itself, Chalmers was delighted. On December 10, 1867, he sent an enthusiastic report to the Imperial board, accompanying it with a box of prime ore samples. After spending an estimated £8,669 to develop the mines and install the 50-stamp mill in the first year, he predicted, the property should yield £104,000 profit from ore shipped to be reduced in England, and another £21,666 from bullion produced at the mine. Return on investment to stockholders he estimated at 20 to 30 percent *per month*.

When Lord Poulett read Chalmers' letter to the stockholders at their First Annual Meeting in April 1868, they were understandably fired with anticipation.

But Chalmers was already getting more lessons in California mining. Starting work on a new Imperial Tunnel, he encountered such hard rock that he had to send to San Francisco for a new explosive, "Giant Powder", from Egbert Judson's company. Still the contractor was agonizingly slow.

"I determined to take into my own hand," Chalmers wrote London, "and am now preparing to run it night & day."

Concerned about claim jumpers, Chalmers built an office and residence across the river from the tunnel and, as he reported, "hold possession by personal occupation."

Frank Jones, who owned shares in the Imperial, saw all this and wrote his brother in London that Chalmers had taken men from the tunnel to work on his house. Still smarting from their clash with Chalmers, Captain Jones passed word to the Imperial board of directors of this and other alleged acts of mismanagement. When these charges were relayed back to Chalmers, he laboriously answered each one and confronted Frank Jones by mail in February 1869. But Jones, while not denying that he was the source of the complaints, answered:

"I propose to let you and my Brother do your own quarreling."

Still working behind Chalmers' back, Frank Jones secured affidavits against Chalmers from disgruntled workers whom Lewis had fired. Getting wind of this, Chalmers was furious.

"I have now commenced to deal with F. Jones," he wrote London, ". . .and I shall make him eat dirt before I have done with him."

Chalmers plunged deeper into the mountain, assuring London that the work "will make a Second Comstock of the Imperial." Seeing the dirt fly, the Alpiners were equally sanguine. One of them wrote the *New York Tribune* that the Imperial Tunnel "will throw the Comstock far into the shade."

But by early November 1869 the Imperial Tunnel had been driven more than 1200 feet, still without striking one of the ore-bearing ledges. With most of its capital spent without results, the Imperial Silver Quarries Company, Ltd., went into bankruptcy. The 50-stamp mill on the dock in San Francisco was sold by the port authority. In April 1870 the company's creditors, including Chalmers himself, swooped in for the remaining assets. It looked as though Lewis had not redeemed his past, after all.

The Jones brothers were, of course, relishing Chalmers' troubles. Frank referred to him as "Lord" Chalmers, perhaps starting with this sarcasm the cognomen that became popular throughout Alpine County. But their triumph was premature; they had not counted on the resolve or the persuasive powers of Lewis Chalmers.

In the fall of 1869 a new London firm, the Exchequer Gold and Silver Mining Company, appeared under the same leadership as in the old Imperial—Lord Earl Poulett and Henry Syme—with a capitalization of $300,000. Chalmers had shifted his interest to Silver Mountain some nine miles south of Monitor on Silver Creek, a tributary of the East Carson. There he bought an eight-stamp quartz mill and adjoining saw mill, and the Buckey No. 2 Mine, renaming it the Exchequer.

This time the enterprise needed no long tunnel through barren rock. The ore was in sight within the mine, ready to be extracted and milled. And the location had the added advantage of being miles away from Monitor and the critical eyes of Frank A.S. Jones.

From London came Lewis' brother, Captain John Chalmers, to manage the mine. In the spring of 1870 he arrived at Silver Mountain and set about his work—erecting a building for the miners, grading a road to the mill, and opening the shafts, drifts and winzes of the Exchequer. By November, mule teams pulling high-sided wagons were hauling ore the five miles to the mill.

But delays still dogged the Chalmers brothers. The little stamp mill could not work the recalcitrant Exchequer ore at a profit.

Undaunted, Chalmers bought the IXL mine, the pioneer of them all, and formed still a third mining venture, the IXL Gold and Silver Mining Company, Ltd., with essentially the same London backers. With a new outlay of $171,000 approved in London, the Chalmers brothers pushed deeper into the earth of Silver Mountain.

To raise more funds, Chalmers took stage, train and ship to London in late January 1872. Spreading his persuasive powers in England for several months, he returned in mid-June, purchasing new hoisting works for the Silver Mountain mines while passing through New York. As his housekeeper wrote, he arrived in Alpine "With Lots of Money and some very rich Presents."

While his brother returned to Scotland, Lewis promptly paid all debts of the Exchequer, bought several more Silver Mountain mines, and took stage for Virginia City to hire more miners at premium wages. Enticing carpenters at $6.00 per day, he installed the hoisting works, then shifted them to construction of an enlarged stamp mill. By March 1873 he was in Grass Valley hiring still more miners. Crowed the Silver Mountain *Alpine Chronicle*, "British gold will produce silver."

Except for shutdowns through two Alpine winters, the improved mill ran for two years. Lewis continued to spend British gold on new furnaces and more stamps for the mill. Managing an empire of 200 miners, he seemed on the brink of redeeming his past.

Led by the Chalmers spirit, the whole of Alpine County was rekindled. In Monitor the famous Tarshish mine was shipping ore, while its furnace was still fighting the rebellious rock. In Mogul, the Morning Star Mine, with its heavy copper content, was the most steadily profitable of Alpine mines. In Silver Mountain the *Chronicle* reported:

"The sound of hammer and saw. . .is heard in all parts of our town."

With money beginning to flow again, the usual hard cases showed up and the usual rascalities were afoot. As early as 1864 the Monitor & Northwestern Mine had been mining the pockets of its investors with stock assessments. Little of the proceeds were used for working the mine. Most were transferring to the pockets of the operators. At one stockholders meeting a lead brick painted with silver was on display to help milk the investors. Not until the spring of 1874 were the scheming operators turned out in favor of new management. But it was too late; within a year the property was sold by the Sheriff.

Nor did Alpine County, which had generally been a more sober and steady district than most, escape the appearance of the reckless breed. In 1874 one Ernst Reusch shot and killed a man; when his lawyer got a change of venue to Mono County, the outraged Alpiners formed a vigilance committee, stopped the undersheriff and prisoner, and hanged their man to the bridge over the East Carson some two miles upstream from Markleeville. Noting the trial of another such murderer in Reno, the *Chronicle* editor wrote:

"Alpine would not be put to much expense in such a case."

But ordinarily the Alpine scene was sylvan and pastoral. Each spring two big events drew attention from the primary mining activity. With the annual flood of the East Carson, the sawyers supplying the Comstock with mine timbers jammed the river with logs—sometimes floating leisurely in quiet glades, sometimes hurtling through rapids while the canyon rang with the crack of timber. And every June, drovers from San Joaquin Valley herded as many as 100,000 sheep over Ebbetts and Carson passes to reach the green meadows of Alpine.

"Our mountains," reported the *Alpine Chronicle*, "are getting lousy with sheep."

It was an idyllic if rugged life—the long winters that blanketed everything in white made bearable by hearth and chimney, the summer months punctuated by joyful picnics, impromptu horse races, hiking trips to nearby peaks, and grand parties in which the women set bounteous tables and enjoyed the surplus of eager dancing partners.

Alpine County Museum
In the 1860s and '70s, Monitor's one street ran for a mile down the narrow canyon of Monitor Creek, while the mountainside behind rang to the blasts of miners digging tunnels and shafts.

Alpine County Museum
Silver Mountain was the biggest town in Alpine County in 1863-4. Hillside above town is mostly denuded of pine trees for timbers in mines and fuel for furnaces and stamp mills. Today, only the walls of the jail remain.

George A. Day, Fraserburgh, Scotland

Lewis Chalmers, the "lord" of Silver Mountain, may have looked something like this portrait of his father.

Alpine County Museum

Antoinette Laughton came to Silver Mountain as Lewis Chalmers' housekeeper and became "Lady" Chalmers.

As for Chalmers, he was still spending money on mines, mills, roads, buildings and payrolls from Monitor to Silver Mountain. In June 1876 he improved the building that was his home at the Exchequer mill and furnace. As the *Alpine Chronicle* declared, "it is now as comfortable as an old English mansion."

From this passing observation may have sprung the local term "Chalmers mansion" that still persists among Alpiners. Viewed today, standing a few feet from the towering brick chimney of the old Exchequer furnace, the two-story frame building can hardly be called a mansion, though it is larger than it looks, since much of it extends down toward the creek. But in its day, mansion or not, it was the showplace of Alpine society. It was the closest Chalmers could come to his lost Witch-hill in Scotland.

Almost from the beginning Chalmers had maintained a housekeeper. Now, Mrs. Kelly, his current houselady, was leaving for San Francisco; she recommended a friend from Shaws Flat, Tuolumne County. When the new housekeeper stepped from the stage in Silver Mountain, Chalmers and all the townspeople were enchanted. She was a strikingly attractive girl of twenty years. Disillusioned by a poor marriage that had ended in divorce, Antoinette Laughton arrived with a baby boy. She was, in effect, a child of the Gold Rush, born in California of French and German parents. An acquaintance who visited her as a girl later said,

"She was a nice looking lady. . . .she had a nice little figure and she had pretty arms. She had a kind of 'Roman' nose and round face, and she was all right."

Temperamentally she was sensitive, high-strung, subtly conniving, and understandably suspicious of men's motives.

Quickly Antoinette settled into the rustic and rigorous life of Silver Mountain. From the beginning she attracted Alpine's unattached men, who offered more attention than she desired. One was Brusaw, who drove the stage from Markleeville into Silver Mountain. Another named Crippin was jealous of Brusaw, and so intense was his feeling for Antoinette that he laid in wait under the brush along the road below town with his Colt cap-and-ball revolver. When the stage rattled by he shot at the driver. But between Brusaw's big overcoat and a wallet full of papers, the ball was spent before it struck him. Dropping below the seat, the driver held onto the ribbons while the horses carried him safely up the hill.

As for Crippin, he was so disconsolate that he went into an empty building in the lower part of Silver Mountain and shot himself. According to one observer, "I think he was the first one they buried out there in 'Boot Hill'."

It was not long before Lewis Chalmers, a widower for at least a dozen years, took more than an employer's interest in Antoinette. Though more than fifty years old and over thirty years older than his housekeeper, he retained his personal charm and imposing presence, and must have been considered the greatest "catch" in Alpine County.

But though Nettie—as he was soon calling her—became more than a housekeeper to Chalmers, she was not quite his woman. Amply aware of her own charms, she knew how to play Chalmers like a harp. She complained of the work, and Lewis hired a Chinese man to do that. Bored with Alpine County, she claimed the mountain climate was harming her health, and wanted to visit San Francisco.

In 1877 Chalmers paid for her and her baby to call on Mrs. Kelly in San Francisco. But then he filled himself with fears of whom she might be seeing in the city of wicked temptations. In June she moved into the new and opulent Palace Hotel, then the glittering center of San Francisco society. Alpine was soon buzzing with what Chalmers called "nasty rumors". On the 19th he sent a frantic wire:

"Money scarce. Better come Home at once."

Then he sent enough funds to pay the fare back to Silver Mountain. With it he wrote of his anguish, his loneliness—"the evenings long" without her. By late August she had moved from the Palace to the Baldwin, still complaining about her health but hinting that she was coming back. The night of September 3 Chalmers waited anxiously for the weekly stage from Carson City, but it rolled past without stopping at his house. Next day he sent more money for her return, promising her "no work to do so long as I can afford to pay a chinaman." Then he waxed playful.

"The sooner you come to bossie, the better it will suit me. . . . Hasten home, and let me nurse my wee pettie well and strong again and then, no more bad girlies."

And he closed with the admonition, "Never again doubt the truth and love of yours truly."

By October, Nettie had returned, but their lives were darkened by new financial tribulations. For years Chalmers had lavished company funds on mines, mills, furnaces, roads and buildings, including his own "mansion". The long and severe winters that blanketed Alpine in white had limited the production of bullion to little more than half the year. And in the good months, much running time had been lost while new furnace processes and stamp batteries were added.

Chalmers had not recognized the essential link between income and expenses. A certain red-ink period for development is expected in any mine, but investors cannot accept this year after year. Unfortunately, Chalmers mistook great activity for good business. So high was his confidence in his own persuasive powers that he believed, when the money would run out, he could always get more. At best, he had simply been bitten by that virus endemic among miners—unlimited hope. And this curious faith in the pot at the end of the rainbow led him on to disaster.

By 1877 Chalmers had not only used up the funds authorized from London, but had sunk the Exchequer deeper into debt. On May 29 the creditors secured judgments against the Exchequer, IXL and Isabelle mine properties. The sheriff got ready to sell them at auction on the steps of the courthouse, which by then had been erected at the new county seat, Markleeville.

With the miners owed weeks of back pay, work was stopped at the Exchequer. Outraged, they descended on Chalmers at his home in mid-June. They would, they said, tie a rope around his neck and set fire to the hoisting works and mill. For two nights he confronted a band of heavily armed men. But on the promise of being paid next week, they withdrew. By June 27, with no pay in sight, Chalmers wrote his friend Syme:

"There are lynching threats."

In this crisis, blame was freely exchanged. In July the Exchequer stockholders, meeting in London, attacked Chalmers and called him a "bad manager". Chalmers blamed the company for not providing enough funds to sink the mine 600 feet deeper. Some of the directors blamed the Exchequer furnace for not yielding enough bullion. Chalmers blamed "the ore and the ore only"—as though $4.00-per-ton ore from the mine he had purchased was no fault of his.

"Everyone feels sore and are doing their best to ruin me," he wrote. "I am ruined."

But the doughty Scotsman was not without devices. His attorney got the sheriff's sale postponed week after week while Chalmers tried to raise the funds. He offered to sell the mines to visiting investors. In July he played host to Adolph Sutro, who would soon be coming into new riches as his famed tunnel into the Comstock Lode was nearing completion. Though Sutro wasn't buying, Chalmers took the opportunity to ask him for a job.

Working with Chalmers to salvage the Exchequer, his friend Henry Syme sent a mining expert from London to confirm its value and justify more funds. But besides proving himself an obnoxious guest, the man badmouthed Chalmers across Alpine County, wrote a negative report to London, and returned home. Chalmers, always the fighter, resolved to counter with another trip to England.

Near the end of the year, in the middle of wintry storms, he bade goodbye to Antoinette and sailed from San Francisco. Arriving at London in January 1878, he sought out his old mentor, Henry Syme. Back in Alpine the atttorney, N.D. Arnot, had secured a large reduction in taxes due and, in the name of his wife, had bought the properties at sheriff's sale, presumably to hold them for Chalmers.

Besides working to keep the mines, Lewis was also worried about keeping Antoinette. Fearing she would tire of the dreary wintry life alone at the Exchequer mill, and that other suitors would captivate her, Chalmers wrote from London suggesting she wait in San Francisco with her friend, Mrs. Kelly. But Antoinette, knowing he would be among old friends in England, affected to interpret this as an attempt to get rid of her. Back to London came a letter implying she had taken up with a passing suitor and might leave with him. Frantic, Chalmers wrote back on March 19 imploring her to stay "at least till my return," and promising to come back "at the very earliest opportunity."

"I am so much accustomed to disappointments now," he wrote, "that I am astonished at nothing—but this intimation of yours was not what I expected of you."

Crossing his letter was a more positive note from Antoinette. Having lost sleep fretting about her, Lewis now renewed his plea for her to go to San Francisco, away from "your old persecutors and any one else who may come along."

"I will be home in a little time and will try and make up by increased kindness for all you have had to put up with during my absence even if unfortunately you have made a mistake...I will forgive and forget, and will be as kind to you as ever....I am your friend...do not throw yourself away."

Antoinette responded by moving to her friend's home in San Francisco. From there a cable told Chalmers that she was "sick, send money". Lewis arranged for his lawyer to give her more funds, and on April 19 wrote more pleadingly:

"I beg of you, whatever you do—don't throw yourself away—at least till we meet—for come what will—even as a beggar I will keep my word and return."

To Chalmers' frantic entreaties Antoinette replied with short notes that brought him even more alarm. She asked what he intended doing for her.

"I intend returning as soon as I possibly can to see after you."

And though he had not proposed marriage, he asked her to stay with him for a time in San Francisco before returning to Alpine. While he had raised some $30,000 in London, he wanted to stay in hiding to avoid creditors while straightening out his affairs. As he prepared to leave England he cautioned Antoinette to tell no one of his return to California.

"You had better not meet me at the wharf—in case you may be followed."

Early in June, Chalmers returned to San Francisco and to Antoinette. For the next five months he dealt at arms' length with his attorney, Arnot, on terms of settlement, first offering 40 cents on the dollar, then half cash and half debentures, to clear the debt. Finally, having sent Antoinette back to Alpine, he felt secure enough to follow in December.

Until now Antoinette had played her cards well. Love-struck, Chalmers was alternately jealous, possessive, fretful—in short, beside himself. When the census-taker came to their door in June 1880, his age was given as fifty (he was fifty-five) and hers as twenty-seven (she was twenty-three). In ingrown Alpine County, where everyone knew everything about everybody, they might not know the full difference in their ages.

If eyebrows were raised and tongues whispering over this attractive pair living under the same roof, the situation was relieved on November 30, 1880. Lewis and Antoinette were married at Grace Church Cathedral in San Francisco. Her wedding dress, the color of "ashes of roses", was stunning in its frills and furbelows.

For the next few years the Chalmers couple was the talk of Alpine County. As at Fraserburgh in Scotland, Chalmers was the leading citizen of Silver Mountain—and the county. Though less popular due to his defaulted payrolls, he nonetheless evoked a certain awe as a charming and cultivated European. To the Alpiners he became "Lord" Chalmers, and Nettie, "Lady" Chalmers. To the happy pair was born a son, in 1881, and a daughter, in 1885. Outwardly, there seemed no end to the bliss of this exotic couple.

In fact, Lewis had not redeemed the past he had left in Fraserburgh. Besides the rebellious ores, his problems were compounded by the dropping price of silver after Congress stopped the minting of it for domestic coins in 1873. From an average of $1.32 per ounce in the New York market in 1872, the price slid to $1.11 by 1884. To save expenses, Chalmers operated the mines and reduction works with a skeleton force.

His faith in the mines undimmed, Chalmers tried to get West Coast experts to come and write favorable reports. But they were less than enthusiastic. Heavily in debt to several creditors, he was able to sell part of the Exchequer holdings to a local resident, John Weiss, who advanced funds to support the Chalmers household. Without him, Chalmers wrote Syme, "I and my family would have starved here."

By December 1883 he was hard-pressed to keep food on the table. On the 10th he wrote a local farmer:

"I would give you a breech loading rifle in exchange for a good fresh cow."

To his friend Syme, who had been trying vainly to raise money from investors in Paris, Chalmers confessed his final loss of faith two days before Christmas.

"I am much afraid, Henry Syme, it is all up and that our 15 years of hard work has been thrown away."

By January 13, 1884, with the mines shut down and creditors demanding a sale, Chalmers advised Syme to abandon the whole business.

"Only I wish you would give me a week's warning," he added, "that I may walk out if possible with a whole head."

A week later he turned the knife in the wound.

"I cannot endure it. Mrs. C. is killing herself cooking, washing, scrubbing & had a fainting spell 2 days ago. She can not stand it, but does not complain when well."

With no word from Syme by the 26th, Chalmers wrote to ask, "Have you thrown up the sponge?"

Getting no reaction from London, Chalmers somehow kept the Exchequer properties afloat under his own name. And he determined to make one more money-raising expedition across the Atlantic. For his family's protection he engaged Dominic Bari, one of his miners who had emigrated from Italy some dozen years before, to become head of the household at the Chalmers home. Bidding goodbye to Antoinette and their children, Chalmers took the stage on the first leg of his last foray to London.

There he found, in his words, "all my old rich personal friends dead or gone away." His mining companies were unable to pay even the wages of the few remaining employees, including Bari, who advanced funds from his own pocket for the care of Antoinette and the children. Adding to the melancholy, the young son of seven years drowned in Silver Creek.

At least once, in 1891, Chalmers sent some English investors to examine the mines and workings in Silver Mountain, but nothing came of it. While he wrote that he hoped "soon to be at the mill to put matters straight once more," he never arrived. Bari and Nettie continued to pay taxes on the home property. In 1898 attorney Arnot got a judgment against Chalmers and bought the mill and home site at Sheriff's sale in 1900. To Nettie Chalmers he was good enough to deed the home and garden.

By this time Bari still lived in the house, though Antoinette's son by her first marriage was now twenty-five years old. The Chalmers daughter, Laura, was a schoolgirl of fifteen. Another Italian miner lived there as a boarder.

At the Exchequer mine, ore cars, drills and foundry equipment lay scattered about, picked over by prospectors and souvenir hunters. At the mill on Silver Creek, the towering brick chimney cast its shadow over the Chalmers mansion and a large pile of ore not worth the firing. Silver

Mountain itself was a ghost town. As early as 1885 a visitor could find only one inhabitant. By this time the quest for a means to reduce the stubborn Alpine ore profitably had been made academic by the dropping price of silver.

Meanwhile, from Chalmers came nothing. Like Madame Butterfly, Antoinette waited while the months passed into years. According to one observer, "The disappointments that were showered upon her and the poverty have partially unsettled her mind." A girl friend of Laura's later commented about Antoinette:

"She had odd spells. She could be very odd."

In 1904 Lewis Chalmers was living in what his obituary called "comparative obscurity" in the north end of London. In February, after being ill for several days, he died of a heart attack at the age of 78. The writer of the long obituary in the *Aberdeen Journal* was unaware that Chalmers had lived in California for eighteen years, or that he had abandoned a wife and child there for another nineteen years. He only knew that in forty years' exile from Scotland, Chalmers had never returned to "redeem my past and claim my position."

So rich a heritage, so promising a mind, so winning a personality—he baffles us with his fate. Yet his rise and fall seem strangely inevitable—a victim of his own ambition and his own charm.

Frank A.S. Jones, Chalmers' old adversary, had fared better in life. In 1874 he took his profits from Alpine mining and, with his wife and daughter, moved to the town of Washington, in the Northern Mines of the Mother Lode. Three years later they settled in an imposing home at Chico. A respected citizen of Butte County, he died at the age of 68 in 1895.

By 1910 Antoinette's children had left Silver Mountain—the son, Henry Laughton, to Markleeville and the daughter, Laura, to live with her husband in Oakland. Now fifty-three, Nettie was listed in the census as head of household. Dominic Bari, ten years older, was listed as a boarder, rather than household head. He made a living doing odd jobs.

Around the middle of 1912 Laura disappeared from her home in Oakland. Antoinette had already spent a large part of her life waiting for the return of her vanished husband. Waiting again for her daughter to reappear was too much. On November 8, 1912, she sold the Chalmers mansion to Mrs. R. K. Whitmore of Ceres, California, for $10 down and $400 within a year. To Mrs. Whitmore she gave copies of all the letters written by Lewis Chalmers between 1867 and 1884. After living for a while in Markleeville she moved to a room on Piedmont Avenue in Oakland, where she vainly tried to locate her daughter.

On the night of December 28, 1913, Nettie locked the door, stuffed cloth and paper into the keyhole and the door and window casings. Then she turned on the gas jets of the chandelier and lay down on the bed with a photo of her daughter in her arms. Later, when gas was detected in the hallway and the door was forced open, they pried the photo from her hands. On its back was a note:

"Daughter, for God's sake come home. I have had months of agony such as only a righteous person can feel."

Nettie's remains lie in a family plot near Silver Mountain, alongside those of her boy who had drowned in Silver Creek. Also buried there are the remains of her first son, Henry Laughton, and those of Chalmers' first son, who had visited from Scotland and died in Silver Mountain at the age of twenty. The fate of Antoinette's daughter is unknown.

Throughout Alpine, the mining towns that once rang with the cry of silver have faded to oblivion. At Monitor, slag piles are all that remain where the bustling street paralleled the creek for a mile-and-a-half. At Mogul, only the imposing rock dumps spread themselves below the mouth, high on the mountain, of the Morning Star—the richest producer in the county. In Silver Mountain only the walls of the jail and traces of a few foundations stand among the noble pines. Down the road is the Chalmers "mansion" and the Exchequer mill chimney, silent for more than a century.

On the mountains above Monitor and Silver Mountain, the siren silver ledges still look down, defying men to take them at a profit. On the middle fork of the Carson the charming town of Markleeville lives on—the smallest county seat, in the least populous county, in California. Through it all, the turbulent East Carson swirls and splashes its way through the forest, just as it did when the people of Alpine were measured, not in hundreds, but in thousands.

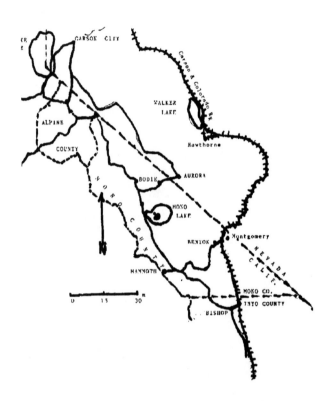

4. THE TOWN WITH NO LAW

>   While other silver camps flared and died, Benton lived high for fifteen years. But not without sharp ups and downs. Pap Kelty preempted the whole town, and almost got away with it. And the two biggest mining companies—the Comanche and the Diana— squared off in a dispute that began with an underground explosion and ended with a grand bilk of Comanche creditors.

In 1861, while gold and silver were on everybody's minds in Mono County, Eugene Conway Kelty settled on pasture land and began grazing stock in Hot Springs Valley. Adjacent was Blind Springs Hill, with one of the biggest silver deposits in the Eastern Sierra waiting to be discovered. But "Pap" Kelty was a cattleman, freighter and stage operator. He was one of the few who wasn't looking for "another Comstock".

Kelty was a man to match this big country of endless meadows and towering peaks. Although jovial with his many friends, he was resolute in pursuing his aims, dogged in defending his rights, capable of great risks, and strong in enduring frontier hardships. Born in upstate New York, he was thirty-nine years old when he joined the Gold Rush in 1849. His ship, bound for the Golden Gate, ran afoul of a storm and cast its passengers ashore 100 miles south of San Francisco. Kelty trudged north through what was then a wilderness until he reached the Bay.

Dead broke, he found a friend in the city who loaned him $350 to start a delivery business with a horse and cart. In six months Kelty had earned enough to take steamer for Stockton and launch a stage business, carrying passengers to the Southern Mines of the Mother Lode.

In 1853 Kelty married Eliza J. Parker, who would bear no less than five sons. Her brother, George Parker, answered the call of gold in Aurora, Nevada Territory, though it was then thought to be in California's Mono County. In 1861 he induced Kelty to take land in the vacant spaces of that county. Herding his oxen and horses from Tuolumne County over the Sierra, Kelty left them in the charge of a staunch frontiersman named E.S.

"Black" Taylor, a half Cherokee who had been the companion of Bill Bodey when he had made the first strike at Bodie in 1859. Kelty went back to his family and business in Stockton. Taylor built a sod and rock cabin with a thatched roof about a mile northeast of the Hot Springs near Blind Springs Hill.

Around the first of January 1862, a man named Sankey killed a Paiute in the neighborhood of the springs and fled southward to Bishop Creek. With the local tribesmen inflamed, Taylor feared that Kelty's stock would be run off. Others warned him to escape, but he decided to stick it out.

At this same time the friction between white settlers and the native Paiutes in Owens Valley had burst into warfare. In a fight around the end of February at what later would become Independence, three whites were wounded and four Paiutes, including a chief, were killed. Their blood up, a party of Indians rode north to the Hot Springs, where the Paiutes were already on the warpath. Together they proceeded to attack "Black" Taylor.

Taking refuge in his cabin, Taylor stood off the attackers for two days, killing ten of them, according to an Indian account. Finally they set fire to his thatched roof, forcing him to run out. Pierced with arrows, Black Taylor paid the ultimate price for his duty to Pap Kelty. The Paiutes burned the cabin and drove off Kelty's stock.

George Parker then headed a party from Aurora to recover the animals, only to find they had been slaughtered. Kelty's first venture in California's last frontier ended in disaster.

By the spring of 1863, hostilities had ceased in Mono County, at least, and Kelty moved again to occupy the land. This time his agent was his brother-in-law, George Parker, who built another cabin, started mowing hay, and drove in horses and cattle.

Also arriving with the end of warfare were the silver seekers. In October 1863, E.P. Robinson and a companion struck silver in the north end of the nearby White Mountains, and the town of Montgomery was born. Around the same time, beckoning outcrops on Blind Springs Hill brought new prospectors. One Cherokee Joe struck pick into galena ore in the summer of 1864.

By early '65 the hill was attacked by a squad of miners. On March 23 they formed Blind Springs Mining District, and among the first claims recorded were the two that would become the bonanzas of the hill, the Comanche and the Diana.

When the news reached mining circles in New York, the firm of Sierra Blanca Silver Mining Company sent out Dr. A.F.W. Partz, expert assayer and mineralogist. In April he located the Elmira, whose first ore samples assayed over $1,000 per ton. At this, the company president arrived to purchase more mines. Dr. Partz promptly founded, in company with others, the town of Partzwick, half a mile north of the Hot Springs.

In October, with men rushing in from Montgomery and Aurora, one arrival wrote, "new buildings are going up rapidly." Among the earliest from Montgomery were Pat Reddy, who would later be Belshaw's lawyer in the Cerro Gordo battle, and his kid brother, Ned Reddy. Together they bought and sold mines as a better way of making a living than digging in them. Already the Diana was turning out bullion with a four-stamp quartz mill, powered by hot water from the springs. By December, Partzwick had a hotel, store, company offices and several saloons.

At this point, there was yet no town at the Hot Springs, though some of the miners working on Blind Springs Hill camped there as the closest source of water. A sod house was built near the springs in the summer of 1865 by Kelty's agent. Early in March 1866 Kelty, arriving from Stockton with his family, moved into the house and began managing his holdings. Catching the silver fever, he started investing in mines and soon was elected recorder of the mining district.

By this time, residents of Partzwick had petitioned for a post office at a place by the Hot Springs that they named Benton in honor of a Rev. J.E. Benton, who later became the postmaster in Oakland. It was obvious that Benton, being next to an ample supply of water and to the district's only mill, would be the main town. Partzwickers began moving their buildings to Benton, and as other miners marched in from other camps, it became a full-fledged town before the end of 1866.

Through the late Sixties the mill was only processing ore from the Diana Mine. All the other miners on Blind Springs Hill had to freight their ore by team and wagon the 140 miles to the Central Pacific Railroad, whence it was carried to the Selby works in San Francisco—rich ore, indeed, at $90 per ton for freight.

Among those lured by such returns were the brothers Albert and August Mack—like Kelty, Forty-niners by the sea route. They were born in the 1820s in Wurtemburg, then one of many small separate kingdoms before Germany was unified. With a third brother, Theodore, they emigrated to the United States in the 1840s. Armed with their industrious German heritage, the brothers Mack knew they would succeed in America. But they had additional traits—willingness to take great risks, and persistence in the face of setbacks.

When they heard the call of California gold, they shipped from New York on the diminutive steamboat *Orus* in December 1848. Some 93 fortune hunters had bought tickets, but on seeing the little craft, all but fourteen refused passage on the fear that she would never reach the Isthmus of

Panama. The Mack brothers took the chance, reached Chagres safely, and crossed the Isthmus by rowboat and foot power. At Panama they caught the steamboat *California*, reaching San Francisco on the last day of February 1849—among the first Forty-niners.

During the Pacific passage, brother Theodore had taken ill, and died at Sonoma in May. Albert and August settled in Tuolumne County, digging enough gold to launch a general store.

Early in 1850, still learning the rough ways of the Californians, Albert bought a horse from a Mexican and rode it into Sonora. He was promptly arrested for stealing the horse, while an angry group gathered amid threats: "Let's hang the damned horse thief!" But Albert was hauled before the famous Judge Richard C. Barry, who ruled the Sonora court with an iron will unencumbered by knowledge of the law.

"This young man don't look to me like a horse-thief," declared Barry.

Putting up $2,000 bond, Albert rounded up five witnesses to his purchase and reappeared in court next morning. Barry acquitted him of the charges, but true to form, made him pay the court costs.

"It was," Albert later wrote, "a narrow call from unnatural death."

Through the 1850s the Mack brothers prospered, adding more stores, a saw mill, a stock ranch and a pack train. Albert, three years younger than August, had a way with people and a flare for politics. In 1857 he was elected a county supervisor. When he was able to send money to bring his sister from Germany to California she arrived in company with a woman friend from Philadelphia whom she had met on route. The new friend, Elizabeth, soon became Albert's wife.

With the gold strike that spawned Aurora, Nevada, the Macks trekked over the Sierra and opened a store in that booming town. Soon Albert, again in politics, was elected an alderman. In the fall of 1864, Dr. A.F.W. Partz, on his way to the new diggings in Blind Springs Hill, came through Aurora. There he became acquainted with Albert in his store.

"In case I find some place down South," he told Mack, "I wish you to put up a store and I will insure you all my trade."

Visiting the new Blind Springs District at the end of 1865, Mack renewed acquaintance with Pap Kelty, whom he had known in Tuolumne County. By September 1866, in response to Dr. Partz's invitation, Mack opened the store in Partzwick. Next year he and his wife moved from Aurora to the new town of Benton. Still unable to resist politics, he was elected Justice of the Peace, the only elective office in town. As for business, Albert Mack soon proved himself a master opportunist.

In 1868 Dr. Partz's principal mine struck a fault and the silver bearing ledge was lost. With no more funds coming from the East, Partz sold his reduction works and other property to Mack for $3,800. Mack proceeded to run much of the ore that the "chloriders", the independent miners, were shipping to Reno. Later Mack wrote:

"My wife became mistress of the fine mansion and we were getting along swimmingly."

Then in 1869 the famous silver strike at White Pine, Nevada, drew many of the Blind Springs Hill miners, and Mack bought some of their claims. One of his acquisitions was the Comanche, which soon became a silver producer famous throughout California.

Using a new process, Mack was able to work the most refractory ores in a small mill he purchased at the declining town of Montgomery, packing the ore seven miles from the mines. After that he acquired a ten-stamp mill, followed by a roasting furnace. And the Diana Mining Company acquired a larger stamp mill; its ruins are still evident behind the town.

By 1872 all the bullion was produced locally, the mines were profiting, Benton was booming and a new wave of miners and merchants rode in. George W. Hightower had arrived from Owens Valley and built a growing business with his blacksmith shop, gun store, machine shop, sawmill and lumber yard. In 1874 came young George W. Rowan, a clerk and bookkeeper by trade, who would later become Benton's leading storekeeper. By that time Benton had two hotels, two general stores, three saloons, four commercial stables, Hightower's businesses, and numerous private homes, including a small Chinatown.

Through the early Seventies, Albert Mack was the big man of the district. In 1870 he gave the main oration, in his German accent, at the Independence Day celebration. Besides the biggest mine in the district, he also owned a general store and the hotel in Partzwick. One acquaintance called him "the legitimate founder of Benton."

By mid-1875 Mack's Comanche shaft was down 650 feet, following a vertical vein of silver ore that was four-and-a-half feet wide. Visiting the mine, a correspondent of the *Inyo Independent* reported that "a real bonanza has appeared in Blind Springs Hills."

Early in 1876 the prosperity of Benton and Albert Mack attracted attention among San Francisco's mining magnates. One was William M.

Lent, previously a California state senator and soon to become the principal owner of Bodie's famous Standard Mine. Another was Gen. George S. Dodge, a heavy speculator in Nevada silver mines and another soon-to-be plunger in Bodie properties.

Both men had been taken in by the Great Diamond Hoax of 1871-2, and were taking no chances now. To Benton they sent professionals to examine mines and mills. On a favorable report, they offered Mack $125,000 for all his properties except the Kearsarge, an extension of the Comanche, which Mack held out of the deal, believing it to be "my best mine."

Taking stage and train to the Bay, Albert and Elizabeth met with Dodge in his San Francisco office in April 1876. Dodge's wife happened to be present at the time. Mack asked Dodge to make out a separate check for $50,000 to Elizabeth Mack, who was a business woman in her own right, investing in Bay Area real estate. Handing her the check, Mack made the playful gesture of asking the others to witness this as a gift from him to his wife. At this, to Mack's amusement, Dodge's wife stepped up to her husband and tapped his shoulder.

"George, this is a good husband," she said pointing to Mack. "Why don't you do the same thing to me?"

Taking her seriously, Dodge answered, patronizingly:

"My dear, I paid this out, I didn't get it—see the difference?"

Everyone was smiling through this awkward moment and, as Albert later wrote, "I felt very proud."

Elizabeth Mack entrained for the Philadelphia Centennial Exposition, soon to be followed by Albert. Then they settled in San Jose, where Elizabeth owned property. For a time they lived in a mansion six miles east of the town, with a grand view of the Coast Range and part of the bay.

"We had all we could wish for," wrote Mack. "A fine double rig, a splendid pair of horses, a billiard room, an English coachman, and lots of friends."

As for Benton, the news that Lent and Dodge had bought into Blind Springs Hill sparked a new rush of miners and speculators from the Eastern Sierra. Dodge and his colleagues incorporated the holdings into the Comanche Mining Company, with a capital stock of $200,000—meaning more money would be invested in mines, mills, and jobs.

Suddenly, hotel and saloon business jumped, and new buildings were going up to the sound of saw and hammer. Main Street was jammed with freight teams bringing goods for new stores and produce from Bishop Creek farmers. Remi Nadeau, who had moved his southern terminus to the nearest rail point at Mojave, could now profitably send his sixteen-mule teams as far north as Benton. They added to the dust, noise and confusion that meant prosperity to the town.

California State Library

German immigrants in the Gold Rush, Albert (left) and August Mack were among the leading mine operators at Benton, Mono County, in the late 1860s and early 1870s. In this portrait, the empty chair honors their brother, Theodore, who died soon after arriving in California.

Demila Jenner

Starting in 1876, E.C. "Pap" Kelty tried to preempt land that included the whole town of Benton. Feeling among the townspeople was so strong that Kelty needed a squad of bodyguards to appear on the street. Ridge behind town is Blind Springs Hill, which produced some $5 million in silver.

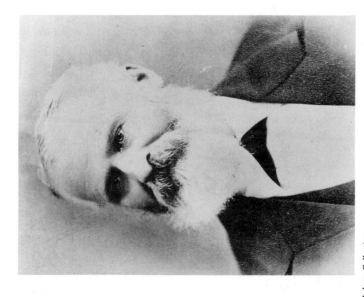

Nadeau Collection
Remi Nadeau's mule teams dominated the freighting business in Eastern California from 1868 to 1882.

California State Library
William M. Lent, the "sharpest of the sharp", was the biggest stockholder in Benton's Comanche Mine.

At this point Albert Mack's departure had left Pap Kelty the leading figure in Benton. According to a later eulogy by a Benton editor, Kelty "was perhaps as widely known and as popular as any man in the state—in Benton...he was simply beloved by all."

From the time he first settled in Hot Springs Valley, Kelty had only exercised squatters rights, without legal title. Neither had anybody else who followed him into the valley. This was partly due to Western informality on legal matters, and partly to the absence of any government land surveys in Mono County. As in many other mining camps, ownership stemmed from possession—"the right of shotgun."

In 1864, through an agent, Kelty had asked the county surveyor to make a survey of the Hot Springs Valley township. It was done in 1865 and filed January 17, 1866, in the nearest Land Office at Stockton. But it was not known in Mono County, and the plat was not transferred when the new Land Office governing Mono County was opened in Aurora, Nevada, in 1868.

Through the following years the town of Benton sprang up and flourished. Buildings were erected along a main street. Town lots were bought and sold. Bob Morrison, the first to move his store from Partzwick to Benton in 1866, got a plat map of the town, but it was unofficial, without benefit of a government survey.

In the summer of 1871, Pap Kelty rode the forty miles to Aurora. Finding no one in the Land Office, he stepped into the street and came back with the local blacksmith, who was doubling as Register of the Land Office. Kelty told him he wanted to file a preemption claim (not a homestead, but an application for fee title). He asked to see the map for the Hot Springs Valley. The Register said there was no such map in his office—"never has been." He had applied to Washington for one, but none was forthcoming.

But from a tract book of previous entries they derived some numbers for Kelty's claim. He filled out a declaration of preemption, paid his fee, and returned to Benton. The filing was not recorded or reported by the Aurora office—hardly unusual for one that was notorious for loose operations. So Kelty still had no legal claim.

In 1873 the Aurora office was transferred to Independence, Inyo County. On February 24, 1874, Kelty walked into the new Land Office and filed a homestead claim for the 80 acres he considered his land. The acting Register told Kelty that the homestead law allowed 160 acres—a quarter section.

"It would cost no more to take 160 acres than 80 acres."

Kelty allowed the added land was "not worth much."

"Bad things," responded the Register, "sometimes become pretty good."

"All right," agreed Kelty.

Then the question arose about anyone else "holding this land." The Register was still accommodating.

"If anyone wanted it," he observed, "why didn't they come and take it, as the land was open on the books of the office."

Clearly the Register did not know that the claim included the entire town of Benton.

As for Kelty, admired in the Eastern Sierra for his lifelong energy and hard work, something inside him snapped. At 64 years old, he was tired from the years of driving freight teams through snowstorms, waterspouts and sandstorms. He knew the claim covered the town of Benton, but said nothing.

For nearly three years no one in Benton knew that Kelty had homesteaded their town. In July 1876, for good measure, Eliza Kelty filed another homestead claim on the same land. Then on December 27, 1876, Kelty rode again to the Land Office at Independence. This time he was ready to "commute" the 160-acre claim to a full preemption title, as allowed by law.

By this time the Register was R.M. Briggs, later the Mono County Judge. Also unaware that the quarter section included Benton, he took Kelty's payment of some $200, recorded the transaction, gave him a certificate of title, and forwarded the documents to Washington for a patent on the property.

Early in January 1877 the people of Benton discovered that Kelty was their landlord when surveyors showed up to confirm his boundaries. The uproar that followed was led by the young clerk, George W. Rowan, and the blacksmith-lumber dealer, George W. Hightower. They immediately rallied the Bentonians to outraged protest. Rowan, given to vehement language, called Pap Kelty a "land shark". With all property ownership in limbo, prices of lots tumbled and they could hardly be sold at any price.

Benton had a lawyer, but its citizens wanted one at the Inyo County seat, close to the Land Office. North from Independence to Benton rode Paul Bennett, senior member of the firm of Bennett & Moffat. On hearing the angry grievances he drew up an equally angry affidavit.

On January 15, 1877, six men gathered in the local attorney's office and signed it. Most of them signed it in a rage without even taking time to read it. Calling Kelty's preemption "fraudulent and void," they asked the Commissioner of the General Land Office in Washington to delay granting the patent until the townsite of Benton could make its own application.

Bennett rode back to Independence and submitted the affidavit at the Land Office.

As it happened, Register R.M. Briggs was in San Francisco on another matter and had placed his legal affairs in the hands of Bennett's law partner. Without referring the issue to the Receiver of the Land Office, Thomas May, the law partner immediately forwarded the affidavit to Washington with no recommendation by the local Land Office. Even after Briggs returned he was unaware of the situation until he received a letter from the U.S. Commissioner ordering him to suspend Kelty's patent until the facts could be discovered in formal hearings.

Briggs' anger was only exceeded by his embarrassment at this "irregularity"; he swiftly wrote an apology to Washington. And on April 1 Briggs notified all parties that the hearing would start April 23, 1877, in Independence.

Immediately Benton plunged into heated preparation. The townsmen were canvassed for contributions to a fund for the legal fees. It soon appeared that sentiment was not unanimous. Pap Kelty had been well liked and respected. One friend said they had "always looked to Mr. Kelty as a leader." Later the same year he was made a member of the local election board. Besides his freighting business, he owned several properties (Eliza ran a two-story hotel) and had an interest in some mines. His four sons (one had died) were taking their places in the life of Benton. The Keltys were woven into the fabric of Hot Springs Valley.

So it was that not every Bentonian readily subscribed to the legal fund. And gossip said that some would testify in Kelty's favor. Savage rancor seized the town of Benton. A decision against the citizens, wrote the *Inyo Independent* correspondent, "will completely bankrupt the majority." Pap Kelty kept a group of men—estimated at twenty in number—around him whenever he ventured onto the streets.

George Hightower was making the rounds for subscriptions to the legal fund when he walked into Joe Hutchins' saloon. When Joe refused, Hightower said that if Joe would not contribute he would hurt him in his business. But the only effect that Hutchins noticed was that Hightower, who had been a frequent customer, stopped coming in.

Writing to Jim Shaw, who had moved to Bishop Creek, George Rowan informed him that he knew Shaw and his brother Frank would testify for Kelty. The Bentonians had asked Washington for a stay pending further information, he wrote, and "if this thing was followed up it would bring Mr. Kelty and his witnesses before the U.S. Court for perjury."

Such was the feeling in Benton when the six contestants, together with their witnesses—a large part of the population of Benton—rode down to Independence for the hearing on the appointed day. But Register Briggs was absent on another case in Kern County. After waiting two or three days in Independence the disappointed contestants rode back to Benton.

Then the Register set June 3 for the hearing. Back to Independence rode the contestants. This time Kelty and his witnesses were not there. But, pressured by the irate Bentonians, Register Briggs started the hearing on June 4.

It was the first of many days of testimony. The Keltys themselves never appeared, contenting themselves with depositions after the hearing. After losing most of their old friends, the Keltys did not relish meeting them face to face. But, having taken his hand, Kelty would stay the game.

In the crowded quarters of the Land Office, the evidence was heard by Register R.M. Briggs and the Receiver, Thomas May. The hearing was conducted like a trial, with a display of courtroom punctilio, although Briggs followed the rules of evidence with considerable creativity. By now Kelty's legal battery included the champion attorney of the Eastern Sierra—Pat Reddy. He attacked the testimony of the Bentonians with frequent objections, which were usually overruled.

From June 4 to the 27th the hearings were held in Independence. Final testimony was taken in Benton on July 23 and 24.

Witnesses for Kelty said he was the first on the ground when he settled on March 6, 1866, in what became Benton. Witnesses for the Bentonians said there were others there then.

Lawyers for the townspeople invoked the legal provision that a townsite took precedence over individual entries. Lawyers for Kelty insisted the Bentonians had never sought to establish legal title.

Since agricultural use was required by the Homestead Act, witnesses for Kelty said the land was suitable for farming and that he grew vegetables on it. Albert Mack, who had known Kelty for twenty years, testified with apparent exaggeration that his friend had two or three acres under cultivation. Even Eliza Kelty took a hand and stated that, from the cows pastured there, she "made a sufficient amount of butter for the use of the family, and some years butter to sell."

Witnesses for Benton said the land was useless for agriculture without irrigation (the only spring in the vicinity was owned by the Comanche Mining Company north of Kelty's claim). At one point lawyer Bennett asked George Rowan whether or not, without irrigation, "will vegetation enough grow to make gruel for a sick grasshopper?"

"Without irrigation," answered George, "there will not anything grow but sagebrush and desert-weed."

Lawyers for Benton argued that a homesteader had to file within thirty days of settling on the land, or if no survey had been made, within three months after the plats were available. Kelty's homestead filing was delayed eight years after his settlement. Meanwhile, the town of Benton had appeared.

Kelty's lawyers said this rule prevailed only if another applicant intervened, and the whole town of Benton had been asleep. Witnesses for Kelty said he had not kept his claim a secret from the townspeople. But their witnesses contended that in repeated discussions about town lots and streets, he had never claimed his ownership.

Finally, an unexpected windfall came to the Bentonians. According to law, one could only use the homestead privilege once. George Rowan wrote to the Land Office in Stockton asking whether Kelty had ever done so before his Mono County filing. Back came letters of May 17 and June 23 revealing that Kelty had filed a homestead claim on a quarter section within Stockton's jurisdiction on November 21, 1863. Included in the correspondence was a copy of his entry certificate.

Failing to do the required labor, Kelty had found that claim challenged by another in 1870. When he ignored a hearing in the Stockton Land Office, his homestead filing had been canceled, November 19, 1871. This previous entry helped to explain why Eliza had later made a new homestead claim in her own name on the Benton property.

Making matters worse for Kelty, he swore in his deposition that "he has never availed himself of either the preemption or homestead laws of the United States until he claimed the land in contest in the year AD 1874." Whether or not his previous claim had been wiped from the slate when it was canceled is a moot question.

In response to Rowan's little bombshell, Reddy argued that it was not proven that the two Eugene C. Keltys were the same person, and that the Stockton information had been presented after the hearing had been closed. Such arguments might have prevailed in a court of law, but not in the Independence Land Office.

With the last depositions taken and the last briefs filed by the end of 1877, the Bentonians waited for the decision. After four months, George Rowan learned from a "reliable person" that all the testimony, exhibits, depositions and briefs had never been sent to Washington. At this the exasperated Rowan wrote to the Independence office demanding to know the status, and a second letter to the Commissioner in Washington, casting off all civility:

"We have been shamefully dealt with by unprincipled lawyers and Land officers who have perjured themselves and are bought and sold like sheep."

Then Rowan gave his version of the report that should have been sent from Independence months before. He embellished it with raw accusation. The people of Benton, he wrote, cannot understand why the papers had been held up in the Independence Land Office, except that money will corrupt office holders at all levels, and "the U.S. Land officers in this District have been found guilty of some of the vilest practices and with such an element to contend with what is a poor struggling Community of settlers to do? How are we to get justice?"

Warming to his task, Rowan fired both barrels.

> How are we to let your Hon know that a community of bonafide settlers are about to be robed by a land shark and soon turned out of our homes which is all we have? Can you answer these questions or is our case to go by default. And are we to be driven with our families into the streets by a land grabber. Is our case to smolder in the Independence office and die?

Spurred by this blast, the Commissioner in Washington shot a letter on May 15, 1878, to Register Briggs. Before this could reach Independence, Briggs and Receiver May had responded to the anger in Benton by shipping all papers, including their hastily written opinion, to Washington. So on May 26 Briggs was able to answer the Commissioner that the papers had already been sent. And he blamed the delay on a shortage of space for writing the opinion!

Meanwhile, copies of the opinion had been given the attorneys on both sides. The decision: "Cash entry No. 153 of Eugene C. Kelty...should be canceled...the County Judge of Mono County...should be permitted to...obtain title to the land in dispute, in trust for the benefit of the inhabitants of said town of Benton."

But the case was not yet closed. So hasty had been Briggs in sending off the papers that they lacked the proper formalities. After the General Land Office sent them back, it was August 1, 1878, before the papers were forwarded again to Washington. And Kelty's lawyers appealed, sending a new brief to the Commissioner of the General Land Office.

For still another year, the Bentonians waited without knowing their fate. The opinion written by the Independence Land Office was known only by rumor. The town continued in exasperating limbo, though not without change.

George Hightower, one of Kelty's opponents, had died in 1877. George Rowan opened a store in Benton and became a deputy assessor of Mono County. In 1879, spurred by new mining activity, Benton had some 23 businesses, including three hotels, five saloons, two breweries, and two newspapers, the *Bentonian* and the *Mono Messenger*. The population had grown from around 100 in the early Seventies to about 600, not counting the many miners who lived by their claims on Blind Springs Hill. And the object of the whole uproar, Pap Kelty, died April 14, 1879—surviving Eliza by a few months. A large part of Benton's population rode down to Bishop for the funeral.

Finally, on November 20, 1879, the General Land Office in Washington confirmed the decision and awarded the land to the Mono County Judge "in trust for the inhabitants of the town of Benton." Kelty's entry was officially cancelled on March 9, 1880.

By this time Register R.M. Briggs was the Superior Court Judge of Mono County. On March 22 he notified the Bentonians that they could get their titles if they would raise the $200 for the land, plus surveying and court costs. The *Bentonian* urged them to raise the money and "at once secure title before there may occur any intervening obstacle."

Still the Bentonians were complacent. Fourteen more years went by before they applied for patents to their properties. By 1910 the General Land Office still had not refunded the money Kelty had paid for his preemption. In approving payment, one official wrote:

"The townsite claimants were more dilatory than Kelty in asserting claims, and of the whole mess, it appears to me that repayment can legally be made."

"Dilatory" was the kindest word for almost all concerned in the Kelty case—the various Land Offices, the Bentonians and Kelty himself. Most telling was the behavior of Land officials who granted Kelty his preemption without troubling to find the town of Benton on the site. But they were not the crooks that George Rowan had charged. Far from being controlled by lawyer Pat Reddy, Briggs had overruled almost all of his courtroom maneuvers, and in the end had decided for the Bentonians.

Actually, Kelty's legal case was as strong as the Benton case, though his plot to seize the property of his neighbors was deplorable. In deciding against him, the Land officials were swayed more by equity than by law. Cynics might also observe that Briggs could hardly turn the voters of Benton into the streets when he had his eye on the Superior bench of Mono County.

Though Benton property values had suffered from the start of the Kelty threat, the mines and mills prospered—with some setbacks—well into 1878. Under the Dodge and Lent group, the Comanche mill was turning out silver bullion night and day. Wrote the correspondent of the *Inyo Independent*, "every stage is laden with the precious metal." In 1878 a professional report on a key Comanche property, the Tower Mine, showed an abundance of rich reserves. The Diana shaft, in ore all the way to its 400-foot depth, was feeding the mill with bounteous ore. Up on the surface, Bentonians were going their happy ways. In August 1876 the same issue of the *Independent* reported a grand ball and a theatrical performance in Benton.

At the annual meeting on October 1, the Comanche stockholders elected five directors of the corporation and confirmed the Anglo-Californian Bank of San Francisco, a British owned company, as the treasurer.

Most conspicuous among the directors was William M. Lent, who had joined General Dodge in buying the Comanche from Albert Mack in 1876. Born in New York City in 1818, Lent had been one of the first to heed the call of California gold. Late in 1848 he boarded the sidewheel steamship *Oregon* for the long voyage around Cape Horn. On April 1, 1849, the *Oregon* was the second steamer to reach San Franciso in the Gold Rush.

At the corner of Montgomery and Washington streets, Lent started as a commission merchant. But his interests were wider—pursuing investments from Mexico to Montana. In the early Comstock excitement he advanced funds to help develop several mines—including the Savage, Mexican and Yellow Jacket—that became bonanzas. Unlike most San Francisco speculators, he insisted on inspecting a mine before investing. When approached, his reply:

"Tell me what you've got; let me see for myself and I'll buy your mine."

In the Fifties, Lent entered politics in the David C. Broderick faction of the Democratic Party, serving a term as a state senator. But he would make his career in mining, not politics. Between the late 1850s and the 1880s Lent was, as one chronicler wrote, "one of the best known big mining operators in the West." First on horseback, then by stage and palace car, he visited most of the camps west of the Rockies.

Soft-spoken and friendly, Lent was hailed in all the camps as "Uncle Billy". As one admirer wrote, "his word was everywhere accepted as a certified check." His generosity was legion. At the end of a long stage ride over the Sierra he would invariably slip a $20 gold piece to the driver. He was liberal in giving hot tips on mining stocks to investors and miners alike, and was known to buy a stock on the rise for a friend out of town.

Yet there was a certain method in Lent's liberality. The stage driver would always assure him a good seat the next time he boarded a crowded coach. And the miner fresh from the face of a tunnel would, if he saw Uncle Billy, confide to him the discovery of a blind lead.

Actually, Bill Lent had two sides—the genial and honest friend beloved from the Bay to the Comstock, and a hard-headed operator when it came to turning a dollar.

"Uncle Billy was as sharp as the sharpest, " wrote one observer.

Just how sharp would be seen late in 1878 as the Lent group imposed its policies on Benton's Comanche Mine. While outwardly the company seemed more solid than ever, matters were not so pleasant under the surface. The superintendent of the Diana Mine suspected that the Comanche company, through one of its properties known as the Kerrick Mine, had tapped into the rich vein owned by the Diana. The Comanche was, he believed, making off with many tons of Diana ore.

William H. Sears.

To test this, he ran an adit northward from the Diana shaft for several hundred feet to the point of suspicion. There he found, not the Comanche stope, but a huge mass of debris, caused by the explosion of many cases of giant powder. Its purpose, concluded the Diana superintendent, was to hide the evidence of ore theft by the Comanche. The obstruction was so massive that, as he later wrote, he "did not have the heart to clean it up."

At once the Diana sued the Comanche for $150,000, the claimed value of the ore. It came at a time when the Comanche owed some $30,000 to various creditors and, in back wages, to its miners and millmen.

Heading the Comanche defense was attorney William H. Sears, a leading California political figure. Born in Connecticut in 1830, Sears was descended from 17th Century Puritans. At 18 he served as a seaman on a ship carrying relief to Ireland in the potato famine. In 1851 he joined the Gold Rush via the Isthmus, settled in the Northern Mines, and took up law. He was elected as a Republican to the state assembly in 1861 and within two years was chosen speaker of that house. There he worked to repeal a law prohibiting blacks from testifying in court. After 1865 he practiced law in San Francisco, and in 1868 chaired the California delegation to the Republican Convention in Chicago that nominated Ulysses S. Grant for President.

Thus Bill Sears was a man engaged in public service, including some noble causes. Another side of his character, and the character of the Comanche directors including Bill Lent, would now appear as Sears represented them in their moment of distress.

To answer the claims of their unpaid employees and the Diana Mine, the Comanche directors hatched an elaborate scheme. Speed was essential, since at any moment the creditors could descend on them.

On November 26 the board authorized a loan of up to $20,000 from their treasurer, the Anglo-Californian Bank.

On November 27 they signed a note to the bank for a loan of $14,971.45 in gold coin, for one day.

On November 28 the Comanche officers endorsed the note and delivered it to Anglo-Californian. The bank did not advance a new loan, but Comanche already owed a similar amount through its regular overdraft account.

On November 29 the bank started an action in the U.S. Circuit Court to collect the loan, obtain a judgment and attach all the Comanche property.

Thus in four days the Comanche directors deliberately put the company in jeopardy.

The judge then summoned William M. Lent, who was not a company officer, but the most prominent member of the directors, to appear in court. Representing him there was Bill Sears, who did not oppose the bank's action.

On December 3, 1878, the millmen arriving for their shift at the Comanche mill in Benton found the U.S. Marshal had taken possession of the works. They could only gather around and read the posted writ of attachment from the U.S. Circuit Court, pending a judgment in favor of the Anglo-Californian Bank. The Comanche, Tower and Kerrick mines were likewise seized.

Too late the Comanche creditors and employees discovered the truth. The Comanche had gone into bankruptcy and all its property was, until further notice, in the hands of the U.S. government and safe from creditors. But this was still merely Phase I of the Comanche scheme.

On December 20, 1878, the judge declared Comanche to be in default on the "loan".

On December 26 the court entered a judgment against Comanche for $15,873.06, including court costs.

Many creditors filed liens against the Comanche, and one of them hurried to San Francisco to get a settlement. In the city, one of the Comanche's largest stockholders, Robert Ellon, suddenly awoke to the danger. Rushing to the Comanche offices, he demanded that the president bring suit to set aside the judgment. This was refused. Instead, on December 27, the day after the judgment, Bill Sears waived the usual ten-day stay of proceedings and authorized immediate execution of the court order.

The public sale of the Comanche property was set for January 25, 1879. Nineteen creditors, including Albert and August Mack, brought suit against the Comanche on the day before the scheduled sale.

But it was a futile gesture. Not only the Diana, but miners and merchants all over Benton, were victimized. Wrote the *Inyo Independent* correspondent, "every person in the town has lost either in cash or business." Miners and millmen had been counting on paychecks to satisfy bills. Merchants were "swamped with bills against men who would be glad to pay but cannot." This meant simply that "men who are perfectly honest must take a walk"—in short, skip town.

Stage II of the Comanche strategy unfolded in Benton on January 25. From San Francisco came Bill Sears, acting for the Comanche trustees, who intended to buy the entire holding in the public sale and start over with a fresh corporation, unencumbered by debt and clear of lawsuits.

One William Balch, who had previously been placed in charge of the properties by the U.S. Marshal, was now hired by attorney Sears to bid on them, representing the San Francisco group. But when the bidding opened Sears retained control. By prearrangement, Balch watched the toothpick that Sears had in his mouth. If it was in the right side of the mouth, Balch was to continue bidding; if on the left side, he was to wait for a while; when the toothpick came out of the mouth, Balch was to stop bidding on that particular item.

The morning of January 25 was cold and drizzly. But the Bentonians turned out to watch, grasping their overcoats and standing for shelter under the Comanche mill roof. When the auction was over, Balch had bought almost everything, including the mill, offices and all three mines, for $16,432.50.

The real purchaser was J.C. Classen, one of the Comanche stockholders. He then deeded the property over to what was understood to be the same group that had controlled the Comanche. William M. Lent personally paid off the overdraft debt to the Anglo-Californian Bank, thus relieving it of any loss in the transaction. Within three weeks the enterprise started up again under a new name, Vulcan Mill and Mining Company.

By this time the elaborate ploy was well understood by the Bentonians. On the streets and in the saloons, accusations were rife. The Comanche, wrote the *Inyo Independent* correspondent, "took a short cut to swindle its creditors." How shameful, he wrote in another issue, "that a company consisting of individuals whose wealth is counted by millions, should be allowed, without even the taint of dishonor, to rob the men, who delved and toiled for them, out of their hard earnings."

On July 24, the stockholder who had tried to get the Comanche to fight the judgment brought his own action. Robert Ellon sued the old Comanche directors and officers, the Anglo-Californian Bank, even A. H. Poole, the U.S. Marshal. He charged that they "combined together to cheat and defraud" the stockholders. And he asked the judge to set aside the previous judgment and sale, and return the property to Comanche.

It was now Marshal Poole's job to subpoena each of the defendants. He served papers on all but Bill Lent, who was nowhere to be found. Poole was told that Lent was living in New York City. When the various other defendants appeared in court, Lent was represented by lawyer Bill Sears. Each of the defendants simply denied all of Robert Ellon's charges.

They did not, and could not, deny the swift succession of events, from the note in favor of the bank to the demand for the property, that occupied only four days' time. As for Robert Ellon, they had already taken care of him. On April 4, 1879, they levied a stock assessment of $1.00 per share. As the company had owned no mines or mills since the public sale on January 25, the assessment had no business purpose other than to freeze out the "outside" stockholders. Assuming Robert Ellon knew of the assessment—it was advertised in the *San Francisco Bulletin* and the Benton *Mono Messenger*—he could hardly be expected to pay $7,200 (a dollar for each of his shares) to a non-operating company that existed only in name.

This, of course, is precisely what the trustees had in mind. Thus when Robert Ellon sued them, they responded to the court that he was not a stockholder, had failed to answer an assessment, and therefore had no standing to sue.

Ellon threatened to pursue the suit but never prosecuted. The case stayed on the books until it was dismissed on July 18, 1887. Bill Lent and his friends had foiled their creditors in what today would be called a scam.

But they were not to reap any big rewards. The path of the new Vulcan Mine was rocky at best. In March 1879 the miners and millmen struck for more than a month's back wages owed them. The mill of the Vulcan ran only intermittently. Under pressure from the Diana, which was still pushing its case, the Vulcan company ceded the Kerrick mine in compensation for the pirated ore. In February 1880 the latest shutdown of the Vulcan mill brought new charges of "freeze-out" from the correspondent of the *Mammoth City Herald*:

"If a man steals because he needs money, they punish him, but these ghouls steal because they have money, and are honored for it."

By this time even the local editors—usually any town's biggest boosters—were complaining of dull times.

"In a business point of view," the *Bentonian's* editor wrote in February 1880, "Benton was never in a more apathetic state than it is today. There seems to be absolutely nothing doing."

As for Bill Lent, he had moved on to rich investments in nearby Bodie. Bill Sears, elected a state senator in 1880, was later rewarded for his service to the Republican Party. He was appointed collector of duties at the port of San Francisco in 1884, and was serving as Collector of Internal Revenue for Northern California when he died in 1891.

Between Pap Kelty claiming the whole town and the Comanche welshing on its workers and creditors, Benton's nerves were frazzled by 1879. Unusually peaceful for a mining town, Benton found itself full of quarrel and fracas. Where occasional gunplay had been the sport of transient ne'er-do-wells, it was now the leading citizens who resorted to violence.

One was Captain Jim Powning, who had come to California's Northern Mines in 1852 from England. By 1873 he was in Benton associated with Albert Mack, earning his captain's title as a mine foreman, and by 1876 was with the Diana Mine. In John Kremkow's brewery on the night of October 11, 1879, Powning was quaffing his share of the brew when he got into a fight and came out second best. His pride wounded, he left and soon returned with his six-shooter. Immediately he opened up on the crowd at the bar, drawing return fire. In the fusillade of perhaps sixteen shots, one man was shot in his rear and Powning was wounded in the thigh. When the shooting was over John Kremkow, who was a huge man of some 240 pounds, emerged from behind a two-and-a-half gallon beer keg where he had tried to take shelter.

Another leading Bentonian was James M. Millner, who owned one of the stores in partnership with John A. Creaser and was variously interested in Blind Springs Hill mines. On October 26, 1880, he and his brother went into the saloon of Dan Ashley, with whom they had previously feuded. After some heated words, Ashley drew his revolver and fired point blank. The Millners both pulled out their six shooters and fired back—all three parties emptying their guns within a few feet of each other. When the smoke cleared Jim Millner was wounded in the arm, and had been saved from a bullet striking just above the heart by a package of letters in his vest pocket. Besides another man who had rushed in to see the excitement and was shot in the leg, no one else was hit. Reporting the affair, a neighboring newspaper chided the *Bentonian* editor for failing to report the event because he was "ashamed of the bad marksmanship."

Still another notable was George Watterson, who with his brother owned a leading store in Benton. Early in January 1881 he exchanged gunfire with Ben Alverson, a pioneer saloonkeeper who had been one of those opposing Pap Kelty's claim. Watterson was wounded in the arm, but recovered. Two months earlier his brother, the local postmaster, got into an altercation with the justice of the peace. Both drew guns, but through the triumph of sanity, did not shoot.

It was clear that Benton's light was going out, not with a whimper, but with a bang. Yet the gunplay of Benton's respectable citizens—their bullets burning the air by the score—brought no more suffering than a few recoverable flesh wounds. Mercifully, they were not the best shots in Mono County. And though Benton had a jail, a deputy sheriff and a justice of the peace, none of these shootings brought an arrest.

There was, in fact, a still greater shame in Benton's last flicker. For several years the law against selling liquor to Indians was openly flouted. Almost every day, Benton's Main Street played host to the antics of drunken Paiutes, both men and women.

This had not been unusual for decades in much of the West, but Benton was uncommonly afflicted for a town of a few hundred souls. Some of the liquor peddlers were transients making money on the Indians' misery; in two years' time, only two were convicted.

But by 1880, one of the two breweries was regularly selling beer and whiskey to the Indians, who held carnival on Benton's streets. Everyone in town knew it, and the only one who did anything was the editor of the *Bentonian*, whose campaign against it ended when his paper died in January 1881. The evil traffic was harder to kill. As the *Mammoth City Herald's* correspondent wrote:

"Benton, like Necessity, knows no law."

By this time, with the mills still idle and the price of silver dropping, the Bentonians threw in their hands and moved on to other games.

George Rowan rode over to Mammoth City, where he launched another general store and a sawmill.

Captain Jim Powning trekked the length of the Eastern Sierra and pulled up in Calico, on the Mojave Desert.

George Watterson rode down to Bishop Creek, where two of his nephews would later lead Owens Valley's fight for water against the city of Los Angeles.

Albert Mack tried again in the late Seventies to turn a fortune in and. near Benton with his Kearsarge Mine and his new sawmill enterprise. Failing this time, he returned to Elizabeth in San Jose, went back to the drug business, and died in 1908 in San Francisco at the age of 85.

One who stayed was Jim Millner, of Creaser &

LUMBER! LUMBER!

ALBERT MACK & CO.'S SAW MILL.

LOCATED ABOUT FOURTEEN MILES from Benton and ten miles from Adobe Meadows, Mono County, California, where we are prepared to fill all orders for

PINE OR TAMARACK LUMBER

on the shortest notice. The lumber, whether for finishing or ordinary purposes, is as good as the very best obtainable in this county.

Price at the Mill,..............$30.

Also, Dressed Lumber, Tongue and Grooved Flooring, Rustic, Door and Window Frames got out to order.

Office and Lumber Yard at Bentoh, Mono Co.

1-tf   A. MACK & CO.

Millner's store. Acquiring the Comanche and other properties, he had Blind Springs Hill rumbling to the blast of powder and the mill at Benton wheezing and clattering once again.

Millner's success was boosted by the coming of the Iron Horse. In 1880 the Carson & Colorado narrow gauge began inching southward across Nevada from Mound House on the Virginia & Truckee Railroad. By the end of 1881 it reached Belleville, drawing all wagon traffic from as far south as Owens Valley. In 1882 the rails reached the new Benton Station, four miles east of Benton under the shadow of the White Mountains and the ghost town of Montgomery. With shipping costs slashed, Millner sent hundreds of thousands in silver bullion over the slim rails in the 1880s.

But during that decade most of the mines on Blind Springs Hill encountered a vertical fault that abruptly cut off the rich silver veins. Despite exploration, nobody ever found the rest of the leads. And in the Nineties, the price of silver plunged further—to an average of 63 cents per ounce in 1894. By the turn of the century Benton's population was down to fifty. Intermittently, other companies have tried to coax a profit from Blind Springs Hill, with mixed success.

Today, a general store, a gift shop, a residence or two, a few abandoned buildings, and the meager ruins of the Diana mill greet the wayward traveler.

Unlike many of the silver camps east of the Sierra, Benton never quite died. In her prime, from 1866 to 1880, she had fought off greed, treachery, complacency—and still survived. Through those years, as the six-horse stages raised dust through Benton's streets on their way to Nevada and the railroad, the boots of the Wells Fargo coaches carried some $5,000,000 in silver bullion.

5. MINING THE STOCKHOLDERS

It takes a gold mine to work a silver mine.

— Old Spanish proverb

While it was the irrepressible miner who tamed the Far West, he quickly learned that developing his mine required capital. To find gold- or silver-bearing ore is one thing. It is quite another to extract it with giant powder, equipment and labor; to secure the mine with timbers; to buy the machinery for mill or furnace; to haul that to the mill by mule team; to reduce the ore to bullion, and then to market that bullion—again by mule team.

The miner could borrow the funds, but a bank had to be assured that both he and his mine were good risks. Besides, in the California of the 1860s, capital was in short supply. The far-flung Western frontier, isolated from "the states", was still pulling itself up by its bootstraps. It took time for a new society to accumulate more wealth than it was consuming. Consequently, interest rates were discouraging. Though lower than in the 1850s, rates still remained around 2 percent *per month* in the 1860s. The miner could not carry this load.

Obviously, the solution was to sell all or part of his mine to others. Since mining district laws allowed each claimant so many feet of a ledge, he could sell a piece of his mine by the foot. While the discoverer could still work the mine, he would share the profits with his silent partners according to the number of their "feet". They could also share in further costs, if the discoverer came back to them in need of more money to develop the mine and works.

For this kind of risk the investor needed convincing. In every mining camp the streets and saloons had their share of miners' pockets bulging with ore samples—the stock in trade of the prospector. Of course he quoted the results of assays on these specimens—the numbers running into several hundreds or even thousands of dollars per ton of ore. But his specimens were selected samples—the best of the best—sometimes with buttons or streaks of silver showing to the naked eye. No wonder the West laughed knowingly at Mark Twain's definition of a mine:

"A hole in the ground owned by a liar."

In this atmosphere a new procedure quickly arose to provide the capital needed for the Far West's mining industry. The miner would sell his claim to one or more investors, usually men of means in San Francisco.

They would then incorporate and sell shares (no longer "feet") in a mining company. They would hold a stockholders meeting and elect a board of directors and officers. With the proceeds of the stock sales they would hire a superintendent and develop the mine.

Basically, this procedure facilitated the opening of underground resources in the West. They could not have been opened without it.

The system made the most of that human trait called hope—hope that the mine would be profitable, hope that it would prove a bonanza—perhaps another Comstock. There was no limit to the hope conjured in the mind of a prospector or investor. Hence the mining business grew out of, and was fueled by, the Westerner's willingness to take a chance. There was only a fine line between mining and gambling—and everyone knew it.

The Comstock itself was the inspiration. Hundreds of millions were at stake. No sooner would the Comstock fall into a time of borrasca when a blind lead in one or more mines would yield another bonanza. So eager were investors to lighten their own pockets, and so many mining companies were offering to take their money, that the need was soon obvious for a clearing house bringing together sellers and buyers.

In September 1862, some 37 investors combined to form the San Francisco Stock and Exchange Board. For years following, the board served as the catalyst for mining stock transactions. Starting with a small room in the Montgomery Block, the Board outgrew one location after another until 1877, when it moved into a five-story, $450,000 edifice on the south side of Pine Street, between Montgomery and Sansome.

Inside, the room for stock transactions was, in the words of one observer, "like the banqueting hall of some great potentate." A gallery for onlookers, attired like the others in their top hats and Prince Albert coats, encircled the room. At one end a crimson canopy hung from a gilded cornice. Before this was a platform, three feet higher than the floor, holding desks of the "caller" and his secretaries and clerks. On the floor in front was a circular rail enclosing the seats of the board members, used by them or their agents.

At first, members gave their orders by rising at their seats, with a fine imposed if they left the seats. But so many brokers jumped up and ran to the rostrum, despite the fine, that the Board gave up trying to keep order and waived the penalty.

The opening of a session was announced by a gong. At this, lively conversation turned to silence. An onlooker described the scene:

> The moment the caller names a stock for which there is a demand, all the brokers rise from their seats and rush to the space in front of the bellowing. They pull and tug and jostle each other as if in anger, till tired, when they resume their seats.caller, and yell the number of shares they will sell or buy, and the conditions, gesticulating like wild men—their eyes glaring, and hoarse with

Stirred by this passionate drama, the clients surrounding the railing were "surging and heaving in sympathy." And with most men smoking cigars or pipes, as another chronicler put it, "the more excited they became the larger would be the volume of smoke arising to the ceiling." Meanwhile, in and out of this maelstrom flitted hundreds of private messengers and uniformed telegraph boys, taking notes to absent investors across the country and even in Europe. If silver was king, here was its throne.

By the late Seventies, the Board had 100 members, with the price of seats growing from an original $100 to $43,000. Among them were such mining magnates as Jim Flood, William Sharon and Senator John P. Jones. The Board listed 184 stocks in Nevada mines and 56 in California, including New Coso and Defiance in Darwin, the Comanche in Benton, the Modoc Con at Lookout, and the Mammoth at Mammoth City.

It also listed ten stocks for mines in other states, plus thirteen municipal bonds and eighteen stocks of railroad, bank, insurance and other companies, including Egbert Judson's Giant Powder Company. With these latter comprising only seven percent of total stocks listed, it was obvious that mining was easily the dominant industry in the Far West. California still had a long way to go toward a mature economy.

In 1874 the San Francisco speculators got news of the "Big Bonanza" in two Comstock mines, the Con Virginia and the California. It was described as the world's greatest treasure house of silver and gold. The excitement brought feverish activity in existing Washoe shares and spawned a generation of new mines and stocks. By early 1875 the San Francisco Board was swamped with buy orders. The morning session usually ended before half the listings were called, and often many stocks had still not been called in the afternoon.

Before the doors opened at 11 am, investors and their brokers would crowd the north side of California Street near Montgomery. Large sales were made among them right on the street. With passage completely blocked, a path was opened every morning by a policeman.

To relieve the pressure the Board opened an early session from 9:30 to 10:30. Still the Board building, from California Street through a long corridor to the exchange room itself, was jammed with shouting investors. It was almost impossible for a broker with a fistful of orders to get through. To bar the riff-raff, the Board began charging a monthly entrance fee of $5.00, issuing a ticket good for thirty days.

It was clear that another exchange was needed. As early as 1872 brokers and investors had launched the competing California Stock and Exchange Board to help handle the volume.

In April 1875, a singular incident at the San Francisco Board triggered a third exchange. E.J. "Lucky" Baldwin presented himself at the door without his ticket. The doorkeeper stopped him. No amount of angry vituperation would avail. The enraged silver tycoon turned away, announcing he would open a new exchange of his own.

Organized in mid-May, with seats selling at $5,000 apiece, the Pacific Stock Exchange opened for trading on June 7 in the Halleck Building, at the corner of Sansome and Halleck. Among its forty original members—known to the irreverent as "The Forty Thieves"—were Lucky Baldwin, president; General George S. Dodge, vice-president; and such other financial giants as George Hearst and William M. Lent. Business was so booming that next year they constructed their own building on the east side of Montgomery Street, between California and Pine, extending back to another entrance on Leidesdorff Street.

Clearly rivaling the San Francisco Board, the Pacific Stock Exchange was a study in Victorian splendor. The ground floor vestibule was lavishly fitted with mosaic flooring and imitation marble walls. From the rotunda with its high dome, the visitor entered the board room itself, ornamented with European frescoes, polished hardwood, candelabras and red damask curtains. A richly appointed anteroom offered the social amenities, and one of the galleries was especially decorated for the ladies. The bidding for stocks easily rivaled the frenzy in the San Francisco Board.

By 1877 these exchanges, though beset with repeated booms and busts, were driving the California economy. In that year alone, combined stock sales exceeded $115 million. It seemed that everybody, from washerwomen to delivery boys, had a position in mining stocks. One could buy what the brokers called "A.O.T."—"Any Old Thing"—and promptly sell it at a profit.

Naturally the free-wheeling character of the business, and the huge profits that were both imagined and sometimes real, led many to try and beat the system. The deception of showing the best ore samples, begun at the lowest economic level by the prospectors themselves, was child's play. Though many mines were managed on the soundest principles, public confidence was undermined by a whole panoply of tricks—some entirely legal, since there were few laws and no government agency controlling the stock exchanges.

One of the most primitive ploys was "salting"—artificially impregnating the ore at the mine face with silver or gold dust, sometimes by blasting the particles in with a shotgun. A variation of this approach was to display at stockholders meetings a bar of "bullion"—actually a lead bar painted with silver or gold.

Much subtler and harder to penetrate was the wording in the prospectus of the corporation. The ore body averaged hundreds of dollars to the ton. The ore body was several feet thick and ran for miles. Wood and water were close by in abundance. Roads were in place or being built. Railroads were within easy reach, or were being built. The mine was near a well-known bonanza, in the same mineral belt, with the same type of ore.

Another corporate practice was to borrow money in order to pay dividends—thus boosting the stock price so the insiders could profitably sell. An easier device was simply to create the rumor that a rich blind lead had been struck in the mine, thus sending the stock skyward.

The mine superintendent could be a key figure in such manipulations. He could choose to process high grade ore or low grade ore, if instructed by the directors, and the published results of a month's run would affect the stock price.. A self-serving superintendent could, if a rich strike was made in the mine, keep it secret until he could get to the nearest telegraph office and buy a large bloc of shares.

Still another game practiced on a discoverer by the purchasing corporation was the inflated par value of the stock. The prospector was often paid partly in cash and partly in shares of stock in the forthcoming corporation. On paper, the company would be capitalized at a generous value per share, usually $100. But when the shares were offered for public sale they brought as little as $2 or $3 apiece. By 1878 the par value of California and Nevada stocks totalled $300 billion—a number roughly 150 times the current national debt.

But this was often just the beginning of the game. The shares were subject, not only to possible dividends, or "divvies", but also to assessments of from 25 cents to $1.00 per share, which were justified as necessary to continue developing the mine or building the mill or furnace. Only if the stockholder paid his assessment within a certain time could he keep his stock. If he failed to pay, his shares were forfeited to the company for sale at auction. Often the insiders were on hand to buy up the shares at a sacrifice price. This process, if pursued deliberately to eliminate outside stockholders, was known as the "freeze out".

The possibility of assessments was well understood and accepted in the business. Even repeated assessments, without dividends, were not unusual, but they soon resulted in driving down the stock price. Assessments were the subject of wry humor among the boys, who called them "mud", or "soap", or "Irish dividends".

Certainly, abuse of assessments could wound the confidence of investors in mines as a whole, or mines of a certain locality. One mining man from Darwin noted that if he tried to sell shares of a claim to someone on Pine Street, the man's hand instinctively went for his pocket.

"Inyo county mines don't pay divvies," would be the stern reply.

Without question, exploiting a mine required time and capital—a fact that many excited investors did not understand. But too often, assessments were spent on elaborate roads, office buildings and other appurtenances before opening the mine.

The prudent superintendent, with one ear cocked to the expectations of stockholders, would start with some mining, if on a small scale, even shipping the first ore for treatment to a nearby mill, to try putting the company on a paying basis. But superintendents were subject to interference from San Francisco. Assessment after assessment had the effect, even if unintended, of freezing out the outside stockholders, including the discoverer.

Nonetheless, the mining industry as a whole did pay good profits. This was due to a relatively few mines that were extremely rich and well managed. In the first ten months of the banner year 1875, dividends from all mines listed on the San Francisco Board totaled more than $67 million, while assessments totaled $39 million—a net income to the investment community of more than $28 million (assessments were thus 58 percent of dividends).

By the year 1878 the situation had deteriorated. Dividends totaled only $18 million, while assessments were nearly $14 million (77 percent of dividends). The fabulous mines of Nevada were still tipping the balance in favor of dividends. But in California, assessments exceeded dividends. More than ever, it took a gold mine to work a silver mine.

One cause of this decline was the demonitization of silver in 1873 (Congress removed silver from the list of metals to be coined for domestic circulation). From $1.30 per fine ounce in 1873, the average price of silver on the New York market slid to $1.12 in 1879. This 14 percent drop simply brought a corresponding drop in mine profits, many of which were already too thin.

So far as the San Francisco stock exchanges were concerned, the problem was aggravated by a provision in the new California Constitution of 1879. If a stock were sold at a price so low that it cut into the margin advanced by the investor, he could sue the broker and recover all of the margin. Thus brokers discouraged margin buyers, which narrowed the volume of mining stock transactions.

As early as 1880 the end of the silver boom was foreshadowed on the San Francisco exchanges. The decline of the Comstock, the drop in silver prices and the so-called anti-business California Constitution of 1879, were taking their toll. Stock prices were dragging. One of the three exchanges—the California Stock and Exchange Board—went out of business. One visitor from Bodie was shocked at the market's "entire lack of life" and sadly recalled "the good old days."

"Pine street, the grand centre of mining stock operations, shows no sign of the excitement . . . known in previous days."

Thus the venture capital on which new mines depended had all but dried up. Except for Calico, where most of the mines were privately financed and innocent of the stock market, California's silver era had played out by the early 1880s.

TRUE BLUE MINING COMPANY.

Mammoth City, Lake District, Mono County, California.

A limited number of the shares of the working capital stock for sale at the office of the Company.

OFFICE—405 California Street, Second Floor, San Francisco.

HENRY R. MILLER, Sec'y

6. THE JINGLE OF NADEAU'S TEAMS

The empire created by the Silver Seekers was strung together—made into a fluid, mobile community 400 miles long—by the men with wheels and teams. Different stage men ran their passengers into Eastern Sierra from north, west and south. But marketing the silver product and returning with life's necessities was, from the late 1860s to the early 1880s, a task dominated by one freight outfit with more than a thousand mules run by a Canuck named Nadeau.

In the fall of 1868, when Belshaw and Beaudry were both turning out silver-lead bullion at Cerro Gordo, they knew those shiny ingots would do no good piling up at the furnaces. They needed a reliable freighter who would contract to haul them over 200 miles to Los Angeles, where the coastal steamers would forward them to the Selby refinery in San Francisco.

There, Thomas Selby would pay for the silver and lead, which in turn would enable the bullion kings to meet the payroll in Cerro Gordo, pay for the freight and make a profit. Calls for competitive bids to market Cerro Gordo's output came from the the office of their partner, Egbert Judson, at 402 Front Street, San Francisco.

One of those quick to respond was a French Canadian, Remi Nadeau, who had freighted with ten- and twelve-mule teams northward from Los Angeles for several years. Believing this was his big chance, he resolved to beat his competitors to San Francisco.

Nadeau shunned the coastal steamer, perhaps because he couldn't afford the fare, or couldn't wait for its periodic call at San Pedro. But he had a long-legged sorrel named Hippy that was good for a long ride. Though he limped slightly in one hind leg, Hippy was celebrated for speed and endurance.

According to the story told by his daughter, Mary Rose Bell, Nadeau mounted his steed and pounded northward out of Los Angeles. On the first night he found one of his rivals asleep alongside the road. Giving him a wide berth, Nadeau raced on. After four hundred miles and little more than four days (as his daughter recalled), he pulled up at Judson's office in San Francisco. He was coming out of the office with the contract in his hand when the first competitor arrived.

This heroic ride rests only on the memory of his daughter, who was a fountain of Nadeau legends when I was a youngster. But the contract was to be the turning point in his life. At age forty-seven, he now faced his most noteworthy career.

Contemporaries described Nadeau as a short, stocky Frenchman, who talked in true Gallic fashion—rapidly, and with his hands as well as his tongue. One acquaintance called him "an agreeable, energetic little French gentleman."

Nadeau was born in 1821 at Kamouraska, located down the St. Lawrence River from Quebec. He was descended from Ozani-Joseph Nadeau, who had emigrated to New France from the province of Angoumois, in west central France, in 1665. Marrying Marguerite Abraham, he had settled as a farmer on Isle d'Orleans, in the river below Quebec city.

Six generations later, Remi was born in Kamouraska, one of a family of fifteen. Leaving the farm, he took up the profession of millwright, moved to Hopkinton in New Hampshire, and in 1844 married Martha Flanders Frye, descendant of a colonial New England family. They moved west to Chicago, then to Faribault, Minnesota, where they raised a family of seven and Nadeau built flour mills on contract.

Still restless, Remi caught the Pike's Peak gold fever in 1860 and struck out alone with a wagon and pair of oxen. He left Martha with four children (the oldest only fifteen); three had died, including a girl of fourteen.

Somewhere on the Platte River trail, according to Nadeau's daughter, one of his oxen died. He found another man who was in the same trouble, but who was headed for California. With Gallic resignation, Nadeau yoked his ox with that of his new partner and headed for the coast—far more distant than planned from his Minnesota family.

Wintering in Salt Lake City, where he built one or two flour mills, he pushed on over the old Mormon Trail, arriving in the fall of 1861 at Los Angeles. There he met Prudent Beaudry, pioneer Los Angeles businessman, fellow French Canadian, and brother of Victor Beaudry. Borrowing $600 from Prudent, he bought a wagon and six mules to enter the freighting business.

On the trail southward to the coast, Nadeau had been impressed with the rising trade between Salt Lake and Los Angeles. Each winter, deep snow in the Sierra Nevada closed the central route to California. Beginning in 1854, winter trade had been opened between Salt Lake and Los Angeles. By 1861 the onset of the Civil War, with guerrilla warfare setting Missouri and Kansas aflame, the trade back to the states was also imperiled, and the winter traffic to Los Angeles was a necessity for the Mormon capital.

Braving possible Paiute attacks, the teamsters rolled down the Sevier River, through the pioneer Mormon settlements at St. George and Las Vegas, along the Mojave River, across Cajon Pass, into San Bernardino, and on to Los Angeles. The usual procedure was for Salt Lake merchants to visit San Francisco and choose their wares—groceries, dry goods, hardware, ammunition—from the shelves of wholesalers and forwarders. Then they contracted for delivery by steamer to San Pedro and by ten- or twelve-mule teams to Salt Lake at rates around seventeen cents a pound. By 1862 some Salt Lakers were traveling directly to Los Angeles to do their buying, gathering for safety in trains of up to sixty wagons each.

Remi Nadeau threw his teams into this traffic, freighting nearly 700 miles and some thirty-five days across the Great Basin. Beginning in 1864 the gold towns of Bannack and Virginia City in the Montana country, locked in for the winter months when the upper Missouri River froze over, drew the Los Angeles traffic still further north. Nadeau and other freighters teamed 1,100 miles and sixty days to those camps—one of the longest wagon routes in a country where such distances were usually covered at least partway by riverboats.

The trade peaked in the winter of 1865-66, when more than 1,000 wagons called at San Pedro and headed northward laden with goods. By the spring of 1868 the Union Pacific and Central Pacific Railroads, rushing to meet each other from East and West, began to draw the trade. No more did the outskirts of San Bernardino and Los Angeles twinkle with the campfires of the teamsters and ring with their ribald songs.

Foreseeing the invasion of the Iron Horse in the Great Basin, Nadeau had already looked elsewhere for freight business; one observer says that he spent time in San Francisco. In any case, by 1866 he had settled in Los Angeles as his permanent home. The Salt Lake trade had left him experienced and hardened in the teaming profession.

Among his competitors in that traffic were Phineas Banning and Don George Alexander, who had pioneered the up-country trade since the 1850s. Apparently Nadeau had some difficulty with them. He delivered his down-freight at San Pedro to Banning's rival, John J. Tomlinson, whose wharf and warehouse stood at Timm's Landing.

In 1868, Nadeau sent funds to his wife and family for their passage to California. According to his daughter, this was the first time Martha had heard from him in seven years. With her four remaining children (the oldest now married), the brave wife took steamer to the Isthmus of Panama, crossed on the little railroad, and arrived by steamboat at San Pedro. Nadeau refused to let the family come ashore by Banning's regular lighter service to Wilmington. Instead, the doughty Frenchman chartered a boat and brought his family ashore to San Pedro without Banning's help.

Starting in December 1868, Nadeau's teams hauled the silver-lead output from Cerro Gordo to San Pedro along what came to be known as the Bullion Trail, or Nadeau Road. The 230-mile trek to San Pedro meant a dozen days of blistering heat in summer and choking dust and sand all year round.

Down the eight-mile descent of the Yellow Grade from Cerro Gordo, wheels were chained in place and several spans of mules were tied behind to help hold the wagons back. The teams reached Owens Lake near what is now Keeler and followed its northeast shoreline to the adobe town of Lone Pine, avoiding the heavy sands along the south flank of the lake. Then they rolled southward, between the Sierra and the lake, whose sparkling waters at that time stretched for twenty miles to the southern end of the valley.

On they pressed past the way stations at Haiwee Meadows and Little Lake. Then they dropped into the upper Mojave Desert, camping in the open where there were no stations, and stopping at the watering places of Indian Wells and Coyote Holes. Through Red Rock Canyon the teams were doubled in strings of twenty-four mules while the wagon wheels sank to the hubs and the brake blocks dragged the sand.

Southward lay a three-day, 38-mile stretch of waterless, sand-rutted road, always the most dreaded portion of the trip. But at the end stood Willow Springs, an adobe tavern eleven miles east of the present Rosamond, where the Cerro Gordo teamsters met around their campfires and broke the stillness of the desert night with their laughter.

Then they pushed across Antelope Valley, with its forests of spiny Joshua trees, and in springtime its fields of orange poppies and purple lupine, frightening herds of antelope that bounded gracefully across the sand. After twenty-eight miles the mules dipped their heads in the pond at Barrel

Springs. At their backs lay the conquered Mojave, before them the brush-covered Soledad Pass.

Along the dry river bed of Soledad Canyon, marked only by a one-saloon mining camp of the same name, wheels crunched through deep sand that closed over the rims and half buried the turning spokes. At length, swinging southward opposite the mouth of San Francisquito Canyon, the teams pulled up at the stage stop and tavern at Lyon's Station.

Here the teamsters, traveling in twos and fours, unhitched one set of wagons and doubled their teams for the grueling climb over San Fernando or Fremont Pass. Upward they lurched, the chock blocks dragging after each right hind wheel, ready to hold the wagon when the mules lost momentum and stopped for a breath. Then the teams were unhitched again and returned downhill for the other set of wagons

Over this section, teamster John D. Cage drove Nadeau's prize "Yellow Hammer" team of sixteen matched buckskin mules. Knowing Nadeau's weakness for fine animals, freightman John Delameter had driven the team past his home in Los Angeles. Sure enough, Nadeau had seen it out his window, rushed outside and bought the team for a fancy price.

At the summit of San Fernando Pass, a cut had been carved in the mid-1850s to facilitate trade with the Kern River gold camps, then deepened further to accommodate the Butterfield stages, and carved still lower in the 1860s for the Owens River trade. By 1870 the traffic was so heavy that a sprinkling cart was employed to patrol the narrow slit and dampen the dust.

Down the south slope creaked the wagons, rolling into San Fernando Valley and stopping at Lopez Station, now under water near the dam of the San Fernando Reservoir. Across the barren valley they crawled, through miles of cactus and sagebrush inhabited by lizards and coyotes. Near the summit of Cahuenga Pass they stopped at Eight-Mile House, then descended into the brush-covered hills that are now Hollywood, and swung into Los Angeles along Sixth Street.

With lead bells jingling, wood and leather creaking, blacksnake whips popping, muleskinners shouting and cursing, mules snorting and coughing, Nadeau's teams turned up Spring Street and raised dust through the business district. Another day took them to San Pedro, where the bullion was loaded onto lighters to be transferred to the coastal steamers.

After September 1869, when Phineas Banning's railroad was completed from San Pedro to Los Angeles, the bullion was unloaded at the Commercial Street platform of the depot. Then the teams repaired to Los Angeles Street's wholesale houses to be loaded with return merchandise. Bales of hay, casks of wine, sacks of potatoes—everything from a frying pan to a crate of live chickens—headed for Owens Valley behind Nadeau's teams.

When Nadeau's three-year contract was about to expire on December 1, 1871, Belshaw insisted on reducing the freight rate for bullion. Nadeau refused his terms. Stepping forward to get the Cerro Gordo contract was James Brady, superintendent of the Owens Lake Company's smelter at Swansea. Together with the output of his own furnace, Brady now controlled the entire bullion shipment of Cerro Gordo. In June 1872 he launched the little steamer, *Bessie Brady*, to carry the ingots across Owens Lake and save three days in the wagon haul.

Through most of 1872, Nadeau kept his teams on the bullion trail as an independent freighter. In November he contracted to market the output of the flourishing borax camp of Columbus, Nevada. With him in the venture were his two sons, Joseph and George. Thirty-four of Nadeau's teams left Los Angeles to haul on the route between Columbus and Wadsworth, on the Central Pacific Railroad. One Angeleno had some fun with the editor of the *Los Angeles Express:*

> I am told Nadeau will remain, and will—horrible thought—possibly keep one team here to run short daily trips in and out of town, to make two items daily for each of the papers. Now I am frantic over that thought, and if Nadeau dares to keep that team here, I warn him fairly he must take the consequences.

Meanwhile, Brady's subcontractor for hauling the bullion was unequal to the task. Several thousand bars were lying at the smelters and at Cartago Landing, on the west side of the lake. Belshaw's and Beaudry's operating funds and profits were tied up in those gleaming bars of silver and lead.

At this point, the Southern Pacific was laying track down the San Joaquin Valley below Merced, reaching as far as a new terminus at Tipton, only fifty miles north of Bakersfield. Once the rails bridged Kern River and reached Bakersfield, that town could challenge Los Angeles for the Owens River trade.

Ready to reach for the Cerro Gordo freight contract was Julius Chester, leading businessman in Bakersfield and freight agent for the S.P. at Tipton. Taking stage for Cerro Gordo in October 1872, he secured the next year's freight contract from Belshaw and Beaudry.

But as the bullion traffic swung over Tehachapi Pass, Chester and the Bakersfield freighters were beset with a host of obstacles. Winter rainstorms ruined the roads, sending one wagon over the side on the Yellow Grade, and rolling another over and over on the trail below Owens Lake. In the spring of 1873 a fatal horse disease, the epizootic, raged through the region and paralyzed traffic over the pass for several weeks. Finally, when the Southern Pacific resumed tracklaying, it stopped at a new railhead, Delano, still north of Bakersfield and the Kern River crossing. It was too far north for the Tehachapi route to compete with Los Angeles. Julius Chester's freighting venture collapsed

In seven months the Bakersfield teamsters had moved between 600 and 1,000 tons of bullion, but had allowed 1,200 more to accumulate in Owens Valley. By the middle of May 30,000 ingots lay piled like cordwood at Cartago and Cerro Gordo, while the three furnaces turned out 300 more per day. At Swansea and Cartago, miners and others had piled the pigs into brick walls, stretched canvas over the tops, and were living in the silver-studded shanties. But to Belshaw and Beaudry, this bizarre evidence of Chester's misfortune meant disaster unless something could be done to market those ingots.

At this juncture, Remi Nadeau was having problems of his own at the dry lakes of Columbus. An early summer hot spell was preventing the borate crystals from rising to the surface of the ground. Recovery operations were suspended and were not likely to resume until fall. In late May 1873 Nadeau pulled off his teams and headed for Owens Valley.

Hearing that Nadeau was passing by, Belshaw and Beaudry rode down the Yellow Grade and hailed him near Owens Lake. Squatting in the sand while the mules stomped and wheezed behind them, they told their troubles to the white-bearded Frenchman.

Nadeau agreed to return to the Inyo trade, on condition that they join him in forming a new freighting company and would put up $150,000 to build a line of stations along the Bullion Trail and buy new wagons, mules and harnesses. Only with these improvements, Nadeau maintained, could he promise to clear those mountainous piles and move all the bullion they could produce.

Belshaw and Beaudry promptly accepted Nadeau's terms. Together with their San Francisco partner, Egbert Judson, they joined Nadeau in forming the Cerro Gordo Freighting Company. It would dominate Eastern Sierra roads for the next nine years.

Nadeau continued south with his teams, bringing the jingle of bells and the rattle of chains into Los Angeles early in June. The *Los Angeles Star* editor was so jubilant that he welcomed Nadeau with a poem in his honor, written in French.

"It will give a wonderful impetus to trade here," declared the *Los Angeles Express*, "and times will be lively and money plenty."

Feverishly, Nadeau organized his new empire, putting eighty teams, each with fourteen mules and three high-sided wagons, on the road by October 1873. From Owens Valley to Los Angeles he built company stations at thirteen- to twenty-mile intervals—a day's haul apart. Where existing stations were not well spaced, he built his own at the nearest spring. At each station a married couple, earning $30 to $40 per month, provided beds and board for the teamsters and provender for the mules.

Over this route Nadeau operated his teams like a stage line, with schedules close to an hourly basis. Two teams would head southward each day, to be unhitched and fed at one of the stations. Next morning they would trade wagons with the teams that had arrived from the south and head back northward over the same route with a load of merchandise. Thus while the wagons rolled through between Los Angeles and Owens Lake, the teamsters and mules plodded back and forth between the same stations day in and day out. It was a system unique in Western freighting.

At the north end the first station was at Cerro Gordo Landing, on Owens Lake. From here the *Bessie Brady*, newly acquired by the freighting company, carried the bullion across the waters to Cartago. Southward from the lake the company established new stations at Rose Springs, Nine Mile Canyon, Coyote Holes and Red Rock Canyon. North of the present Mojave at Forks-of-the-Road, where the route over the Tehachapi Pass branched westward, Nadeau had to build a five-mile pipeline bringing water from the nearest spring.

The rest of the Mojave Desert was spanned with company outposts at Willow Springs, Cow Holes and Barrel Springs. The grueling haul through Soledad Canyon was eased by stops at Mill Station and Mud Springs, while Lopez Station brought the teams into San Fernando Valley.

Abandoning the grade of the Cahuenga Pass, Nadeau switched the last lap of the route by cutting a road across San Fernando Valley through the thick cactus and brush jungle in the trough west of the Verdugo Hills, across the Los Angeles River east of the pueblo, and into her traffic along Aliso Street. At his headquarters in the city, Nadeau took over most of the block bounded by Fort (now Broadway), Hill, Fourth and Fifth streets for his barns, corrals, stables, and blacksmith and repair shops.

By the fall of 1873 Nadeau's teams were beginning to catch up with the furnaces. At the beginning of 1874 Belshaw and Beaudry were confident enough to install larger equipment and increase their total capacity to 400 bullion bars per day—twice the output of 1871.

Gradually the shanties of silver-lead bricks tumbled before the advance of the freight mules. In the middle of August 1874 the last ingots were cleared and the teams were matching the pace set by the smelters.

By this time the Iron Horse was taking a hand in shortening the wagon route. In Southern California the S.P. extended its tracks from Los Angeles to the new town of San Fernando and began train service in April

Nadeau Collection
From 1868 to '82, Remi Nadeau dominated the freighting business in Eastern California, hauling the silver product of mining camps—first to coastal steamers and later to the railroad. Leaders in this 18-mule team are wearing the usual bells to warn oncoming traffic around a blind bend.

Nadeau Collection
Two of Nadeau's 14-mule teams pose for this photo of his headquarters at Mojave, on the Southern Pacific Railroad. From 1876 to '82, outfits each with three wagons and from 14 to 20 mules served silver camps from this terminus as far north as Benton and Mammoth City and as far east as Calico and Providence.

Thompson & West, *History of Los Angeles County*, 1880

In the 1870s and '80s, Nadeau lived in this home at Fifth and Olive Streets, Los Angeles, where the Biltmore Hotel now stands. Here he had a pair of prize thoroughbreds that were twice stolen and twice returned to collect his rewards. After that he guarded his horses with a tall fence and a bulldog.

Nadeau Collection

Leaving the freighting business in 1882-3, Remi Nadeau built the first four-story building in Los Angeles and the first with an elevator, located at first and Spring streets. It opened as a deluxe hotel with what the *Los Angeles Times* called "an elegance never before attempted in this city."

1874. On the same day Nadeau opened his new terminus near the train depot, cutting two days off the Bullion Trail.

A year later the S.P. tracklayers in San Joaquin Valley had passed Bakersfield and halted at the western foot of Tehachapi Pass. There, while eighteen tunnels were being pierced to provide a viable grade for the rails over the pass, the end-of-track would remain for more than a year. This connection, eliminating Los Angeles and the steamer passage up the coast, shortened the total route to San Francisco by about eight days.

In April 1875, Nadeau moved his headquarters over the Tehachapi to the railhead at what was first called Allen's Camp, then named Caliente, from the nearby Caliente Creek. During the next year the wagon road crossed the surveyed roadbed of the Iron Horse many times. One of Nadeau's muleskinners, P.N. Arnold, was urging his team across the tracks when a work train roared out of a tunnel and took out two spans of mules.

Finally, when the Southern Pacific finished conquering Tehachapi Pass in August 1876, Nadeau transferred his terminus to Mojave, where it remained for the next six years. A month later the S.P. completed its connection with Los Angeles at Lang Stration, in Soledad Pass.

Despite its name, the Cerro Gordo Freighting Company was not content to serve Cerro Gordo alone. At every new silver strike, Nadeau was on the spot to get the contract for freighting in mill and furnace machinery, and then to win the contract for hauling the bullion to market. By assuring payloads on the down trip, he could reduce freight rates against his competitors on the up trip. And with adequate capital he was quick to buy the watering holes and build the roads (on which he could take toll) over the way to the new discoveries.

Before the Cerro Gordo Freighting Company was born, Nadeau had reached with his teams to other distant camps. In 1872 he had freighted to Ivanpah, in eastern San Bernardino County, and to Mineral Park, across the Colorado in Arizona Territory.

While still headquartered in Los Angeles, Nadeau ran his teams to Panamint in 1874 and Darwin in 1875. At the same time he hauled borax for John Searles from Borax or Searles Lake. In 1876, out of Mojave, he freighted to the new camp of Lookout, in the Argus Range overlooking Panamint Valley. The Mojave terminus enabled him to reach as far north as the Mono County towns of Benton and Mammoth City in the late Seventies and early Eighties.

In 1878, when the historic old oasis of Resting Springs came alive with a silver strike, Nadeau beat out his San Bernardino competitors by blazing a road through the southern end of Death Valley—the first to haul goods across that forbidding region. From Owl Holes his road descended eastward into the valley, crossed the Armagosa River to Saratoga Springs, then proceeded through Ibex Pass northward to Resting Springs.

Through the Seventies and early Eighties, Nadeau dominated the freighting business in Eastern California, where the Cerro Gordo Freighting Company was spoken of simply as "the Company". The *Kern County Gazette* called Nadeau "the boss teamster of Southern California," while the *Inyo Independent* called the Cerro Gordo Freighting Company "one of the heaviest institutions in this or any other country."

Originally Nadeau's teams had ten mules each, drawing one large wagon with a four-ton cargo capacity. But he found that using bigger wagons and more mules yielded a disproportionately higher cargo. By late 1870 he had thirty-two teams of twelve mules and two wagons each. By the following spring he was experimenting with still bigger outfits of fourteen mules and three wagons, hauling more than eight tons. A key advantage was that increasing the length of teams did not increase the number of teamsters on the payroll.

Nadeau was not the inventor of large mule teams. In the Comstock excitement of the early 1860s, the long grade into Virginia City required teams of as many as twenty mules. In his early career Nadeau probably saw such teams and may have operated some. Early in 1875 a twenty-mule team and two eighteen-mule teams from the Comstock passed through Owens Valley, and were hailed as the largest ever seen there.

But Nadeau was the first to apply such teams in the California desert on a large scale. By 1874 he had eighty teams, most of them with fourteen mules and three wagons each. In 1875 he had 100 teams, all of fourteen mules, which were increased to sixteen before the end of the year. These outfits regularly hauled ten tons of freight—the capacity of a narrow-gage boxcar.

In the late 1870s most of the teams had eighteen mules, and by 1880 some had twenty and even twenty-two mules each. The length of the teams varied, of course, with the weight of the load and the difficulty of the route. Actually, the wheelers were usually draft horses, which could handle the heavy tongue and take the shocks of the wagon better than mules.

Such outfits hauled three blue-painted, high-sided wagons. The lead vehicle had rear wheels six feet in diameter, while the trailers, known as "back actions" or "tenders", were smaller. Generally the outfits traveled in pairs, to help each other in case of trouble, or to double the teams up a steep grade. Each span of animals had its own designation. The pair ahead of the wheelers was known as the "swings" or "pointers". Beyond them were the "sixes", the "eights", the "tens", and so on to the front pair, known as the "leaders".

The leaders each had a set of bells (five to seven in number) attached to the top of the collar. Since the teams plodded along at about two miles an hour, these bells were a warning to oncoming traffic, which might suddenly burst upon them around a blind bend.

Each team was managed by a driver, or "muleskinner", who rode a saddle on the near (left) wheelhorse. Since reins become impractical for a team of more than eight, he guided them through a long rope, known as a jerk-line, running through harness rings to the halter of the near leader. A jerk on the rope meant a right turn, while a steady pull meant a left.

The driver also had a long blacksnake whip which he could pop with remarkable accuracy. The expert skinner would claim he could knock a fly off the rump of a lead mule without touching the skin. Rarely did the popper of the whip actually touch the animal. Usually, the crack of the blacksnake over his head was enough to stir him on.

Still another weapon of the driver was a collection of rocks in the small bin at the front of the lead wagon. These he aimed with superior marksmanship at recalcitrant mules. One of Nadeau's teamsters, "Pegleg" Smith, would apply still more persuasion, if necessary. Leaping from his horse, he would run up to the offending animal and hit it with his wooden leg.

> **$50. REWARD.**
>
> THE ABOVE REWARD WILL BE PAID BY the Cerro Gordo Freighting Co., to any person who will furnish at their office in Caliente positive proof that any teamster or employee of the C. G. F. Co., has tapped barrels of liquors, or opened cases of liquor, or in any way illegally meddled with any freight or property in the charge of the company, or sold without authority, wagon sheets, or any other property belonging to the company, or in their charge; provided, such proof is furnished as above, before the company have settled with the employee complained against. R. NADEAU,
> n28 7m3 Superintendent

Accompanying these inducements was a collection of oaths that for sheer originality was unrivalled in the annals of profanity. The repertory of a muleskinner, with its marvellous juxtaposition of words, was a literary form that never has received due recognition. Ever since those freighting days the ultimate compliment to any oath is that it would "shame a muleskinner."

The driver also handled the brake on the lead wagon, and often walked alongside the team. Riding in one of the rear wagons was the driver's helper, known as the "swamper", who handled the rear brakes, kept the rock bin filled with missiles, helped harness and unharness the mules, cooked the meals if they camped in the open, and otherwise assisted the muleskinner.

These knights of the freight teams were not noted for their genteel ways. When Nadeau once said that it was very difficult to get good teamsters, he was referring to many requirements: the ability to handle a long string of animals on bad roads in the worst kind of weather; the responsibility to care for the animals and the freight as though they were his own; the honesty to keep from stealing cargo items and selling them; the fortitude to resist sampling the barrels of brandy and wine that went north from Southern California to thirsty mining camps; and the agility to stay out of trouble when in Los Angeles or any of the equally tough mining camps which they served. Nadeau had to run notices in mining camp papers that his teamsters were not authorized to sell goods from the wagons. And the same papers, as well as those in Los Angeles, reported instances of skinners getting into barroom fights, suffering wounds in a knifing or shooting scrape, landing in jail, escaping from jail.

Once a month Nadeau drove a buggy up the bullion trail with bags of gold coins to pay his station-keepers and teamsters. The muleskinners normally earned $50 a month, plus board and lodging at their home station. There is no question that they earned ever dollar.

Subject to sandstorm and cloudburst, the desert roads were often obliterated, and at best were full of ruts and chuckholes. On downhill grades, there was real danger that the heavy wagons would bolt out of control and carry teams and teamsters over a cliff. On the Yellow Grade from Cerro Gordo down to Owens Lake, one of Nadeau's skinners invented a double brake with blocks before and behind the main wheels, clamping them like a vise.

As in most desert freighting, mules were preferred to horses because they were hardier, less nervous, and more easily managed in the harness. They were also preferable to ox teams because they were faster and could go further without water.

Nor did the mules fulfil the popular notion of stupidity. In a fog or sandstorm, when the outfits strayed off the road the mules knew it before the drivers. They would stop and circle, refusing to move until they could see again. When a mule would go lame or otherwise give out on the trail, he was turned loose and would invariably head straight for the nearest station. With such intelligence they could be trained to perform extraordinary feats in the harness.

The pulling power of the mules was exerted along a chain, really an extension of the wagon tongue, that ran between each pair to the leaders and was attached to all the doubletrees. Each pair of mules saw to it that the chain extending ahead was kept tight, thus assuring that the pair in front was pulling its share. Melvina Lott, Nadeau's niece and the wife of his wagon boss, told of the "beautiful sight to watch" when a team started up.

The two leaders would start first, the next pair wouldn't budge until the chain between them and the leaders was taut. After this second pair had taken a few steps the third pair would start, then the fourth, and the fifth, until every animal in the team was pulling the load evenly.

When going around a bend, the spans in the lead could not make their strength felt along a curving chain. The tendency would be to pull the rear spans and the wagons off the road and across the corner. This might mean ramming the wagons into the hillside or dragging them off a cliff. Thus on passing such a bend the lead mules would step out of the trail and swing as wide as possible to keep the wagons going straight up to the point of turning. As the lead span passed the turn the following pairs had a special duty. As they approached the turn, the inside mule would jump across the chain and pull with his partner at right angles to the turn, so as to keep the wagon in the road and prevent it from cutting across the bend.

Managing the teams and muleskinners in this hard-bitten operation was Nadeau's wagon boss, Austin E. Lott, a tall, sandy-haired Southerner who had served as a postboy in the Confederate Army. Lott was a demanding boss with no patience for incompetence; he expected his teamsters to be "at least as smart as the mules." His outbursts of temper, complete with a consummate command of profanity, earned him the nickname "Tor-Nadeau".

In 1877 Lott was in Mojave when he was accosted by a Mexican named Jocko, who threatened his life, threw a stone at his face, and made the mistake of drawing a pistol. Lott pulled out his double-action revolver and fired five quick shots. At the inquest the wagonmaster was exonerated on grounds of self-defense.

Lott was camped with two outfits south of Little Lake in 1871 when they were joined at the campfire by a horse thief who was fleeing from the Inyo County sheriff. Knowing his identity, Lott let him bed down with them that night. After breakfast next morning they presented a double barreled shotgun at his chest. Tying him with a rope, they put him on top of a wagon and proceeded on their way. But, stupidly, they left the shotgun in the same wagon. The fugitive untied himself, grabbed the weapon, and bidding them goodbye, disappeared up Sand Canyon. When the sheriff arrived Lott was embarrassed to tell him their story. But the officer followed the robber up to the snowline of the Sierra and captured him.

Nadeau himself had his encounters with outlaws on the wild and lonely roads of Eastern California. From his daughter, Mary Bell, and his niece, Melvina Lott, are passed the stories about the arch-bandit, Tiburcio Vasquez.

For three successive winters between 1872 and '74, Vasquez and his lieutenant, Cleovaro Chavez, had their hideaways in the mountains around Soledad and Tejon Passes. On one of his monthly trips up the Bullion Trail with the payroll for his employees, Nadeau came across a wounded man lying beside the road. Helping him into the buggy, Nadeau took him to the next freight station. There Nadeau ordered the station-keeper and his wife to care for the man until he recovered. It turned out to be Vasquez, who left the station swearing never to rob Nadeau's teams.

Accordingly, Nadeau was again driving his buggy with sacks of gold coin for the payroll under the seat. Around dusk he saw a band of horsemen approaching. Hoping to elude them, he drove into a clump of trees. When the riders came up, the leader—Vasquez—saw him and called him to come out. When Vasquez recognized him as Nadeau, he let him go.

Another time, Vasquez and his gang were waiting in ambush in the narrow cut at the top of San Fernando Pass. Up the steep north grade a mule team was laboriously climbing. When it hove into sight through the narrow slit, Vasquez recognized Nadeau's blue-painted wagons. Astride the near wheeler was Nadeau's youngest son, George. Vasquez called off his men and let the team pass.

Once, according to Melvina Lott, Nadeau's wife Martha asked him why he did not tell the authorities after any of his Vasquez encounters.

"Freighting is my business," he replied, "and as long as my freighters are not bothered by Vasquez, Vasquez is not bothered by Nadeau."

This attitude was typical of Nadeau as a Frenchman and as a single-minded businessman. However, these Nadeau-Vasquez stories have to face one clear contradiction: Vasquez and Chavez robbed Nadeau's station and teamsters at Coyote Holes, only three months before Vasquez was captured.

Through the 1870s, Nadeau's northbound wagons were hauling an amazing variety of goods to the mining camps—blasting powder, mining and milling machinery, grapes, nuts, wine and brandy, corn, potatoes, bales of hay, billiard tables and mahogany bars, water pipes, even boats for the waters of Owens Lake. Occasionally an outfit would take on a passenger, who necessarily would be in no hurry, at a flat rate of $5.00.

To feed his legions of mules, Nadeau leased thousands of acres at what are now Culver City and Beverly Hills in order to raise hay and barley. And he added to local prosperity by huge purchases of provender from farmers in the Los Angeles basin and Owens Valley.

In 1874 the Inyo trade from Los Angeles was practically half as large as the city's exports through San Pedro. And about one-third of that San

Pedro export was Cerro Gordo bullion. Declared the *Los Angeles News* in February 1872:

"To this city, the Owens River trade is invaluable. What Los Angeles now is, is mainly due to it. It is the silver cord that binds our present existence."

Beginning in 1880, Nadeau's freight empire was under threat. The mines of Inyo County were faltering. And from Nevada came the neigh of the Iron Horse. William Sharon and D.O. Mills of the California Bank group had founded the Carson & Colorado Railroad Company. South from Mound House on the Virginia & Truckee Railroad, a small army of Chinese was building a roadbed and laying narrow-gauge track across the desert. Skirting the east side of Walker Lake, it reached Hawthorne in the spring of 1881.

At this point the C & C drew the Mono County traffic northward. Nadeau, unable to compete with the lower rates and shorter distance to San Francisco, withdrew his teams from Benton and Mammoth. And he reduced his Inyo service to half a dozen teams.

Then on January 1, 1882, the tracks reached Belleville, near the mining camp of Candelaria. With a further reduction in rates and time, this pulled the Eastern California business as far south as Bishop and Big Pine.

As for the rest of Inyo County, the mines at Darwin and Lookout were not producing enough bullion, and the towns of Lone Pine and Independence were ordering too little up-freight, to support Nadeau's wagon empire. Teams would be waiting at Mojave for days at a time for shipments from the Southern Pacific.

On December 27, 1881, Nadeau's agent, Austin E. Lott, anounced the Cerro Gordo Freighting Company was pulling its teams from the Bullion Trail. Ten days later he was at Independence, as the newspaper put it, "making about his last cleanup of coin—creating a general rustle and scaring the impecunious." To meet existing contracts only two more wagon trips were dispatched for Inyo. In January 1882, the C.G.F.Co. bade farewell as the last team of twenty-two mules raised dust through the main street of Bishop.

As early as the spring of 1879, Nadeau had captured new business in Arizona Territory. Early in June he dispatched seven teams across the Southwest to the famous Vulture Mine near Wickenburg, where they hauled ore from mine to mill. Then in the fall of 1880 Nadeau won a five-year contract to haul the ore of the Tombstone Milling and Mining Company from mine to mill. For the next year-and-a-half he spent much of his time managing his business in wild and wicked Tombstone.

On November 12, 1880, he paid $400 for a ninety-nine year lease on property several hundred feet west of town between the extensions of Fremont and Safford Streets. On April 21, 1881, he secured another adjoining piece for $200—all for use as stables and corrals for his mules. The sellers were Wyatt Earp, James C. Earp, Virgil W. Earp, and R.J. Winders.

Quickly Nadeau began sending some thirty-five more teams to Arizona. By mid-May the mules and blue-painted wagons were raising dust through Allen Street, carrying ore to the mill.

Stirred by the silver excitement in Tombstone, Nadeau paid $2,000 for a fourth interest in a group of Tombstone mines on October 27, 1881. On the same day Tombstone witnessed the funeral of the three Clanton and McLowry men killed by the Earps and Doc Holliday in the fight at the O.K. Corral.

This Indenture, made and concluded this Twenty-First day of April A. D., 1881, by and between V.W. Earp, W.S. Earp, J.C. Earp and R.J. Winders of Tombstone as parties of the first part, and The Cerro Gordo Freighting Company as party of the second part WITNESSETH: That said parties of the first part, for and in consideration of Two Hundred DOLLARS in lawful money of the United States to them in hand paid, the receipt whereof is hereby...

But Nadeau's venture in Arizona was short-lived. Apparently at the instance of his partners, M.W. Belshaw and Egbert Judson, Nadeau sold the teams and wagons in Arizona for something over $100,000 in February 1882, and the Cerro Gordo Freighting Company was disbanded in May.

Nadeau himself continued to freight in San Bernardino County, which became the center of silver excitements in California in the 1880s. Early in 1882 he began hauling silver ore from Calico's famed Silver King Mine some forty miles to the Oro Grande Company's mill on the Mojave River. Later the Oro Grande company, which by then owned the Silver King, built a closer mill near Daggett on the south side of the Mojave River only seven miles from Calico, thus greatly shortening Nadeau's haul and his freight profits.

Then in November 1882 his teams delivered the mill machinery for the Bonanza King Mine at Providence, near Mitchell Caverns. By this time the Southern Pacific Railroad, laying track east from Mojave, had reached Daggett, six miles south of Calico. Then it pushed on to meet the Atlantic and Pacific iron at the Needles on the Colorado.

Nadeau's empire, invaded by the Iron Horse from north and west, had now faded like a desert mirage. In 1883 he sold the rest of his teams to his wagon boss, Austin Lott, who continued to freight between Calico and Daggett.

By this time Nadeau's energies had already turned to other challenges. In 1880 he launched the first sugar beet enterprise in Southern California. From Europe he imported sugar beet seeds, a $100,000 sugar beet factory, and experts to run it. Besides planting several hundred acres himself on land purchased east of Florence, he distributed much of the seed to farmers in Los Angeles County. By the fall of 1880 over 1,100 acres had been planted on his own and other farms. Nadeau planted another 2,800 acres in sugar beets around the mouth of Ballona Creek.

"That this enterprise will prove a success, there is no doubt," wrote J. Albert Wilson in Thompson & West's 1880 *History of Los Angeles County.*

However, the sugar refinery failed to work. What sugar Nadeau did produce was, as friend Harris Newmark wrote, "bad at best, and the more sugar one put in coffee, the blacker the coffee became." But Nadeau was undismayed.

"I'll make a winery out of this outfit," he told a friend.

In November 1881, Nadeau began setting out a million cuttings of wine grapes—mission, zinfandel and charboneau—and converted the sugar beet factory to a winery. By the spring of 1884 he had put out more than 2,000,000 cuttings on 3,600 acres of land east of Florence. It was, according to more than one chronicler, the largest vineyard in the world. As for the wine, it was a better product than the sugar. It looked as though the wine venture would succeed.

Next, Nadeau took up 31,000 acres in what would become Inglewood and, still operating on the grand scale, sewed the largest barley crop in the world. But grain farming depends on rain, not irrigation. One of Southern California's recurring droughts dried up his fields and his ambitions as a barley king.

Meanwhile, Nadeau had been pursuing an equally grandiose venture in Los Angeles. In 1878 he paid $20,000 for a plot at the southwest corner of First and Spring streets. Since the Plaza was still the center of business activity in the late Seventies, such a price for business property that far south was considered foolish. An old adobe stood on Nadeau's property and a block further south was a horse corral. Harris Newmark told him he was "crazy".

But Nadeau, soon suffering losses in his sugar beet venture and stretching his resources with the world's largest vineyard, was undeterred. Early in 1882 he had made a killing in barley speculation, buying 80,000 sacks from Newmark and profiting hugely when the price shot up. With this windfall he planned a 1500-seat theater that would be the showplace of Los Angeles. To San Francisco he sent an architect to study the theaters there and draw up elaborate plans.

In July he changed his mind and resolved to build a four-story business block—the first south of San Francisco. With that many floors it needed a passenger elevator—another first for Los Angeles. By the end of July his workmen were tearing down the adobe on the lot and preparing the ground.

During construction, Angelenos shook their heads at "Nadeau's Folly". When it was nearing completion in July 1883 he invited some visiting friends to see how the elevator worked. It was an Otis hydraulic model, a cage enclosed inside another cage—a fancy Victorian ornament for the main lobby. With everyone safely on the platform, Nadeau pulled on a cable and miraculously, they started upward. At the top, another tug impelled them downward, past three flights of stairs, without mishap. With this technical marvel, Remi Nadeau was now a long way from those plodding mule teams.

At first Nadeau planned his grand edifice as a business block for stores, professional offices and private residences. In June 1884 he began renting the three upper floors to tenants. But in May 1886 he leased the entire building to a local hotelier, who opened it as the Nadeau Hotel with a grand banquet and ball on July 5. The ten-course meal included choice of ten viands, five kinds of pie, seventeen other desserts and such specialties as *Green Turtle a l'Anglaise* and *Pyramid Paté de fois gras a la Strasbourgoise*. And the furnishings—the lobby was called "the hall of a million mirrors" and the dining room was fully paneled and chandeliered. Altogether it had what the *Times* called, "an elegance never before attempted in this city."

From then until the turn of the century the Nadeau was the reigning hostelry of Los Angeles. Its fancy omnibus met guests at the railroad depot. From a porch over the sidewalk a band offered concerts every Saturday night. For the sporting crowd, poker games in the saloon could last for ten days. In the dining room the cuisine was famous. In 1886 Charles Dwight Willard wrote his mother from his hotel room:

"They dine at night at the Nadeau and set an elegant table."

For years the staff talked of the celebrities in the bridal and Presidential suites—of dancer Lotta Crabtree, who inaugurated the "Star Room" with imported wallpaper just for her; of actress Anna Held, who astounded Angelenos with the daily delivery of dairy cans for her famous milk baths; of the time special provisions were made to enable actors Edwin Booth and Lawrence Barrett to share the Star Room because neither could play second fiddle to the other; of champion John L. Sullivan, who once pounded the table in the bar and roared:

"I can lick anybody in this place!"

Among the famous who graced these halls were presidents Benjamin Harrison and Theodore Roosevelt; King Kalakana and Queen Liliokalani of Hawaii; stage celebrities such as Sarah Bernhardt, Nellie Melba, Adelina Patti, Lily Langtry and Lillian Russell; and such varied guests as Mark Twain and Wyatt Earp.

By the early Twentieth Century the Nadeau was eclipsed by more modern hotels. Half a century after it was built, it was razed to make way for a parking lot, then for the *Los Angeles Times* building.

Though Nadeau had capped his career with the world's largest vineyard and Southern California's leading hotel, his fortunes had actually declined since 1879. The period coincided with a drastic change in his private life.

Martha Flanders Frye Nadeau, the New England daughter of colonials and patriots, was unhappy and told him so. Nadeau had, after all, left her in Minnesota for seven years in the 1860s, with little or no word from him. In California she found herself alone again for long periods while Nadeau was off on business. There was, perhaps, another reason for her indignation.

Born in Warren, Pennsylvania, in 1832, Laura Mehitabel Hatch was eleven years younger than Nadeau and had married twice before she sailed for San Francisco in 1862. With her were her three children—Frank Egleston, son of the first husband, who had died, and two daughters, Cora and Gertrude Jones, by her second husband. She had divorced him before 1875, when she moved to Los Angeles and took a home on Main Street near Fifth, some four blocks east of the Nadeau residence. There was some talk in the Nadeau family that Remi had met her in San Francisco.

Though nominally at least a Roman Catholic, Nadeau sued Martha for divorce in June 1879—a rare and shocking step in those Victorian times, and practically unique among his friends in Los Angeles society. In the complaint his attorney claimed that throughout the marriage of thirty-five years, "and particularly during the year last past," Martha had treated Remi cruelly—"continued scolding, and use of abusive, unkind and unfeeling language, and insulting epithets spoken in the presence of their children." If this was due to his possible attentions to another woman, the complaint was silent. As a youngster I visited one of Nadeau's nieces, Rose Seymour, who was then around eighty years old.

"I don't blame Remi Nadeau for divorcing Martha.," she said. "She was always nagging him."

Martha, the proud daughter of New England patriots, refused to appear in court within the required ten days. As witnesses substantiated Remi's claims, the judge granted the divorce at the end of June. He awarded to Martha three large pieces of property including several thousand acres of the Nadeau ranch, which lay between the Los Angeles and San Gabriel rivers, and all the farm equipment and animals. Nadeau was awarded the rest of the holdings, including the home at Fifth and Olive streets and the plot at First and Spring where he would later build the hotel.

As for Laura, one observer later called her "exceptionally beautiful in form and feature." She loved fine clothes and jewelry, and on her marriage application with Nadeau, gave her age as forty-four, rather than the actual forty-seven.

In 1879 her marriage to Nadeau, only three and a half months after his divorce, must have raised eyebrows and set tongues wagging in Los Angeles society. If they had not known each other previously, one can easily see how the wealthy Frenchman, suddenly the catch of Los Angeles, was a target of opportunity for the beautiful divorcee.

In any case, Laura threw herself wholeheartedly into the marriage. As the mistress of the home at Fifth and Olive, she proved a different wife from Martha. She accompanied Nadeau on his business trips, including several months in Tombstone. She had twin portraits of Nadeau and herself painted by the popular contemporary artist, Albert Jenks. She was active in the social life of Los Angeles and supported charitable causes. For his part, Nadeau put her son, Frank Egleston, in charge of his vineyard and winery.

That there was at least some acceptance of Laura by Nadeau's children is suggested by the simultaneous appearance in 1881 at Tombstone of Nadeau's youngest son George, Laura's son Frank Egleston, and the husbands of her daughters. Apparently inspired by Nadeau's investment in a silver mine there, each of them staked and recorded a claim of his own.

Meanwhile, Nadeau was feeling the financial pinch of his unsuccessful ventures in sugar beets and barley farming. On January 17, 1886, the *Los Angeles Times* announced that Remi Nadeau "is financially entangled." He had mortgaged the vineyard, the Nadeau block, his home at Fifth and Olive, and four other properties. A San Francisco company was attaching them all for an unpaid amount of only $18,500. It seemed that everything had gone wrong since he had left Martha.

Evelyn Cawelti Collection
Martha Flanders Frye of New Hampshire married Nadeau in 1844 and brought the family to California in 1868.

Nadeau Collection
Laura Hatch married Nadeau after he divorced Martha. He deeded Laura the Nadeau Hotel just before he died.

Courtesy Colorado Historical Society
The flamboyant Johnny Moss promoted mines throughout the Southwest and in 1869 laid out the town of Ivanpah.

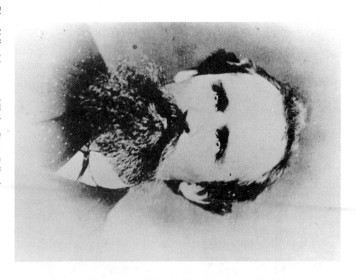

Mojave River Valley Museum and Thelma Prater Collection
One of four brothers who developed the Ivanpah mines, John McFarlane died defending the property in 1881.

Nadeau held things together through 1886 while his property values shot up in Southern California's Great Boom of the Eighties. In November he secured funds totaling $225,000 through a new mortgage on the Nadeau Hotel from Jasper Harrell, a noted businessman and rancher in Visalia. This was more than ample to satisfy Nadeau's other creditors.

But by this time he had a far more serious problem. For much of 1886 he had been suffering from kidney, heart and liver problems, diagnosed as Bright's Disease, a form of what has since been called acute nephritis. According to his children, the pain from the disease caused him to use liquor excessively for relief. They also claimed that his wife, Laura, continuously badgered him to sign a will to shift the inheritance toward herself and her children. By the last days of December 1886 she was, claimed his children, barring them from seeing Nadeau at his home, or when she did let them in, allowing no privacy.

On December 30, 1886, he conveyed to Laura the Nadeau Hotel, with the provision that its mortgage would be paid as soon as possible. For her part, Laura renounced any other claim on the Nadeau estate. Two days later, January 1, 1887, he signed a will that assured the hotel mortgage would be paid before any other distribution to heirs. It named as executors his wife and his friend, lawyer Samuel Caswell. And it left $5,000 each to his four children, Laura's three children, and his own brothers and sisters, most of whom were in Canada.

All of this would, when made public, confirm the fears of the Nadeau clan. Apparently Nadeau had no previous will, depending on California law that distributed estates to spouses and children, but not stepchildren or siblings. With the hotel, the crown jewel of the estate, given to Laura outside the will, California law for a normal distribution was further subverted.

Within a week of signing the will, Nadeau took a worse turn. Through the second week of January he sank lower. On the night of January 14 he was visited by his eldest son, Joseph F. Nadeau, who found him improving. As Joseph was leaving he said he did not like to leave his father "for the night". Nadeau assured him "It would be all right"—he would "be well soon." Early in the morning he got up from his reclining chair and lay down on the bed, complaining of nausea. In a moment he was dead.

With this event the acrimony between the Nadeaus and Laura went public. The long and eloquent obituary in the *Los Angeles Times*, apparently written with the help of the Nadeaus, named his surviving children but gave no mention of his wife, Laura.

After she and Caswell filed the will for probate on January 19 the Nadeau children launched one of the most celebrated civil contests in the Los Angeles superior court. The will, they claimed, was invalid—made under duress at a time when Nadeau was of unsound mind, only two weeks before his death.

In April they settled with Laura for $50,000 apiece. But Laura later reversed herself and claimed she was ill at the time and had not read the agreement. The issue would soon become irrelevant.

Though most of Nadeau's properties were still mortgaged, the estate benefited by the boom of the Eighties as real estate values peaked in the summer of 1887. But the bursting of the bubble sent property values crashing below the money owed on them.

Part of the vineyard was sold for a real estate subdivision before the collapse, but after that Laura could not bring herself to sell anything at a loss. The estate was unable to raise the funds to pay either debts or interest. Though Laura herself paid $10,000 in interest on the hotel mortgage, the property sank into default and was acquired by Jasper Harrell through foreclosure in 1889. Laura demanded that the estate pay her for the mortgage money that had been promised in the will, but there were no funds for this or anything else.

Nature herself took a hand with the great vineyard. A plague of those arch-enemies of French and California vines, the phylloxera plant lice, destroyed many of the vines. In the flood of 1889 the Los Angeles River roared through the vineyard and over the roof of the winery, covering the fields with sand and scattering $100,000 worth of wine and brandy barrels for miles. For days, passersby shot holes in the barrels and helped themselves. The vineyard, too, was sold at foreclosure.

When the estate was finally settled at the end of 1892, Laura and Caswell received a few thousand dollars due them for expenses and executors' fees, and the Nadeaus got nothing. For them and for Laura, it was bitter ashes after a million dollar dream.

Financially, Martha Nadeau had the lion's share after all. Divorced before Nadeau had plunged into his ill-fated Los Angeles ventures, she reigned in comfort at the old Nadeau ranchhouse on Central Avenue near Florence. Not long before she died in 1904, someone took a picture of her sitting in a rocking chair in her home. In her hand she holds a gold-headed cane once given by her brother to Nadeau. On the wall above is Remi's picture.

Laura, living in a modest home on South Hill Street, was not ready to be shelved. In 1894, according to gossip printed in the *Los Angeles Times*, she ran an advertisement in a matrimonial newspaper for a wealthy husband. From Grand Rapids, Michigan, an answer came from Lorenzo D. Ballou. In the preliminary correspondence each party claimed to be worth $100,000. With the wedding date fixed in July 1894, Ballou arrived by train and was immediately taken by Laura to the bank and shown $25,000 worth of diamonds in her safe deposit box.

Whether or not this chain of events was more than gossip, it is true that the couple were married two days later. It is also true that, before many months, they were divorced.

Still the romance of Laura Hatch Egleston Jones Nadeau Ballou was not ended. In April 1900, at the age of sixty-eight, she married a man named McKay. But she was single again when she died in December 1909. She owned fifteen pieces of property from Malibu to San Pedro, and it took an appraiser two legal size pages to list her jewelry.

In her lengthy obituary, complete with photograph, the *Los Angeles Times* reporter recalled her beauty and wrote that at age 77, she "was well preserved." Her name was given as Laura M. Nadeau, "widow of Remi Nadeau, a Los Angeles pioneer." No mention of her other four husbands, before and after Nadeau. She had the last word over the Nadeaus who, in their father's obituary, had left her out.

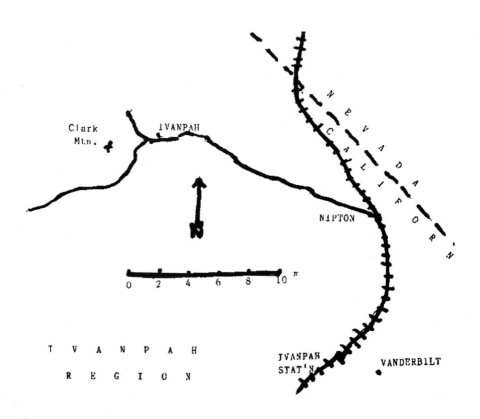

7. THE FIGHTING MCFARLANES

> South of Death Valley and north of nowhere, mines had to produce rich ore to overcome the distance to market. At Ivanpah and later at Providence, the ore was indeed rich. The four McFarlane brothers were in the vanguard of those arriving to take it—braving waterless wastelands, sandstorms, Indians and those interminable distances. As isolation spawned the wildest and wickedest camps in California, survivors also had to get past the worst that other men could do to them—as John McFarlane discovered too late.

John Thomas Moss was thirty years old when he ventured into the east end of California's Mojave Desert. With him he brought a reputation as a buckskin-clad frontiersman who discovered gold mines. From Iowa he had crossed the plains to California in 1857, became a Pony Express rider, and spent a good part of his life among Southwest Indians, especially in Arizona Territory. There in 1862 he discovered and sold numerous mines in El Dorado Canyon and next year was one of the locators of the famous Moss Ledge in the San Francisco Mountains south of the Grand Canyon.

Equally at home with Indian tribesmen or San Francisco capitalists, the "irrepressible Johnny" was of medium height, lean of build, handsome with his moustache and well-trimmed goatee, with wavy chestnut hair that dropped to his leather-fringed shoulders. He never turned a shovel in any of his claims, but was expert in exploiting the pockets of investors.

Flamboyant, fast-talking—Johnny had a reputation for tall tales and legendary binges. Yet George Hearst and William Lent were among the mining magnates who paid attention to Johnny's stories of discovered riches and profited from them. A writer in the *Alta California* called him "the mining 'Kit Carson' of the Coast."

Captain Moss, as he liked to be called, was in Mohave County, Arizona Territory, when an Indian brought him a sample of copper ore in 1868. Moss rode westward with him across the Colorado and, in what he called the Clark Mountains, found not only copper but silver as well. The location was as remote as one could get and still be in California—185 miles from San Bernardino to the west and 45 miles from the nearest settlement eastward on the Colorado. No roads, no wood, very little water—the place lacked everything—except mineral riches.

Very well, Johnny Moss took ore samples and rode westward the 260 miles to the coast. In San Francisco he spread his wares before his friends and with them formed the Piute Company on April 13, 1869.

Outfitting in Visalia, Johnny rode back into the Mojave Desert at the head of a prospecting party that included James H. Crossman, a Forty-niner known as a mining expert. In a vast territory covering the Clark District and northward into Nevada, they located about 130 claims. Then they shipped several tons of ore to San Francisco.

The assays were so promising that Johnny and his friends incorporated the Piute Company and issued 50,000 shares at a par value of $100 each. The prospectus called the enterprise "second to no other in the mining history of the Pacific Coast."

With capital in his pocket, Johnny laid out a town called Ivanpah at the only spring in the area. Several miles to the west in the Clark Mountains, he set a band of Paiutes digging in the main mine at 50 cents per day plus board. Then he explored a roadway to San Bernardino and arranged for a pony rider to bring in mail from the nearest California settlement, Camp Cady on the Mojave River.

Armed with his flare for publicity, Johnny never went through San Bernardino without calling on the editor of the *Guardian* and flashing his rock specimens. Assays of ore shipped to San Francisco were running up to $800 per ton.

Between this and the recognizable names of Piute Company officers, Ivanpah became the silver excitement of 1870. The San Bernardinans awoke to the opportunities and sent out parties to improve the road. From a Piute Company official in San Francisco came congratulations to the San Bernardinans; the mines, he wrote, "have never been exceeded in value by any discoveries on this coast." And he hinted at construction of a narrow-gauge railway. As early as August, with 300 men already braving 100-degree heat in Ivanpah, more were riding in from camps in Nevada and California.

Among the earliest were the four McFarlane brothers, experienced miners from the Kern River gold diggings. Arriving in February 1870, they were caught in a desert rainstorm and took refuge in a cave on the slope of Clark Mountain. There brother Tom picked up a rock and found it rich in silver. The McFarlanes staked their claim and stayed to become the masters of Ivanpah.

The four brothers came easily to frontier life. One of their ancestors was Louis McFarlane, a scout in the Lewis and Clark expedition that first explored the Northwest to the Pacific in 1804-6. Their home town was Allegheny City, near Pittsburgh, at a time when western Pennsylvania was little past the frontier stage.

Moving with their parents to Iowa in 1844, the brothers headed for California as they grew to manhood. At age 19 Tom got a job driving six yoke of oxen from Fort Leavenworth to Salt Lake. Arriving in California in 1859, Tom joined two of his brothers in the Kern River mines. There Andrew, the eldest brother, discovered the famous Long Tom Mine, which the brothers sold for $20,000.

Now in 1870, Tom and Andrew, together with a third prospector, pooled $450 and rode eastward across the Tehachapis and the Mojave Desert to the silver strike at Ivanpah. After Tom's discovery they located other mines—the Monitor and Beatrice 1 and 2—and formed the Ivanpah Mining Company. They produced some bullion by crushing ore in Mexican arrastras and smelting it in the vasos.

But most of the Ivanpah ore was sacked and shipped by wagon and steamship to San Francisco. What the Selby Works could not handle was sent as ballast in sailing ships around the Horn to the smelters of Swansea, Wales. Rich ore, indeed, that could return a profit on such a system.

Soon Tom and Andrew were joined by a third brother, John McFarlane, who became the Clark District's expert on minerals. Living with his brothers in a large tent near one of their mines, John collected more than 200 specimens, identified them, and could tell the place where each had been found. As host to inquiring visitors, he was an amiable and accommodating guide. As one arrival of 1871 put it:

"John really looked what he is, the very ideal of a modern miner, a man whose well balanced head and mind show. . .[the] will to undertake and the will to execute; he is as he well deserves to be the master spirit on this side of the mountain."

But withal, John was a man of physical courage, and had a violent streak when aroused. In April 1880 one Andy Laswell shot and killed a man over a gambling dispute. Laswell took refuge in his cabin and warned he would "make a 'lead mine' of anyone" who tried to take him. There being no law in Ivanpah, it was John McFarlane who met the challenge. In his blacksmith shop he forged some handcuffs and marched up to Laswell's cabin. If Laswell did not come out, John hollered, he would "blow the place up." Laswell came out, was handcuffed and taken under guard to San Bernardino for trial.

At first the going was slow for the McFarlanes. The expense of getting Ivanpah ore several hundred miles to market certainly discouraged any but the most persistent miners. Water to some of the mines had to be packed by burros an average of six miles from the Ivanpah spring. Lacking hoisting machinery, the ore was raised up the shafts by hand-cranked windlasses. It was hauled by mule team 260 miles along the Mojave River, through Cajon Pass, San Bernardino and Los Angeles to San Pedro. Then it was shipped by steamer another 400 miles to be smelted in San Francisco.

All of this cost, for one typical mine, $435 per ton. Ore worth less

was left on the dump till a mill and furnace could be built. But much of the Ivanpah ore ran more than $1,000 per ton—yielding a handsome profit.

By August 1871 Ivanpah, despite its paralyzing isolation, was a town of fifteen buildings, mostly made of adobe or rock with shake roofs. Counting the mining operations on both sides of Clark Mountain, 300 miners were at work by October 1872, and still shipping ore—not bullion—630 miles to market.

But in 1873 the McFarlanes changed this. That spring the boiler, bellows and machinery were hauled in by mule team from San Bernardino, and the furnace was turning out silver bullion before the year's end. Crushing by arrastra was still limiting output, so early in 1875 the McFarlanes hauled in a five-stamp mill from the nearby New York Mountains.

Two other mills soon started up in the Ivanpah mines, and the town became a lively oasis in this remotest part of California. At least one was hauled in by Remi Nadeau's mule teams. Besides the clatter of the stamp mills and the roar of the furnaces, the desert stillness was shattered at night by the shouts of revelry from the two saloons.

When L.M. Wilson opened a second hotel, the Occidental, it was nicknamed the "Accidental" because it was "an accident if you get anything to eat, and an accident if he gets any pay for it." In 1880 the camp even had, for a short time, a newspaper laboring under the title, *Green Eyed Monster* —named after one of the local mines.

With such prosperity came the usual lawlessness. One observer wrote:

"Everybody had money, and consequently nearly everybody was drunk, or trying to get that way. Fights were the order of the day."

In 1876 another arrival wrote in his diary:

"Many teamsters and miners were drunk. One fight."

A visitor of 1880 found two drunks lying in the street and another trying to find a gun to shoot someone.

"Oh, what a mess," he wrote. "Deliver me from such a place."

Giving Ivanpah a more distinct flavor was the presence of numerous Paiutes, of whom the *Inyo Independent* declared:

"There are some pretty wild Indians out in that region."

At first their peaceful behavior belied this description. Some of them worked quietly in the mine of Johnny Moss, whom they had known and trusted for twenty years. Soon they were also hauling water on pack mules from Ivanpah to the mines. In August 1871 a traveler at Camp Cady was introduced to Pachoca, chief of all the natives in the Mojave. He was, wrote the visitor, a "very respectable gentleman," and "said to be a good Indian."

In October 1876 some white men killed two Indian horses. In retaliation, Indians burned a white man's cabin and killed his Indian wife and child. Fearing a general outbreak, the Ivanpah men sent out letters calling for help. It was Tom McFarlane who rode the 200 miles to San Bernardino for arms and reinforcements. But by the time he returned with both, the so-called "Ivanpah Indian War" had subsided.

By 1876 a fourth McFarlane, William A., had joined his brothers as superintendent of the Ivanpah Mill and Mining Company. That spring they sold out for a reputed $200,000, with Bill and John McFarlane staying on as superintendent and mill foreman. Somehow the company faltered and was sold at sheriff's auction in January 1877. Changing hands still again in 1879, the Ivanpah Consolidated Mill and Mining Company was headquartered in San Francisco, though still managed at Ivanpah by the McFarlanes.

By 1880 Ivanpah and its mines on Clark Mountain were humming along gloriously. Ore running $300 and $400 to the ton, which had been put aside as "low grade", was now milled at a good profit. The Ivanpah Consolidated was stoping a vein half a mile long with at least five tunnels and numerous shafts. By this time it had shipped half a million worth of bullion. Chloriders—the independent miners who brought their ore to the mills for custom treatment—were flocking to the mills and prospering.

Within a few months the Ivanpah Con was beset with trouble. Though the miners were paid $4.00 a day, the management in San Francisco began paying them in company-issued scrip, which could be exchanged for U.S. currency at the company's offices in San Francisco. This was a sham, since no miner could afford the time or cost to travel to San Francisco. And the scrip itself could only pass for currency at the company's Ivanpah store, operated by a nephew of the president in San Francisco. The stated object was to save the cost of shipping gold coin from the city more than 600 miles to Ivanpah, but it was also aimed at forcing trade at the nephew's store.

The miners grudgingly accepted this scheme. But the issuance of scrip as a substitute for currency was subject to a 10 percent Federal tax, which the company did not pay. The government sued and got a judgment against Ivanpah Con for $1,480. When the company still failed to pay up, a deputy Internal Revenue collector named E. P. Bean was sent from San Francisco the 630 miles to the farthest corner of California to attach the Ivanpah Con's mines and mill.

Bean was an experienced, cool-headed official, aware of both the extent and limits of his authority, and determined to fulfill his mission peaceably if at all possible. Stupidly, the San Francisco owners had sent word to their Ivanpah office not to surrender the mill.

On the afternoon of May 16, 1881, officer Bean stepped from the stage at Ivanpah and went immediately to the office of the Ivanpah Con.

Near there he encountered J.B. Cook, an established Ivanpah citizen who had been one of the founders of the short-lived *Green Eyed Monster*.

A former employee of Ivanpah Con, Cook was aware of Bean's purpose and proceeded to challenge his authority in threatening words. Bean told him he was "a U.S. officer", had come "with the authority of the government" and "must discharge" his duty. The hotheaded Cook calmed down for the moment.

Next morning Bean notified Superintendent Bill McFarlane that he was seizing the mill, then posted notices and formally took possession. According to law, Bean and McFarlane each chose an appraiser to fix a value on the property. In the course of the appraisers' work they needed to know the value of the quicksilver, used in the ore treatement, that was on hand at the mill. There, accompanied by Bean, they encountered the redoubtable John McFarlane, the mill foreman, and asked him the value.

Flying into a rage, John grabbed a hammer, attacked them with insults, and ordered them out of the mill. As they withdrew, McFarlane seized a double-barreled shotgun and followed them with a torrent of abuse. Outside the company store, he aimed the shotgun at Bean and threatened to shoot. Risking his life, Bean jumped at McFarlane and shoved the gun away. He was there, he told John, "as a U.S. officer to take the property," and would do it if he lived. If they killed him, "a force sufficient to take it would be sent" and it was "useless to resist."

At length McFarlane quieted down and apologized. On Bean's request he blew steam off the boiler and closed down the mill.

Bean's next move was to appoint watchmen to guard the mill in eight-hour shifts. Knowing nothing of existing tensions in camp, Bean appointed Cook and a man named Fred Hisom, who was in some difficulty with Cook. Known as "a peaceable, quiet young man," Hisom did not usually carry a gun, but on Bean's advice borrowed a revolver as he took the night shift at the mill.

Cook was playing cards in an Ivanpah saloon when he learned that Fred Hisom was guarding the mill. He jumped up, declaring that the "damned son of a bitch" would "not stay there," that he would "drive him out." Passing Bean and John McFarlane, he shouted the same threat on his way to the mill.

Alarmed, Bean followed, with McFarlane behind. When Bean entered the mill, its interior lighted only by two candles, both Cook and Hisom were facing each other with guns drawn. But Cook had the drop on Hisom, the hammer at half cock.

Rushing up, Bean grabbed Cook's revolver with his right hand and swung at him with his left. The two men were struggling when McFarlane came into the mill and stepped up to help Cook. Hisom, fulfilling his duty as guard, moved in and told McFarlane not to touch Bean. The moment was charged with disaster—four men contending, two revolvers drawn.

McFarlane, armed only with a knife, lurched back to the door where his shotgun leaned against the wall. Grabbing it, he aimed it at Hisom, shouting he would "blow your brains out."

Still Hisom did not fire. Instead he rushed at McFarlane, dodging as he ran, and threw the muzzle of the shotgun upward just as it went off. The charge flew over Hisom's head. Before McFarlane could fire the other barrel, Hisom pushed him against the wall, making him drop the shotgun. While they grappled, McFarlane jammed against Hisom's side, drew his knife and cut a savage gash in the back of Hisom's head. At this, Hisom pushed his revolver under his left arm, pressed McFarlane's body and shot three times. One of the bullets found McFarlane's heart. He fell, groaning and pulling Hisom with him in a death grip.

At this point Bean had knocked down Cook. Seeing McFarlane's fate, Cook—whose ferocious temper had triggered the tragic episode—now pleaded for his life. Bean took Cook's revolver and the three left the mill. Within a few steps two shots were fired by someone in the dark, at the same time that Bean fortunately slipped and fell, one bullet whistling just over his head.

After having his head wound stitched, Hisom gave himself up to a deputy sheriff, who now represented the law in Ivanpah. He escorted Hisom and Cook for several days over the 200 miles to San Bernardino for a hearing. Hisom was freed on grounds of justifiable homicide, while Cook was tried on lesser charges. John McFarlane's body was buried at Ivanpah, and a few months later was moved to San Bernardino at the request of his widow. Grief-stricken to the point of imbalance, she spent the summer with relatives in the East.

Meanwhile the Ivanpah Con, either out of poverty or perfidy, temporarily suspended mining operations without paying back wages due for several months. Led by Bill McFarlane, the miners and millmen sued the company in November 1881 at the Superior Court in San Bernardino. After winning a judgment, McFarlane bought all the property of Ivanpah Con at sheriff's sale in February 1883, thus returning the mines and mill to the McFarlanes. They promptly paid the government's $1,480 claim and under the management of Bill McFarlane went on to produce a total of $3,000,000 in silver.

The Southern Pacific Railroad, building between Mojave and Needles, reached Goffs Station by 1883, only 40 miles southeast of Ivanpah. The cost of marketing bullion dropped enough to keep Ivanpah going a few more years.

Nadeau Collection
The McFarlanes sold out to the Ivanpah Con. Co. in 1876, but William McFarlane continued to supervise the works.

Mojave River Valley Museum
The Ivanpah Co.'s failure to pay Federal tax on this scrip brought on the fight in which John McFarlane was killed.

California State Library
When Sen. John P. Jones of Nevada bought into Panamint he launched California's biggest silver stampede.

California State Library
Sen. William M. Stewart of Nevada opened the Panamint mines and the pockets of San Francisco investors.

As late as 1885, Bill McFarlane was still running the Ivanpah Con mill. The next year Tom McFarlane was operating an Ivanpah mine under lease, and in 1890 two mills were still running in Ivanpah. But the falling price of silver, from an average of $1.15 an ounce in 1880 to 94 cents in 1889, was as devastating to Ivanpah as to other Southwest silver camps. When the price of silver crashed to 78 cents in 1893, the Ivanpah saga was over.

Bill and Tom McFarlane settled in San Bernardino in the mid-1880s, though for a time Bill, married to John McFarlane's widow, operated at the gold camp of Vanderbilt in the New York Mountains. Andy McFarlane had gone prospecting to the northwest and had repeated his Kern River gold strike by discovering the Providence mine at what became the silver camp of Providence. When he later rambled to Jerome, Arizona Territory, looking for a new strike, the *San Bernardino Times* was ready to call him "the most successful prospector on the Pacific slope."

Returning to the Ivanpah region, Andy joined the gold rush to Vanderbilt, where he was appointed deputy county recorder. Old age found him unrewarded for his lifelong search for silver and gold. He was living in a boarding house in San Bernardino when he fell down the stairs, fractured his skull and died in May 1905.

Johnny Moss, who had first discovered and promoted Ivanpah silver, continued to divide his time between California, Arizona and Colorado until he died in 1900.

Silver in Ivanpah naturally spurred prospectors to look for mineral throughout the eastern Mojave Desert. It was Johnny Moss who shared in discovering silver and copper in the Avawatz Mountains in 1871. Four years later, William and Robert Brown, the same brothers who had been early locators at Darwin, found silver and lead ores six miles south of the old pioneer oasis of Resting Springs, some 35 miles east of the south end of Death Valley.

Early in 1876 came Jonas Osborne, a mine superintendent from Eureka, Nevada. He bought the Brown brothers' claims and several others including one auspiciously called the Gunsight. Then he found several Los Angeles investors and joined in creating a stock company in May 1877.

At first, Osborne built a furnace at the Brown brothers' location, where the town of Tecopa sprang from the ground. San Bernardino saw the new district as another Ivanpah, and its streets came alive with wagons bound the 180 miles to the new excitement.

A year later, with several hundred men on the ground, Osborne bought a stamp mill in San Francisco and engaged Remi Nadeau's Cerro Gordo Freighting Company to haul it from Mojave. This time Osborne had the mill erected at Resting Springs, so that Tecopa was largely abandoned for the new townsite. Soon Resting Springs consisted of saloons, hotels and stores—one of them operated by the Cerro Gordo Freighting Company.

Life in this treeless wasteland could test the hardiest souls. Just getting to Resting Springs through 100-degree heat and piercing sandstorms was a challenge. Twice the pony rider bringing the mail from San Bernardino was driven to delirium, losing both his way and the mail. Through the deep sand of the dry Armagosa River bed, stage passengers had to get out and walk, with the wind blowing up the sand until, at one camping spot, it partly filled the coach.

"I will bet there was a peck in the Stage that drifted in since morning," one traveler of 1880 wrote in his diary.

And when he arrived in Resting Springs he was shocked at a society unrestrained by Victorian convention:

"White men. . .playing cards with Squaw Indians others Bathing with them and some drinking with them in the Saloons."

As for Osborne's mill, though promoted as the latest invention, it refused to work. The frustrated superintendent bought another, conventional ten-stamp mill, which again was hauled in by Nadeau's teams. By August 1879 it was pounding away on ore from the Gunsight, turning out bullion that was loaded onto Cerro Gordo Freighting Company wagons bound for Mojave and the railroad.

But Resting Springs' isolation in the southeast corner of Inyo County was the implacable enemy—freight costs denied any profit at the current output. To increase production Osborne made the same mistake committed by Lewis Chalmers in Alpine. He began a 1,000-foot adit, all of it dead work, to tap the lode hundreds of feet below the outcrop. The cost, without any compensating income, brought one stock assessment after another, until Osborne quit and sold his own shares.

A new superintendent ran the mill one shift per day on ore from the surface shaft until the adit struck the lode and provided more. But by this time most of the people had left town, leaving only some 60 men in and around Resting Springs; the pony mail was reduced from tri-weekly to once a week.

"This camp has not been worth speaking about the past three months," one resident wrote in June 1880.

At the same time Resting Springs was threatened from another quarter. A hard case named Rockwell, known in camp for what one citizen called "cunning deviltry", was yet stupid enough to steal a mule from the neighboring Shoshone tribesmen. While usually peaceable, the Shoshone in that region were not necessarily friendly. Pursuing Rockwell, they caught him in the process of making jerked mule meat. Taking all his belongings, they then turned him loose. That night, still stretching his luck, the wretch sneaked into the Indian village. There, using a miner's pick, he killed four Shoshone—men and women—and escaped in the dark, headed eastward toward the Nevada line. Into the Springs came a delegation of outraged Shoshone, demanding that Rockwell be captured and brought to justice. If not, they would attack Resting Springs. As there were an estimated 500 Shoshone warriors in that section of the Great Basin, the threat was taken seriously among the 30 or so miners in the town itself. They sent one stalwart, accompanied by an Indian, on Rockwell's trail. And by the San Bernardino stage they called for help from the outside world—rifles, ammunition and soldiers.

> **INDIAN TROUBLES**
>
> **500 Shoshones Threatening an Attack on Resting Springs.**
>
> **WHITE MAN KILLS FOUR INDIANS**
>
> Nathan Berry, the driver of the Resting Springs stage, arrived from that camp this morning, and brings the alarming intelligence that the Shoshone tribe, numbering about 500 warriors, have threatened the miners with complete annihilation. The provocation seems to

In San Bernardino, fears were rife for the beleaguered miners. One veteran trader who knew the Shoshone advised that the whites would do better to "raise the devil than get into trouble with them."

The response, though slow due to Resting Springs' isolation, was determined. A detachment of troops was sent from San Francisco by rail to Mojave, then eastward across the desert. And by May 8, a stagecoach rolled out of San Bernardino with guns and ammunition.

But the danger of another Indian war on the Mojave had faded by the time help arrived. Though Rockwell was not captured, at least by the Resting Springs pursuers, the Shoshone retired peaceably. The best reckoning is that they themselves had found Rockwell.

By this time a richer silver strike had been made some fifty miles south of Ivanpah, in the Providence Mountains. Early in 1880, Richard Gorman and P. Dwyer rode south from Ivanpah and found a mineral belt, with scattered silver-bearing deposits, some 600 feet wide and several miles long. When they returned to Ivanpah in April with ore samples assaying up to $5,000 per ton, excitement ran through the camp and the rush was on.

Among the early arrivals were Charles W. Hassen and Andrew McFarlane, who was the most experienced prospector of the four brothers. A mile north of the first discovery they located the Bonanza King Mine, which would become indeed the bonanza of the district.

Back went Andy to Ivanpah, where brother Bill McFarlane assayed sixteen samples with an average of $383 per ton—more modest than the earlier assays but also more credible. The rush to Providence redoubled, and by June more than a dozen mines were opened. Shipping ore to the Ivanpah mill, Andy McFarlane was soon writing to a San Francisco friend that Providence "is going to be the best District in S B Co beyond a doubt."

But more prospector than miner, Andy sold his interest in the mine and moved to other beckoning mountains. In January 1882 the remaining owners discovered a rich blind lead, which enabled them to sell out to a new corporation of Eastern investors, the Bonanza King Consolidated Mining Company, for an amount said to be $200,000. One of the owners, Thomas Ewing, arrived to charge Providence with new life.

With 150 men in camp, Ewing began sinking the shaft and tapping ore bodies with three shifts around the clock. Since water was available from nearby springs, he ordered a steam hoisting works from San Bernardino. From a quarry on the site, stone was cut and fitted together for the walls of offices and workshops. And from San Francisco he bought a ten-stamp mill, which the Southern Pacific hauled to Mojave.

In the summer of 1882, Chinese rail crews were laying track eastward, with the line scheduled to pass a few miles south of Providence within a few weeks. But with 2,000 tons of rich ore waiting on the ground, Ewing could not wait for the railroad to save costs in transporting the mill machinery. Instead he spent far more with Remi Nadeau, whose mule teams could haul the machinery from Mojave to Providence in ten days. When the mill started up at the beginning of January 1883, turning out $2,000 worth of bullion daily, Ewing made up for the added cost in a few days' time.

By mid-September the Bonanza King had produced $445,000 in bullion, with another half million on the ore dumps. It was already surpassing Andy McFarlane's prediction as the best silver mine in San Bernardino County. As the *Calico Print* declared, it was "the richest silver mine in the state." In its first eighteen months the ten-stamp mill produced nearly $1,000,000—a record for California silver output.

But in mining, as in gambling, no winning streak lasts forever. In mid-1884 Tom Ewing left as superintendent of the Bonanza King. Among the new management was H.C. Callahan, who proceeded to alienate many of the miners. In February 1885 one of them wrote to the *Calico Print* that they were driven "to do the work of two men." And though there were two general stores and three saloons in camp, the miners were expected to trade and drink at the store operated by Charley Callahan, the superintendent's brother. If they did not—still according to the same account—they were fired. And the diatribe closed with the threat of "tar and feathers" and a "ride on a rail out of camp and the country."

The Bonanza King company declared that the miners could eat and drink where they pleased. But a second report to the *Calico Print* disagreed:

"Miners think it necessary to spend so much a month with 'my brother Charley' or it is not safe."

Early in March the two principal owners, Wilson Waddingham and Samuel Kelley, travelled by train and stage to Providence. They shut down mine and mill, blaming the drop in the price of silver. Thrown out of work, many miners left town, hiking or riding the twenty miles to Fenner, the nearest station on the railroad. A few days later the company announced it would start again, but—still citing the reduced price of silver—at a cut in wages from $3.50 down to $3.00 per day. Advertising for miners in the *Calico Print*, the Bonanza King Mine started up again in mid-March and the mill late in June.

But on July 31, flames engulfed the mill and burned it to the ground. Most of the men took their blankets and left town. The owners collected the insurance and also left town.

Mining and some milling recurred at Providence into the early 1890s, depending on the price of silver, but the Bonanza King mill was never rebuilt. The last effort, spurred by the need for silver during World War I, ended when silver fell again in 1920. The "richest silver mine in California" is marked today by a few finely cut white walls and the scattered remains of the mill and huge ore bin of the Bonanza King.

Across this empty corner of California, where once the silver seekers swept in and out like a desert whirlwind, there is again the eternal silence—unless the breeze across the sagebrush brings to your imagination the clatter from stamp mills and laughter from the saloons.

Miners Wanted

—AT THE—

BONANZA KING

CONSOLIDATED MINING CO.'S

—WORKS AT—

Providence.

20 SKILLED MINERS WANTED.

WAGES $3 PER DAY.

Board $7 per Week.

SAMUEL KELLEY,

General Manager.

March 14, 1885.

8. THE SILVER SENATORS

No camp had more illustrious fathers and a more soaring reputation than Panamint, and deserved them less. When Senators John P. Jones and Bill Stewart of Nevada bought into the silver discovery just west of Death Valley, the rush was on. When they invested a fortune in mines and mill, the boys were sure they were onto a good thing, and for a time Panamint rode high, wide and wicked. As mines and mill closed down, some claimed it had been a stock promotion. But the answer lay in something simpler—bad judgment.

For twenty years in the mines of California, Nevada and Mexico, prospectors were familiar with the figure of Richard C. Jacobs—the stocky build, the sandy hair and whiskers, and what one acquaintance called a "pleasant address". He had the twin skills necessary for a successful prospector—an eye for ore and a tongue for selling claims to others. But through those twenty years that Jacobs had followed the end of the rainbow, he never quite found the pot of gold.

Now, in January 1873, he and his partner, Bob Stewart, were to find a pot of silver. An acquaintance called them "well-bred gentlemen, honest miners and men whose word was good as gold." Grubstaked by W. L. Kennedy, a merchant in Kernville, they had ridden eastward across the Coso and Argus ranges to remote Panamint Valley. There they camped in Mormon Canyon, on the west slope of the Panamint Range. In these highest reaches of the Mojave Desert, Telescope Peak surpassed 11,000 feet and wore a cap of snow in many winters. Beyond to the east, blistering Death Valley sank below sea level, but at the 6,000 foot elevation of their camp the nights were chilly and the campfire welcome.

Among the thick piñon pines and junipers at the head of Surprise Canyon, north of Mormon Canyon, they found outcrops of silver-copper glance that fairly leaped to their eyes. The ledges paralleled the canyon on both sides and were cut by side canyons that exposed them to easy view. As one later arrival wrote:

"Nature herself had done the prospecting."

There were other silver-bearing outcrops—chloride of silver, sulfurets of silver and, as in Cerro Gordo, argentiferous galena. It was a silver treasure-house waiting to be discovered.

On February 1 they put up four claim notices. Ten days later, at the Mormon Canyon camp, they organized the Panamint Mining District and elected Bob Stewart as recorder. Then, carrying their specimens, they rode back to Kernville to get their friend Kennedy, and to get assays, which soon proved to be from $125 to $3,000 per ton. And the lodes, according to their stories, were from two to thirty feet wide.

Out of Owens Valley and the Havilah mines in the Tehachapis rode the first to heed the call. By July, Surprise Canyon was peppered with eighty or ninety claims. Among them were Jacob's Wonder and Stewart's Wonder, claimed in April on the north side of the canyon; and the Wyoming and Hemlock, located in May and September on the south side. These were to be the famous mines of the district. By November, as one arrival wrote:

"Cabins are rapidly going up, mines are being surveyed."

But as usual the pioneers had plenty of hope and no capital. It was time for the promotional phase. In November, Jacobs joined four others including Gustave Wiss, who had challenged Belshaw at Cerro Gordo, in forming the Panamint Mining Company. By December the third of the original triumvirate, W.L. Kennedy, was in Bakersfield and Los Angeles, showing ore samples. With him was "Colonel" Eliphalet P. Raines, a persuasive mine promoter armed with Southern charm and hard experience in such free-wheeling camps as Pioche, Nevada. W.A. Chalfant called him "a man of daring character but limited attainments." Raines had paid $20,000 for one-half of Kennedy's mine.

With Raines and Kennedy's promotion, the rush to Panamint was on—mostly by stagecoach up the Bullion Trail to Little Lake, then by foot or horseback over the Coso and Argus ranges to Panamint Valley.

By early spring Richard C. Jacobs and his partners incorporated their Panamint Mining Company and began selling capital stock with a nominal value of $2,000,000. And up the Bullion Trail, Jacobs started hauling in a ten-stamp mill. By March 1874 more than 700 men were in camp, staking over 150 claims.

"A town has been laid out," noted the *Los Angeles Herald*. "Stores and shops are going up."

But genuine development of the mines still required roads for hauling in machinery for mill and furnace. In Los Angeles, R.C. Jacobs urged the city fathers to finance a road that would leave the Bullion Trail from Indian Wells and reach Panamint Valley by way of Borax Lake and the pass between the Slate and Argus ranges. Most of the route was, he argued, "a good natural road".

With both Bakersfield and San Bernardino eyeing the Panamint trade, the Angelenos moved quickly. By June 1874, Jacobs had a small army of Chinese blasting a roadbed on the Slate Range crossing. Over this back-breaking grade, mule teams were soon hauling his ten-stamp mill. An Owens Valley enterpriser, Bart McGee, carved a road up tortuous Surprise Canyon, and by August 1874 Panamint was opened to wheeled traffic.

One of the earliest to arrive was Dave Neagle, who had killed a man in Pioche and had hurriedly left town. In his violent career Neagle wielded his six-shooter on both sides of the law, leaving a trail of victims from Montana to Arizona. Much later, as bodyguard to U.S. Justice Stephen Field, he would put an end to the stormy career of David S. Terry—the state supreme court judge who had killed U.S. Senator David Broderick in a duel. But now, after reaching Panamint as early as February 1874, Neagle opened his Oriental Saloon and began dispensing cheer to the thirsty miners.

The first to arrive in a buggy was Inyo County's crack attorney, Patrick Reddy, who looked over the claims with a miner's eye. But it was his younger brother, E.A. "Ned" Reddy, who came and stayed, opening with a partner the Independent Saloon. Veteran of three shootouts, Ned joined Dave Neagle at the head of Panamint's hard reputations.

Up to this time Panamint was riding on enthusiasm and hope. It remained for Col. E.P. Raines to strike silver in another quarter. At San Francisco early in 1874 he had first buttonholed John Percival Jones, junior U.S. Senator from Nevada, with the Panamint story. But the $1000 Jones had advanced him was dissipated in a wild night of celebration.

Nonetheless, after a friend sprang him from jail, Raines had nerve enough to follow the senator to Washington and approach him again. And Jones had faith enough to slip him another $15,000 and enter the Panamint venture in earnest. In May the senator's brother, Henry A. Jones, rode to Panamint and secured five of the mines, including Wyoming and Hemlock, on options totaling $113,000.

John P. Jones was just the man to send Panamint soaring. He had a stocky build, medium height, an ample beard, and as one acquaintance put it, "a face at once striking, joyous, genial and commanding." Though he was born in the English county of Herefordshire, he took pride in a Welsh heritage.

"I left Wales at five years of age for the United States," he liked to say, "and brought my whole family with me."

Raised in Ohio, Jones had shipped around Cape Horn to California in 1850 at the age of twenty-one. With his brother he wielded pick and shovel in Tuolumne County, then moved to Trinity County, where he was elected sheriff in 1856.

When an Indian war broke out in 1858, Jones served as a volunteer in a punitive expedition. With fifteen others, he was surrounded by a large war party which killed several men before the fighting ended sixteen hours later.

After a term as a state senator, Jones met a setback in an unsuccessful campaign for lieutenant governor in 1867. But as a miner he clung to his reputation of having a "nose for ore". In Virginia City he launched a new career as superintendent of the Crown Point and Kentuck—mines controlled by the Bank of California group that dominated the Comstock Lode. He soon found that the position was indirectly helping him toward a political comeback.

In April 1869 a fire in the neighboring Yellow Jacket Mine filled the Crown Point and Kentuck mines and shafthead buildings with fumes and smoke. For hours Jones led the effort to rescue his miners. Two days later, in order to stifle the fire in the Crown Point, Jones took the cage down to the 800-foot level and cut the air pipes. Twenty minutes later he emerged, nearly overcome, to find himself a hero among the Comstock miners.

A few months after, Jones discovered a blind bonanza in the depths of the Crown Point. With his brother-in-law, Alvinza Hayward, he quietly bought up enough stock in the mine to wrest control from the California bank group. When the strike became known, his stock soared. One estimate of his new-found wealth was as high as $10 million—making him one of the richest men in the West.

Far from reproaching him for deceiving the California bankers who had controlled Crown Point, the Nevada miners admired him more than ever for outsmarting them. He ran for the U.S. Senate and, in a day when senators were chosen by state legislatures, he spent $500,000 in the campaign "just to set a pace for the next man who runs."

An inveterate mine speculator, Jones had already lost and won two fortunes when he went to Washington in the spring of 1873. There he became known as an able orator and the father of considerable mining legislation. He was a fast thinker and faster talker; his words were fired like a volley of shots, filled with amusing similes and spiced with the picturesque jargon of the mining camps. But he was widely read in the classics and could hold his own with any Eastern senator on subjects ranging from world economics to philosophy. His speeches in support of free silver coinage lasted for hours—one for several days. He was also a penetrating student of human nature. As one observer put it, Jones "could detect real manhood under the gray shirt of a miner as quickly as under a senatorial robe."

When he arrived in the capital for the spring session in 1875 he brought a young bride who charmed Washington society as Jones himself had charmed his colleagues on the Senate floor.

To share the Panamint venture he invited two friends whose names added still more prestige. One was William M. Stewart, who was ending his second term as a U.S. Senator from Nevada by returning to private life. The other was Trenor W. Park of New York, a heavy investor in mining stock, whose career was no less eventful than those of the silver senators.

A Forty-niner, Park had been attorney for the San Francisco vigilantes of 1856, was now a director in the Pacific Mail Steamship Company and president of the Panama Railroad that linked two oceans. His most famous role was as trustee of John C. Fremont's estate in Mariposa County, California. Rescuing it from disaster, he made it the mainstay of Mariposa County's economy. It was a happy resolution for an enterprise that, beginning in 1851, had been involved in a mining stock promotion in Paris in which French investors had lost fortunes.

In 1865, Park moved to New York, where he became one of Wall Street's leading movers. Nine years later in San Francisco, he joined Stewart by train down the San Joaquin Valley, by stage over the Tehachapi and into the Mojave Desert. By horseback they rode up steep and winding Surprise Canyon and into the primitive street of Panamint. Park stayed long enough to inspect the mines, nod his head, and depart for other pressing business. Senator Stewart, an old-time miner who had made his fortune in the silver camps, remained to buy the rest of Panamint and exploit "another Comstock".

Bill Stewart was a lion of a man—in carriage, in stature, even in the tawny hue of his abundant whiskers.

"He had," wrote a friend, ". . . a bearing like that of a lion when he stalks up and down his cage."

Born in 1827 on a New York farm, Stewart was at Yale when the Gold Rush drew him across the Isthmus in 1850. At first he drove bull teams in the Northern Mines and worked with pick and shovel on the Yuba River. It was Stewart who discovered the Eureka diggings, a rich placer deposit worked for many years.

But late in 1852 Stewart took up law, and no sooner had passed the state bar than he found himself district attorney of Nevada County, California. What he lacked in experience he made up in nerve. During his first case he is said to have knocked the opposing attorney unconscious to the courtroom floor. By wits and will he eventually conquered the finer points of the law, expanded his practice, and became attorney general of California.

When Bill joined the rush to Washoe in 1860 his rough-and-tumble methods quickly made him a power on the Comstock. He was once trying a case in a Nevada town when a local bad man burst in to bulldoze the court; Stewart covered him with a pair of deringers, put him on the witness stand, and subdued him with the words:

"If you attempt any of your gunplay or give any false testimony I'll blow your fool brains out."

When the Civil War began Stewart threw his weight in favor of the Union and, at a critical moment, helped to prevent secessionist elements from seizing control of Virginia City.

Meanwhile, he was building a fortune in legal fees from the maze of Comstock mining litigation. He soon invested his savings in a mine and quartz mill and made ready to share more directly in Washoe silver. His position on the Comstock became so dominant that in 1864 it was Bill Stewart, more than any other man, who fathered the new state of Nevada. He led the campaign for statehood, wrote the Nevada Constitution, and went to Washington as the first senator from the new state.

Stewart soon proved himself as much at home in the capitol halls as in the rude courtrooms of Nevada. With his broad-brimmed hat and luxuriant beard he cut a unique figure on Pennsylvania Avenue. He fought for the rights of his Washoe miners against unjust taxation, framed new mining laws of the nation, wrote the Fifteenth Amendment giving former slaves the right to vote, and helped steer it through the Senate chamber.

Like Jones, Stewart championed the cause of silver. When Congress halted the free coinage of silver, the outraged Stewart called it "The Crime of '73!" In that year he built a five-story Victorian mansion on Dupont Circle. Fitted with amenities purchased by his wife in Europe, "Stewart's Castle" became the showplace of Washington.

Less attractive was Stewart's early relationship with the Bank of California ring and the Big Four of the Central Pacific Railroad, which contributed heavily to his campaign war chests. In the case of the Central Pacific, at least, support went beyond political contributions to include gifts of stock and land. For his part, Stewart served the interests of the California bank crowd and the railroad, though without losing his independence.

"He is peculiar, but thoroughly honest," was Collis P. Huntington's left-handed compliment in a letter to Charles Crocker, "and will bear no dictation."

Thus it was three men of near-legendary reputation—Jones, Stewart and Park—who now took hold of Panamint. In forming the Surprise Valley

Mining and Water Company, Bill Stewart bought Jacobs' ten-stamp mill, the Surprise Canyon toll road, a site for a twenty-stamp mill and furnace, and the rest of Panamint's leading mines. Then he imported Chinese laborers to blast roads along the mountainsides to reach mine tunnels and piñon patches.

Soon explosions were jarring the atmosphere every moment of the day, flinging rocks across the canyon to the opposite mountainside, or showering them into the bustling street below. As one resident reported, this "of itself has a tendency to make things lively."

Employing every miner in camp at four dollars a day, Stewart began driving ahead on the Wyoming, Hemlock and other mines. He ordered that all first-class ore—any worth more than $500 per ton—be stuffed into gunny sacks for shipment to Welsh smelters across the Atlantic. Second-grade rock he set aside for reduction in his projected mill and furnace. Some rich Panamint ore was even packed on mules along the trail over the Argus and Coso ranges and up the Inyo Range to the furnaces at Cerro Gordo.

When word traveled through Western mining camps that Stewart and Jones were developing Panamint, the district jumped into prominence as a potential Comstock. The Nevada senators, it was agreed, knew their silver. To the miners it meant that Panamint would be yielding not only jobs at good wages, but new strikes and silver fortunes for those who drove their picks into the right ledges.

At one point Jones had reason to regret inviting Stewart and Trenor Park to the Panamint party. In 1871 the two had been among those involved in promoting the sale of the Emma Silver Mining Company, incorporated in New York with mineral property in Utah. The targeted market had been the group of British mining speculators in London. By 1874 the Emma had proven to be a notorious swindle that tarnished the U.S. Ambassador to London. Though by this time Stewart and Park were out of it, the scandal was fresh enough to cast some doubts in San Francisco on the Panamint scheme. On October 1 the *San Francisco Bulletin* voiced its suspicion under the heading, "Is a new Emma looming up?"

"The shapes of some of the Emma and Mariposa speculators are flitting about in connection with Panamint."

But for most miners in California and Nevada, the Panamint excitement was real enough. It was, in fact, the biggest rush in California's silver era. A correspondent of the *San Bernardino Argus* called Panamint "another Virginia City". And a reporter to the *Inyo Independent*:

"The excitements at Virginia and White Pine were as nought to what there will be here."

In the fall of 1874 the adventurers of California swarmed through Los Angeles, Bakersfield and San Bernardino, converging along dust-marked trails to Panamint.

"The talk is Panamint, and nothing but Panamint," recorded the *Los Angeles Herald*. "The exodus to Panamint continues in undiminished numbers."

In San Bernardino, whose location at the foot of Cajon Pass placed it at Panamint's back door, excitement ran still higher.

"The air is redolent with Panamint," exclaimed the *San Bernardino Argus*. "Everybody talks Panamint—the young and old, men and women."

By October the editor was complaining, "The Panamint excitement has carried away so many men from this valley that laborers are in great demand."

From Los Angeles the Argonauts followed the Bullion Trail, joining the contingents from northern California at the foot of Walker Pass. The fortunate ones rode horseback; others trudged along under the protection of the Cerro Gordo mule teams. East of Indian Wells the traffic was so great that a traveler moving against the tide was always in sight of another silver seeker. At the foot of Surprise Canyon they joined an even greater horde of prospectors who had ridden the length of Inyo County—the Nevada boys from Columbus, Austin, Eureka, Pioche and Virginia City. In their hurry to reach Panamint, some of the Piochers even rode due west 280 miles, braving the rigors of the southern Nevada desert and even California's Death Valley to reach their pot of silver.

Early in November a wagon road from Owens Valley through the Coso and Argus ranges to Panamint Valley was finished by John Shepherd, a pioneer settler who had brought his family to Inyo during the Indian wars. Without delay Panamint's first stagecoach, a four-horse Concord, lurched over the route and was soon making regular trips twice a week, loaded inside and topside with silver seekers.

By the second week in November a six-horse line was bringing Argonauts eastward three times a week across the Slate Range from Indian Wells, where it connected with the Bakersfield stages and eventually by train with San Francisco Bay.

On November 15 a four-horse stage began a weekly run through Cajon Pass from San Bernardino, and nine days later another weekly line opened from Los Angeles to Panamint via San Francisquito Canyon and Indian Wells.

Within a month the bustling camp had been converted from an isolated locality, reached only by horseback and buckboard, to an established terminus receiving seven stages a week along three converging roads.

Travelers on these routes were strangers to luxury in any form. Suffering 100-degree heat and driving sandstorms, most of them had little sleep as the coaches stopped only for a change of horses and for meals, which were minimal. On stretches that were reaonably level the horses whirled along at a full trot, with much swaying and jolting for the passengers. At some steep climbs, such as the Slate Range crossing, the passengers had to walk, and even to push. And going from Panamint back down Surprise Canyon, where the grade was 500 feet to the mile, the ride was exhilarating. At least two wagon teams ended in smashups.

At the end of June 1874, Pat Reddy reported there were yet no stores or saloons in camp. Most arrivals slept in their blankets in the open, though some lucky ones bedded down in a huge tent, the "Hotel de Bum", said to be owned by E.P. Raines.

But by October a row of stores and saloons, many of them built by merchants from Lone Pine and Independence, lined Panamint's main street. Between 700 and 800 men filled the canyon, living in canvas tents, rude cabins and even caves in the mountainsides. Lots were selling from $500 to $1,000, title resting not so much in a written deed as in a well-loaded shotgun. Jones, Stewart and Park had arrived to direct operations. When buxom Martha Camp arrived from Nevada with a bevy of frilled femininity and the nights soon stirred to the harp and fiddle, the boys hailed Panamint as a full-fledged camp.

By the end of November, Panamint's main thoroughfare was nearly a mile in length, extending from Jacobs' mill at the lower end to the Surprise Valley company's store at the other. The muddy street, deeply rutted and strewn with rubbish, was swarming with pack burros and mule teams, jostling each other in constant passage.

Filling the canyon were some fifty buildings, half of them constructed of finished lumber from Owens Valley sawmills, using Sierra timber. The rest were log or rock shanties, whose inhabitants were determined to shiver through the approaching winter to be on hand for the expected boom in the spring. Newer arrivals simply rolled up in their blankets in the stores and saloons at night, grateful to their obliging owners for a space on the floor.

Lining the main street were six general stores, two drugstores, three barbershops, three bakeries and restaurants, a livery stable, a boot shop and a meat market, whose butcher wagon doubled as the town hearse in case of a fatal shooting. A dozen saloons were jammed day and night while the poker pots regularly ran into the thousands of dollars.

Foremost of these was Dave Neagle's sumptuous Oriental Saloon, whose billiard table, black walnut bar and six-by-eight foot mirror were advertised as "the finest on the Coast outside of San Francisco."

The rival saloon in elegance was the still-larger Independent, where one of the two owners was E.A. "Ned" Reddy, younger brother of Pat Reddy.

Still another deluxe saloon, the Dexter, opened early in 1875 with satin-gilt wallpaper and four chandeliers. For the wall behind the bar, owner Fred Yager ordered a convex seven-by-twelve foot mirror (bigger than Dave Neagle's), which was carried from San Diego by steamship, rail and wagon all the way to Panamint. Unloaded in front of the saloon, it was about to be positioned behind the bar when it slipped from grasp and crashed on the floor. The distraught Yager replaced it with five large color lithographs of "the delightful and romantic scenery of the Rhine." A local editor consoled him that these were "a much more pleasing and satisfactory scene than that of a fellow with a dirty shirt standing before a magnifying mirror."

In another section of what was grandiosly termed the Neagle Block was an institution of which even Cerro Gordo, queen camp of Inyo, could not boast—the Bank of Panamint. It was soon the business center of town and the point of arrival and departure for the stage lines.

Other metropolitan necessities were also being established in Surprise Canyon. At the center of town a log-cabin brewery was erected. In November a Sacramento newsman named T.S. Harris arrived with his typesetter and printer's devil. For the first week they camped on the ground, cooked meals over a fire, and raised a canvas tent as shelter for the hand press and type cases. On the 26th they turned out the first issue of the small but robust *Panamint News*.

Finally, up in Sourdough Canyon a cemetery was improvised when two luckless combatants, killed by bullet and knife, required simultaneous burials.

Sourdough Canyon did not lack for residents in this rough and isolated camp—at first lacking any law. One of the early Panamint "bad men" was one Kirby, who had arrived from Pioche and announced he was a "little rooster, and if you don't believe it just draw and toe a line." After a card game dispute in September 1874 he shot Bill Norton in the leg. As Norton returned the fire, Kirby retreated out the door—not quite the "rooster" he had claimed. Then he waited behind the flap of a tent and took aim while the crippled Norton came after him. Witnessing the trap was W.C. Smith, the district mine recorder. As the only elected official in camp, Smith drew his own revolver and got the drop on Kirby, ordering him to throw down his weapon or be killed "in an instant". Kirby dropped his gun and quietly stole out of Panamint. Norton recovered, and Smith was later elected Justice of the Peace.

Another who brooked no nonsense was Jim Bruce, tall and affable, who had arrived from Pioche as early as April 1874. Bruce was, in the words of the *Panamint News* editor, "one of our most charitable and public spirited citizens." Among other enterprises, including professional gambling, Bruce was part owner of the Clear Springs Water Company that supplied Panamint. When it was realized that Panamint was vulnerable to the fires that devastated other mining camps, Jim was a leader in promoting a fire company. And when the miserable main street became too disgraceful for the sensibilities of Panaminters, Jim Bruce took the lead in getting it graded for "a fine appearance".

But Jim Bruce was too human to keep out of trouble in Panamint. On the night of March 22, 1875, he was at Martha Camp's bawdy house talking and drinking with two others, including Edward Barstow, a night watchman of the town. There was an argument and Barstow, who had been drinking heavily, stalked out. Around 2 am he came back looking for Bruce, who had gone to bed with Martha. At her bedroom Barstow ordered her to open the door.

"Go away from here," cried Martha. "You can't come in here."

"If you don't open the door," hollered Barstow, "I will kick it in."

Martha told him not to kick in the door.

"I want to see Jim!" shouted Barstow.

Martha got up and went to the door. From the bed, Bruce told her to light the lamp on the bureau. Then Martha opened the door. Barstow pushed in, six-shooter in hand. Martha screamed and tried to stop him.

"I've got the long-legged son of a bitch at last," Barstow growled, blowing out the lamp. Barstow and Bruce started shooting, filling the air with thunder and smoke. Martha saw the flash from Barstow's gun. Unscathed, Bruce shot him in the right leg and left groin.

A few moments later Jim walked over to Dr. F.T. Bicknell's and asked him to "go and see a man who was shot." The doctor and others rushed to Martha Camp's room. Lying there, Barstow said he "was drunk and to blame." Bruce offered to have Barstow carried to his own quarters, where the wounded man's right leg was amputated. But to no avail; he died that night. At the hearing before Judge W.C. Smith, Bruce was exonerated on grounds of self-defense. He was, after all, "one of our most charitable and public-spirited citizens."

By this time the dynamic Bill Stewart, whose vigor had sown such a riotous camp, had turned to the problems of exporting ore and importing the mill machinery that would make Panamint a full-scale producer. Most logical means of transportation was Remi Nadeau's Cerro Gordo Freighting Company, which already hauled almost every pound of Inyo cargo. But when Stewart approached Nadeau, the shrewd Frenchman took one look at the heavy sands across Indian Wells Valley, the soft marsh of Borax Lake, the backbreaking Slate Range grade, and quoted a freight rate that sent the senator seeking other transportation.

Early in September 1874, Stewart jolted down through Cajon Pass in his buckboard and contacted Caesar Meyerstein, San Bernardino forwarder and wholesaler. Agreeing to deliver upbound machinery at eighty dollars per ton and downbound ore at thirty dollars, Meyerstein had his mule teams plodding northward from San Bernardino by early October.

From the Southern Pacific railhead at Spadra, Caesar's wagons labored 200 miles to Panamint by way of Cajon Pass, crossing the Mojave Desert via such stations as Hennington's Ranch near the present Victorville, Granite Springs twenty-four miles east of the later town of Johannesburg, and Post Office Springs halfway up Panamint Valley where the adobe walls of Ballarat, a later gold camp, are fast crumbling under the elements.

The sudden freighting activity breathed new life into Southern California in the fall of '74. At Spadra, where the Southern Pacific railbuilders east from Los Angeles had temporarily halted, fifteen new stores and saloons appeared overnight, while long lines of wagons crowded about the Southern Pacific depot to be loaded for Panamint. San Bernardino County farmers sold their barley crops at thirty dollars per ton and rejoiced over the best market in years. Blacksmiths and wagoners in San Bernardino were overwhelmed with business, while new frame buildings were rising to accommodate the trade. The streets were jammed with mule teams camping or resting before the toilsome ascent of Cajon Pass.

"Everyone that can," observed the *San Bernardino Guardian*, "is rigging out teams for the Panamint trade."

At first it appeared that San Bernardino had captured the Panamint prize. But Meyerstein's heavy wagons, foundering in the sands of the Mojave, were leaving hundreds of tons piled on the Spadra platforms. By early November so many Los Angeles teamsters had entered the trade that a continuous line of freight wagons was raising dust past Indian Wells toward Panamint.

Even Remi Nadeau of the Cerro Gordo Freighting Company, finding the commerce too rich to disregard, joined the traffic himself. Erecting a station at Salt Springs and another at Water Station north of the present Trona, he began buying all the teams he could find.

By year's end forty sets of his wagons, each drawn by fourteen mules, were operating between Los Angeles and Panamint. At the Southern Pacific's railhead in San Fernando, where Nadeau headquartered his teams, new buildings were rising to handle the raging Panamint traffic.

Thus it was Los Angeles, after all, that was enjoying the surge in prosperity powered by the Panamint boom. Lumber and grain, flour and whisky were pouring northward from the forwarding houses of Los Angeles Street in welcome reminder of the Cerro Gordo excitement five years before. Farmers were selling their surplus grain crops at fancy prices, and lumberyards were deluged with more orders than they could fill.

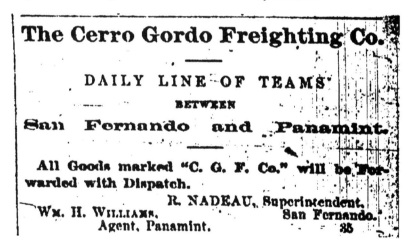

With the coming of winter Jones and Stewart's silver camp reached its shining moment. The Surprise Valley company was operating a half-dozen mines, working two hundred miners in two ten-hour shifts per day. In mid-December, Jacobs' ten-stamp mill chugged into action, and silver bullion joined the ten tons of sacked ore that jolted down Surprise Canyon every day to the jingle of team bells. When the silver senators climbed aboard the stage near the end of the year to appear at the next session of Congress, town lots were bringing $2,500 to $3,000, and the population stood between 1,500 and 2,000 people (some estimates put it at 2,500 and even 3,000). The ponderous duties awaiting the senators in Washington included getting a daily mail service from Owens Valley to Panamint.

It was time now to, in the jargon of the day, "run an adit through the San Francisco Stock Board." On January 28, 1875, the two mining companies incorporated by Jones and Stewart placed their stock on the market. The Wonder Consolidated Company combined footage in Jacobs' Wonder, Stewart's Wonder and two other mines on the north slope, plus Jacobs' mill. The Wyoming Consolidated comprised footage in the Wyoming, Hemlock and five other mines on the south side. Each offered 10,000 shares at $15 per share.

"The mines are so well developed, " according to the newspaper ads, "that the companies can assure the subscribers to these stocks that no assessments will be required."

Four days later the *Alta California* ran a front-page story that, for two-and-a-half columns, equalled any promotion in mining history. Writing from Panamint, the *Alta* correspondent outdid himself with each paragraph in his praise of the richness of the ores, the solid and respectable character of the town and its citizens, and the energy of Jones, Stewart and Park, "whose wealth and enterprise we have admired in San Francisco and on the Coast." The mines grew richer the deeper they went, indicating what mineralogists called a "true fissure vein" that was virtually inexhaustible. And worries about the distance from market were put at rest by a planned railroad from the coast—"the iron horse will go snorting into the new town in less than a year."

With such heavy artillery did Jones and Stewart launch their offensive. Within a few days all the shares were snapped up at the asking price. Many investors were disappointed at not getting any. One Panaminter offered a premium of $20 each for 500 shares, but was only able to get 100.

The $300,000 proceeds from the sale were modest enough for the three enterprisers, who claimed they had already spent $1,200,000 on mines, mills and roads. But the footage sold in the mines was only part of the total footage, and there were hints that more would be available later.

A celebration was in order, and the owner of Panamint's Wyoming Restaurant, Delia Donaghue, organized a grand ball for Washington's birthday. Delia moved out the furniture to make room for the dancers and the platform for the four-piece band, which included a flutist and a harpist. Special tickets in bright blue ink were printed by the *Panamint News*.

However, many of the miners held back, thinking Panamint could not muster enough women for a chance on the dance floor. But when the party began, sixteen ladies were on hand, dressed in their finest.

"Where they all came from is still a mystery," wrote editor Harris of the *News*.

When it appeared there were more women than men, word flashed through camp. The boys jumped into their best clothes and "proceeded to storm the works." The band struck up a merry tune, the harpist contributing "a heap of extra and most excellent twangs." In a moment Delia's floor was bouncing to the stomp of hob nail boots (the ladies were more merciful with their delicate prunella shoes).

Delia's social triumph was followed two weeks later by a surprise party given by the Panamint women for the employees of the Surprise Valley Mill and Mining Company—this time with eighteen ladies present. And on Independence Day, 1875, a grand parade was cheered on its way through town, headed by a marching band consisting of a tuba and a bass drum, and highlighted by young ladies and children riding on the butcher wagon, which was decorated with red, white and blue. At last Panamint had a social life beyond the saloons and Martha Camp's.

Meanwhile, the heavy snows of a harsh winter had sent shivering Panaminters hustling down Surprise Canyon for Inyo County's milder regions. The spring of 1875 found the camp with a scant 600 citizens. When the Bakersfield stage brought Senator Bill Stewart and Trenor Park into Panamint on April 13, they beheld a camp wallowing in that doldrum period between the first stampede and the beginning of heavy production.

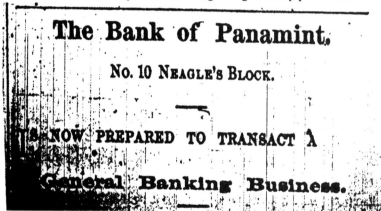

But during their three-day inspection they found the gigantic twenty-stamp mill nearing completion, the ponderous machinery ready for installation after its long wagon haul from Spadra, and full-scale production attainable by early summer. On the morning of April 16, when the two departing visitors dropped their luggage at the stage stop in front of the Bank of Panamint, they were convinced that the camp would not be quiet for long.

Just as they prepared to swing into the coach, the early morning calm was shattered by two citizens in violent argument at the door of the bank. One was Robert McKinney and the other was the same Jim Bruce who had killed Barstow in Martha Camp's bedroom. They were quarreling over a borrowed gun. All at once McKinney reached for his revolver and started firing. Bruce, hit in the left forearm, pulled out his pistol and fired. Some eight shots rang so fast they shook the canyon in a continuous roar.

Next moment the two gunmen lay bleeding on the board walk. Bruce had a second would in the flesh above the hip. McKinney was shot in the hip, wrist, arm and leg—wounds from which he later died.

While the stretcher-bearers came running from the justice's office, Bill Stewart and Trenor Park emerged from their temporary positions behind a stone wall next to the bank and took their seats in the waiting stage. The dignity of Wall Street and Capitol Hill, they discovered, was a useless commodity when bullets flew in Panamint. The senator and the financier rocked down Surprise Canyon with the impression that their town was not quite so lifeless as they had imagined. As for Jim Bruce, he recovered and unintentionally took his place with Dave Neagle and Ned Reddy in the gallery of Panamint gunfighters.

Nor were shootings the only show of lawlessness in remote Panamint. Wildrose Canyon, a few miles north of camp, was the hideout for a gang of desperadoes who made a living robbing Nevada stages. Among them were John Small and John McDonald. In the summer and fall of 1874 they had robbed six stages near Austin, Eureka and Pioche. In one incident they had seriously wounded the driver. At their stomping grounds in the Panamints, they had even sold a mine to Senator Stewart—according to his memoirs.

As described in the Wells Fargo reward poster, John Small was of average height, with light hair and moustache and a round, reddish face. John McDonald, a Canadian, was a slender five-foot-ten, with black hair and beard, long face and sharp nose. Small usually wore a gray suit, McDonald a frock coat and wide-brimmed hat. Small was talkative, McDonald quiet. Wells Fargo offered $4,000 for their capture.

This arrogant pair and their comrades of Wildrose Canyon swaggered the main street of Panamint without fear of the law, boasting that they could not be taken by 100 men. One of them walked into the Wells Fargo office and, looking at the reward notice, growled:

"By God, this poster offering a reward for Small and McDonald does not please me."

With that he tore it down and walked out.

By the end of spring Panamint was almost ready for full-scale production. In June 1875 a cable tramway, supported by great wooden arches and equipped with iron ore buckets, was completed from the Wyoming mine to the reduction works below.

On the 29th the new mill was tested. While a crowd gathered, the boilers were fired, steam was applied to the engine, and as the *News* described it, "the huge monster started gracefully"—its twenty stamps dancing in measured succession. One of the superintendents popped a bottle of wine, which was quickly exhausted.

For the next two months the bandits of Wildrose Canyon took a proprietary interest in the workings, pointedly asking Stewart when the furnace would start producing bullion. At the end of August the furnace filled the air with smoke and began turning out silver bullion worth some $30,000 per ton. According to Stewart's reminiscences, the robbers were on hand when the first gleaming ingots rolled out of the furnace. Normally they would weigh around 90 pounds—the most that an average man could routinely carry.

But the bandits' arrogance turned to chagrin when they found the huge blocks weighed more than 400 pounds apiece (with his usual exaggeration Stewart put the number at 750 pounds). Turning on the grinning Senator, they called him the meanest man in the Panamints.

"Do you think it's right to play that game on *us*," cried one, "—and after we sold you the mine, too."

"You can't expect *me* to be sorry for you, can you?" returned Stewart.

In exasperation—still according to Stewart—the outlaws tackled one of the silver ingots, but their heaves failed to budge it. A few minutes later they brought a husky pack mule. But, with no stomach for robbery, the animal kicked and jumped as they tried to put a silver pig on his back. Finally they abandoned the scheme. After all, where could they cash in such currency even if they could make off with it?

Though this story may be one of Stewart's exaggerations, it was true that the silver bars were cast so heavily just to foil any possible bandits as Nadeau's wagons rolled for days unprotected across the upper Mojave to the S.P. railhead at Caliente. In fact, the *Panamint News* dutifully refrained from announcing the first shipment, as though it were a military secret. And in November the Panamint correspondent of the *Los Angeles Express* reported:

"Some robbers nearly got away with a large shipment of bullion the other day, but on finding that it was in five hundred pound bars, and that they could not carry it very well, they concluded to let it alone."

At this juncture, while the Cerro Gordo Freighting Company regularly carried the ponderous cargo to the railroad unperturbed, John Small and John McDonald robbed a general store of $2,300 and left Panamint for good.

With more than forty tons of Wyoming and Hemlock ore being reduced to bullion every day, Panamint was now a mature camp with full employment, a modest number of families, and fewer of the drifters that had arrived in the first stampede.

While Bill Stewart was opening production in Panamint, John P. Jones was on the coast rounding out the enterprise on a grand scale. If Panamint was to be another Comstock, it needed a railroad. And so in September 1874 Jones rode into Los Angeles and put up $220,000 toward a proposed line from tidewater to the silver country—the Los Angeles and Independence Railroad. Then he visited Santa Monica Bay and bought up three-fourths of the Rancho Santa Monica y San Vicente as a sea-going outlet for his tracks.

What was more, Jones contacted the Eastern railroad magnate, Jay Gould, and suggested linking up with his Union Pacific line for a transcontinental connection. Gould agreed informally to meet him halfway, and the Panamint venture became national news.

All this did not suit the Big Four directors of the Southern Pacific—Leland Stanford, Charles Crocker, Collis P. Huntington and Mark Hopkins. They had their eyes on the Panamint traffic, and were miffed when Jones carefully avoided a connection with their Southern California rails. Worse still would be any linkup that would introduced a competing transcontinental line into California. To the Big Four, such a challenge meant action.

The strategic spot in Jones' master plan was Cajon Pass, where a Los Angeles and Independence survey party had already made preliminary explorations. On January 7, 1875, a band of Southern Pacific men rode into the Cajon to drive their stakes and take possession.

But the L.A. & I. had an alert construction engineer—young Joseph U. Crawford. Getting wind of the S.P. move, he took his own crew and headed for the pass the same day. Somewhere along the sandy bottom of Cajon Creek they passed the rival S.P. gang. Knowing the ground from his previous reconnaissance, Crawford rode straight for Tollgate Canyon—the one spot too narrow for more than one railroad. His men were driving stakes and running lines when the Southern Pacific men arrived. Jones' railroad had won the first round.

Then Huntington of the Southern Pacific tried to stop the upstart railroad with an act of Congress. On January 8 a bill was introduced to give the S.P. an exclusive right-of-way through Cajon Pass. With Huntington's lobbying power, fortified by liberal campaign contributions and other fa-

Eastern California Museum

In 1874-5, California's biggest silver excitement was Panamint, just west of Death Valley. With a population of some 2,000, it boasted a bank, brewery, newspaper, more than a dozen saloons, and a graveyard for men who died with their boots on. Only the ruins of the 20-stamp mill remain today.

Westerners Brand Book No. 11 and Russ Leadabrand Collection

Driving force for Panamint was the Surprise Valley Mining Co.'s furnace and 20-stamp mill, built by Nevada senators John P. Jones and William M. Stewart. When the mill started up in 1875, bandits were on hand to steal the bullion. But Stewart foiled them with silver bars weighing over 400 pounds.

California State Library
Builder of California's first railroad, Lester L. Robinson was president of the big New Coso mining co. at Darwin.

California State Library
Third partner in Panamint was Trenor W. Park, who controlled the Panama RR and the Pacific Mail Steamship Co.

vors, the bill might have passed. But the Los Angeles people heard of it and deluged Congress with such outraged protest that it dropped the scheme. Huntington soon had to calm his alarmed partners in California.

"I do not think they will hurt us much," he wrote. "I will ventilate their safe harbor."

But Jones had only begun to fight. In July 1875 he founded the town of Santa Monica with a grand auction sale of lots. Into the Pacific he built a wharf to serve ships bringing in rails and ties. By December the L.A. & I. rails had reached from the bay 14 miles to Los Angeles, and the Iron Horse puffed into town from a new direction. In Cajon Pass his men were digging a tunnel that would give access to the Mojave Desert—and Panamint.

With the huge mill and furnace turning out bullion bars every day, it appeared that Panamint's flame was brighter than ever. Yet for several months the fire had been flickering. Over in the Coso Range, Darwin's silver ores were more cheaply worked—much of them without requiring a furnace—and many chloriders left Panamint for the newer camp. Many of the business houses that had closed by the spring of 1875, awaiting the startup of Panamint's big mill, had not reopened. Even T.S. Harris shipped his press and type cases by Nadeau's teams over to Darwin, where he launched the *Coso Mining News* in November.

Worse still, questions were arising over the size and richness of the ore deposits. As early as July 1874 one expert who had visited Panamint gave a mixed review in a letter to the *Eureka Sentinel*:

"It is a fine prospect on the surface but I do not believe the mineral will extend to any great depth."

In the following months Stewart and Jones pushed their mines deeper into the earth, extracting some $300,000 in silver by May 1876. But the ore pockets that showed so well on the surface began to fail, and no large blind leads appeared.

Meanwhile, bigger events in California's silver economy were stirring. By August 1875, speculation in Comstock mines—both established bonanzas and wildcat prospects vying for public investors—had reached the breaking point. The Bank of California, headed by financier William C. Ralston and Senator William Sharon, lost the confidence of depositors. On August 27 they staged a run on the bank that caused its collapse. Ralston's body was found floating in San Francisco Bay, raising the assumption of suicide.

In the ripple of subsequent failures that swept the state, the Temple and Workman bank in Los Angeles also went down. Since it had been a key supporter of the Los Angeles and Independence Railroad, that enterprise stalled, and rail construction halted. It would be purchased at a sacrifice price four years later by its adversary, the Southern Pacific Railroad.

Senator Jones, already injured by the crash in the mining stock market, found his grandiose venture toppling about him. In Panamint the Wyoming and Hemlock were scraping the bottom of their lodes. At the end of May 1876 Stewart and Jones decided to shut down the mill and furnace. The steam engine was blown off, the stamps stopped their incessant marching, the bellows were collapsed and the furnace charcoals banked. Wyoming Con and Wonder Con stock sank off the blackboard.

For still another year the mine shafts were sunk deeper in search of new leads, but below the 600-foot level the rock was still barren. On May 8, 1877, Stewart and Jones closed the mines and discharged their employees. By July 1878 the superintendent was hauling the last of portable machinery down Surprise Canyon.

Was the Panamint excitement just another stock promotion, as some claimed? Stock promotion it was, but that effort was underplayed, with only $300,000 worth of stock offered to investors (who were fortunate they could not buy more). Basically, the story does not match other stock schemes familiar at the time.

Jones, and especially Stewart, invested much of their time on the ground at Panamint launching the enterprise. While the total estimate of $2 million that they spent is probably overdrawn, it is reasonable to believe the figure of $1,200,000 that was announced during the stock sale. With Nadeau's freight charges at $40 per ton from Caliente, the cost of delivering mining, mill and furnace equipment was staggering. So was the erection of the mill and furnace, the payroll of miners and millmen, the blasting and grading of roads.

The expense clearly outran the revenue, which consisted of about half a million in proceeds of shipped ore and bullion, plus $300,000 in stock sales. Stewart, Jones and Park probably lost several hundred thousand dollars.

Thus the Panamint fiasco was marked more by mismanagement than by swindle. Erecting a 20-stamp mill in this wilderness, on the hope that the lodes would get bigger and richer as they went down, was the fatal error. Today the crumbled ruins of mill and furnace, which are all that mark the site of Panamint, remain a monument to the folly of the silver senators and a Wall Street wizard.

Trenor Park died at sea on board the steamer *City of Para* in December 1882. So great were his holdings that his death impacted stock prices of the Pacific Mail Steamship Company, Western Union and Erie Railroad.

Far more volatile than Park's fortune was Bill Stewart's, which ran the roller coaster of mining speculation. When he left Washington after his last Senate term in 1905, he was "practically penniless," as one observer wrote. He recouped a little of his fortune in gold mining at Bullfrog, Nevada—only a couple of mountain ranges east of Panamint. He was back in Washington when he died at age 82 in April 1909.

On hearing the news at his home in the Santa Monica he had founded, Senator John P. Jones said that Stewart "could be relied upon to be on the side of right and to stick there through the thickest machinations and the strongest pressure." Three years later, Jones himself had died.

Panamint itself did not succumb easily. In 1879, mill and furnace were purchased by the Leota Gold and Silver Mining Company, which started them up on ore from some sixty chloriders working their own mines. Early in September, thirteen bars of bullion, averaging 90 pounds apiece, were hauled down Surprise Canyon under the escort of miners armed with Henry rifles.

By December the Leota company failed. Employees and other creditors attached the property and fired up the mill again. In four days they shipped $5,000 in bullion by mule team and train to San Francisco. When the proceeds were returned in gold coin by Wells Fargo, a bandit held up the stage north of Indian Wells and made off with the loot.

But this was a small setback compared to the bigger ones before. By 1880, local men acquired the old properties, fired the mill again and turned out bullion. In 1881 another concern leased the business, found new bodies of silver chloride in the Wyoming and Hemlock, and in four months' time 25 bars of bullion went down the grade. In 1882 Panamint still had some fifty people and the mill was still producing silver. A waterspout washed out the Surprise Canyon road, so that shipments in and out were by pack mules. Late in the year the mill burned down. But Panamint itself, still innocent of the fires that devastated most camps, stood like a brooding ghost. Reported one visitor:

"It seemed like a funeral procession of one, to pass up the deserted streets of Panamint, which so few years ago were teeming with a population of at least 3,000, and where millions of money was spent in a misdirected manner."

9. PAT REDDY FOR THE DEFENSE

> The silver deposits at Darwin had been sitting under everybody's noses as they rushed to Cerro Gordo and Panamint. But Darwin had rich ore that was easily worked, and a handy location near Owens Lake and the Bullion Trail. Suddenly it was a magnet for all the characters, good and bad, who had followed silver up and down Eastern California. Leading them was Pat Reddy, miner turned lawyer, who grabbed and exploited the biggest mine. Pat fought for law and order, but he could not prevent the violent climax to one of California's first mine labor disputes.

Early in December 1874 Victor Beaudry, sitting in his assay room at Cerro Gordo, was applying chemicals to ore samples brought to him by Bentura Beltran, a veteran Mexican prospector widely respected in Owens Valley. Beltran had carried the specimens from his Buena Ventura claim near Darwin Wash, a few miles northeast of Old Coso. Results of the assay: $700 in silver to the ton.

Quickly Beaudry took to horse, joined by most of the leading men of Cerro Gordo. Down the Yellow Grade and southward past Owens Lake they rode. Leaving the Panamint stage road, they jogged a few miles through low hills covered with brush and joshua trees and into the Coso Range. After topping the last rise they descended on a bustling scene where men dug holes into the side of what came to be known as Mount Ophir.

Examining the Buena Ventura, Beaudry offered a reputed $7,000 to Beltran and his partners. When they refused, the Frenchman set about buying several adjoining mines.

A few days later, awakened by miners coming into Independence to record their claims, another party headed by lawyer Pat Reddy raised dust for the Mt. Ophir mines.

"Stand by your claims, boys!" chided the *Inyo Independent*.

The rush to New Coso had begun—little more than a month after the first discovery. In October 1874 a friend of Beltran's, Rafael Cuervo, had started picking into promising ledges on Mount Ophir. On the 22nd he located his Promontorio, first to be recorded in the new district. The enterprising Cuervo had moved quickly to stake other silver claims on the same day, together with a nearby spring of water that became known as Cuervo Springs.

In early November he had been joined by Beltran, who located the Buena Ventura on the east side of Mt. Ophir. The same day, November 16, Rafael staked still another, the Cuervo.

By this time the Mexicans were not alone on Mount Ophir. Professor William D. Brown, an expert mineralogist in San Francisco, had visited Cerro Gordo in 1867 and recognized its riches. With news of the Panamint discovery, the good professor reckoned that the country in between must hold undiscovered wealth

With his brother, Robert Brown, he left San Francisco early in October 1874, took the Southern Pacific down San Joaquin Valley, and bent onward over the lower Sierra by stage and horseback to Old Coso. There, Mexicans showed samples of galena ore but were silent on where they were found.

His anticipation whetted, Professor Brown climbed the nearest mountain and surveyed the landscape. To the northeast he spotted the multicolored outcroppings of Mount Ophir. With his brother he rode from the Cosos to the yellow mountain and found Rafael Cuervo's claim to the Promontorio Mine. Here was the pot of silver at the rainbow's end.

Swiftly the Brown brothers prospected nearby and, starting on November 4, staked claims of their own, including the New York, whose vein of ore was forty feet wide and some 2,000 feet long. When assayed at Cerro Gordo and Swansea, the results averaged a respectable $100 per ton in silver, and 50 percent lead.

New Coso fever flared still higher. From a dozen men on the ground in late November the population swelled to 150 by mid-December, and at least 250 more were on the 70-mile road from Panamint. On December 3 the miners already on the ground gathered at the Brown brothers' camp to form a new mining district—New Coso—which generously took in all of the Coso and Argus ranges from the Bullion Trail east to Panamint Valley. For recorder they elected Abner B. Elder, pioneer of Cerro Gordo, who promptly received some 60 claim notices.

One of these was for Crystal Springs, discovered by the Brown brothers, who sold it to Victor Beaudry and Elder; their experience at Cerro Gordo had taught them the power of water control. They promptly began laying pipe from the springs seven miles into the new camp.

When Pat Reddy reached New Coso in mid-December his early miner's training took him to Beltran's Buena Ventura. By this time it had been combined with the Defiance lode, claimed by one John Wilson, and on December 16 the package was sold to Reddy as the Defiance claim for $10,000. Since the San Pedro lead was an extension of the Defiance,

Reddy bought that from Beltran and Pedro Ruperez for $21,000. He now owned what would become the bonanza of New Coso.

With his usual vigor, Reddy drove the tunnel started by Beltran deeper into Mount Ophir, the miners working in two shifts extracting ore averaging $120 per ton. By January 4, 1875, Reddy had filed a location site for his mill and furnace on the mountainside northeast of camp. Suddenly he was spending less time in court and more time in the ground.

By the end of January 1875 the camp had become a town named Darwin. Located just west of Darwin Wash, it was named for the Erasmus Darwin French who had led the first expedition to the Coso Range in 1860. A stage from Lone Pine pulled into Darwin three times a week, then lurched on to Junction Ranch near the head of Shepherd's Canyon in the Argus Range, where it connected with the Panamint Stage from Indian Wells.

The town now had nearly 200 people and some 60 rock or adobe buildings finished or under construction, including a hotel, two restaurants, two butcher shops, two general stores, a livery stable, lawyer's office, doctor's office and six saloons. Among these was the Capital Saloon, first frame building in town, where Pat Reddy's brother Ned shared ownership with John Wilson, the same who had sold Pat the Defiance Mine.

Arriving in style by the new stage line were early investors from Virginia City and San Francisco. One of the first was George Hearst, who had led the rush to the Comstock in '59. His reputation—"one of the best mining experts on the coast," as the *Panamint News* put it—spurred still more excitement.

So did the arrival of Lester Ludyah Robinson, by profession a railroad engineer, by career an investor in large mining, real estate and railway enterprises. Born in central New York in 1824, he had started as a chain bearer in a surveying crew with the New York and Erie Railroad. He had bossed the construction of railroads and bridges in the northeastern states and Canada before he came to California in 1854 to build the state's first rail line, the Sacramento Valley Railroad.

Beset with the financial panic of 1855 and the company's own financial distress, Robinson borrowed $750,000 from a French banking firm and was able to finish the 22-mile railroad from Sacramento to Folsom in February 1856.

After that, Robinson turned to many enterprises in California, including street railways in San Francisco, hydraulic mining in California's Northern Mines, various other mines in California and Mexico, a land and irrigation project in Riverside, and an association with Egbert Judson in the Giant Powder Company.

A lifelong bachelor, Robinson shunned social life, though in a small company of friends, according to the *San Francisco Call*, he showed "brilliant powers as a conversationalist and raconteur." In appearance he was tall and powerfully built, with neatly trimmed gray hair and beard.

"His blue eyes are remarkably piercing and his face is set in a severe even somber expression, which lightens under the genial influence of his smile."

In short, Lester L. Robinson was all business.

"He follows up the slightest detail or incident," wrote another acquaintance; "he can figure up the cost of extracting and reducing a ton of ore to a nicety."

Such were the weapons Robinson brought with him as he swung down from the stage in Darwin's Main Street. He promptly bought two of the most promising mines—the Lucky Jim and the Christmas Gift—and returned to San Francisco to start the New Coso Mining Company.

By this time the Nevada contingent—mostly adventurers from Pioche—were pouring in "on the jump", as the *San Bernardino Guardian* described them. In consequence, miners with claims moved out of town and lived in tents on the ground.

On April 12 Darwin was baptized with its first shooting when Jim Russell, a fighting man from Cerro Gordo, tried to take a horse from Tom Star's stable. There followed a heated argument over the horse's ownership; both men drew revolvers and fired, but it was Russell who hit the ground with lead in his heart and head.

Arresting Star, the Darwin constable allowed him to have dinner at a restaurant before taking him to Independence. While the officer stood guard at the front door, Star cooly departed by the back door, ran to a horse and raced off under a volley of shots. The Kern County sheriff caught him and notified the deputy sheriff of Inyo County. By the time that officer arrived in Bakersfield, Star had been released on *habeas corpus*.

In the summer of 1875 Darwin had 1,000 citizens, 150 buildings plus many tents and shacks, two baseball teams and a boundless enthusiasm. But though plenty of ore was piling up at the mines and furnaces were being built, no bullion had been shipped. Darwin's prosperity was founded entirely on anticipation.

On July 3 Victor Beaudry changed all that. His pipes from Crystal Springs had reached town and his hydrants appeared on the streets. All was ready to supply the boilers for the stamp mills and the water jackets of the furnaces. At 4:00 a.m. on Independence Day, everyone in town gathered at Beaudry's invitation for what one observer called "a remorseless sluicing down" from the hydrants. At Beaudry's tank "the private cuvée flowed as freely as the water," with toasts by Beaudry, Pat Reddy and other dignitaries. At 2½ cents per gallon, Darwin had all the water it needed to get on a paying basis.

Early in August the Cuervo became the first furnace to turn out silver-lead bullion. By October the machinery for Pat Reddy's Defiance was swaying across the Mojave by Remi Nadeau's mule teams from the Southern Pacific's end-of-track at Caliente; the Defiance was fired up in December. And in the same month L.L. Robinson, returning from San Francisco, started his New Coso furnace, located in the Lucky Jim Wash southwest of town.

By the end of 1875 a fourteen-mule team of Nadeau's Cerro Gordo Freighting Company was laboring out of Darwin every day loaded with bullion. It rolled to the nearest point on Owens Lake, where the *Bessie Brady* hauled the silver-lead bars across the water to Cartago on the Bullion Trail. Then the wagons carried the cargo south across the Mojave and over the Tehachapis to Caliente.

Finally Darwin was producing silver, and the flow of money spurred her growth. Town lots were going for $1,000, and as a Darwinite wrote, the "sound of hammer and saw is heard continually throughout the day, and at night the streets are crowded with merchants, miners and all classes." Beaudry provided free water carts to dampen the foot-deep dust in the streets, making the atmosphere bearable for the twenty or so women in the population.

On November 6, T. S. Harris brought his Washington hand press from Panamint to turn out the first issue of the *Coso Mining News*. It carried advertisements for three blacksmith shops, two markets, three general stores, two hotels, two drug stores, three restaurants, six saloons, a real estate office, and most important, a combination bath house and barber shop.

Within another month there were actually fifteen saloons, some of them still in tents. One of the most elegant was Col. J.R. Norton's Centennial, which had a billiard table "of the same pattern as used in the Palace Hotel in San Francisco." Like several other saloons, including Ned Reddy's Capital, it sported plenty of gilding and Bohemian glass. As the *Kern County Courier* observed:

"Superb chandeliers shed their soft radiance through ground glass globes; gorgeous mirrors reflect the motley crew that drift in and out."

The Centennial was, in fact, a trade center and stock exchange. "Here the new discoveries are announced, prospecting parties are made up or come here to report, bargains are made and contracts entered into."

By now the stampeders to Darwin could ride the stage from two directions—the tri-weekly from Nevada through Owens Valley, and the daily from Caliente up the Bullion Trail to Indian Wells, where it branched eastward on the new road through Panamint Junction, then north to Darwin through Mountain Springs Canyon.

As of mid-summer several leading mining companies—led by L.L. Robinson's New Coso—were incorporated and selling shares on the San Francisco exchanges. In August, Pat Reddy incorporated his Defiance, but announced he would wait until he could pay dividends before putting shares on the market. Darwin was riding the crest of its boom, and Pat Reddy was in the saddle.

The frontier lawyer had come to this turn of fortune after thirteen years in the mines of California and Nevada. His father, Michael Reddy, had emigrated from County Carlow, Ireland, and settled in Woonsocket, Rhode Island. Born on February 15, 1839, Pat was one of four Reddy children. In 1861, when the Nevada and Colorado mining excitements were aflame, young Pat shipped for California and arrived in San Francisco that February with, as one chronicler put it, "a common school education and his indomitable energy." He stepped off the ship a tall and broad-shouldered Irishman, with curly red hair, deepset eyes and a smile of anticipation on his face. With too much cheek and not enough meek, he faced the high-flying world of Western mining. Unblinking, he would become, if not its master, one of its movers.

Wilth him was his younger brother, Edward Allen Reddy, then age sixteen. With the same formidable build as Pat, young Ned was even more brash than his older brother.

Together they worked as miners in the Mother Lode, then trekked over the Sierra to the Nevada mines. In Aurora, according to one resident, they were the "terror" of the town. Pat was known as a hard drinker, hard fighter and a good man to have on your side in a brawl. Yet he was kind-hearted and generous to those down on their luck, and fiercely loyal to his brother and their friends.

Besides handling sledge and powder in the depths of the earth, Pat started to seek better means of sharing in the wealth. In September 1863 he and four others incorporated the Garryowen Gold and Silver Mining Company and offered 800 shares at $100 each.

At this time, lawlessness was rampant in the Nevada camps. On October 26, 1863, the editor of the *Gold Hill News* wrote:

"The Territory of Nevada is establishing a reputation for murderous affrays, which vies with that of California in that elder day when it was looked upon as the slaughter-house of the world!"

Eastern California Museum

South of Owens Lake in the Coso Range, Darwin sprouted overnight when silver was discovered in 1874 on Mt. Ophir, shown in background. Pat Reddy rushed down from his law office in Independence and bought the Defiance Mine, the real bonanza of the district.

Eastern California Museum

Burros dominate the scene at this street corner in Darwin, Inyo County, which produced several million dollars in silver. Darwin's prosperity brought five stage robberies in a two-year period. One of the West's violent labor disputes rocked the camp in 1878.

Grace P. Crocker Collection and Mrs. Patricia Crocker Denton
Emily Page Reddy nursed her husband back to health and urged him to study law. Portrait was taken in later life.

Eastern California Museum
Losing an arm in a Virginia City shooting, Pat Reddy took up law and was one of the West's most feared attorneys.

And he predicted that such violence could bring on mob justice. "The warning had better be heeded."

Two nights before, a fatal shooting had occurred in a saloon at Aurora, where the Reddy brothers lived. George Lloyd, known as "the hero of many street fights," was killed in a quarrel with his brother-in-law. At the time, Pat Reddy was in Virginia City, perhaps to sell shares in his Garryowen Mine. On the night of October 27, 1863, he found himself on B Street, which the *Gold Hill News* editor had identified as "the favorite shooting gallery of the town."

In front of the San Francisco Saloon, Reddy got into an argument with Jack Mannix and Tom McAlpin over the Lloyd shooting in Aurora. Mannix, associated with the Headquarters Saloon, was a wild character, though respectable enough to be a member of the invitation committee for the next St. Patrick's Day Ball.

At the height of the quarrel, Reddy was set upon by Mannix and McAlpin. Mannix drew his revolver and shot Reddy in the upper arm.

SHOOTING.—A shooting affray occurred last night opposite the San Francisco saloon, on B street, Virginia. A man named McReady, from Esmeralda, was shot by John Mannix, of the Headquarters saloon. His right arm was amputated by Dr. Bryan. The trouble grew out of the killing of Geo. Lloyd.

Quickly Reddy was taken to a doctor, who immediately amputated the arm at the shoulder. Mannix and McAlpin were arrested and released on bail until Reddy either died or recovered. For Mannix this meant he would be charged with either murder or assault with intent to murder.

Within days Pat Reddy recovered enough to return to Aurora. Mannix and McAlpin were not arraigned before the grand jury until April 22, 1864, when they pleaded not guilty. The case was bound over to the next court session. Meanwhile, Mannix was arrested again for assault and battery, and still again for being drunk and disorderly. Since Pat Reddy was no longer in Virginia City to press charges and testify, the case seems to have faded away in court proceduralism.

During this time, Pat was in Aurora recuperating and thinking about his life. Equally affected by the event, brother Ned resolved henceforth to carry a pistol and shoot first.

In Aurora, Pat hired a widow named Emily Page to nurse him back to health. A cultivated woman in the Western wilderness, she would later contribute literary pieces to California publications. Though she was fourteen years older than Pat, they fell in love and were married on February 14, 1864. They moved to the new silver camp of Montgomery, which was then booming in the White Mountains of Mono County, California.

There, unable to hold a drill and swing a single jack, Reddy bought a town lot in October and opened the Pioneer Saloon. Soon he was elected recorder for the new Blind Springs Hill mining district. Both Pat and his brother Ned bought more property, mostly mines, in the neighborhood. It looked as though Pat Reddy would settle in a business career in the mines.

But Emily saw in her husband a different future. He was alert, analytical, fast-talking and competitive. In her eastern home town of Wyoming, Pennsylvania, her brother was Judge A.N. Conklin, an accomplished lawyer. She urged Pat to take up law, and helped to tutor him. She brought him standard law books, including William Blackstone's *Commentaries on the Laws of England*, James Kent's *Commentaries on American Law*, and the works of the great English jurist, Edward Coke. It is said that she enlisted the help of another Pennsylvanian, Will Hicks Graham—lawyer, editor, politician and adventurer, whose exploits were famous in the California and Nevada mines.

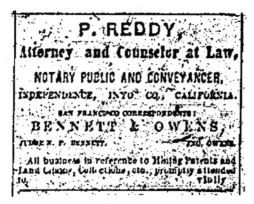

Dedicated to combating the Southern chivalry then trying to dominate California politics, the diminutive Graham was a fighting bantam cock with any weapons, and once bested the filibusterer, William Walker, in a duel. Now trying his luck in the Eastern Sierra, he is said to have coached Pat Reddy in finer points of the law and courtroom tactics.

In 1867 Reddy passed the bar and was admitted to practice law in Mono and Inyo counties. Soon the oak tables of the Bridgeport and Independence courtrooms were shaking under the pounding of his strong left fist. At twenty-eight, Pat Reddy had found himself.

At first, brother Ned had stayed close to Pat's side, taking property near Benton. Besides mining, he worked as a blacksmith. But in 1867 he turned up in California's Southern Mines. On the night of September 3, in the town of Sonora, he got into a fight with one John Noonan. Big and agile enough to handle most anybody, Reddy beat his antagonist, who nursed his grudge until two mornings later. In front of Labétoure's saloon they were exchanging heated words when Reddy pulled his revolver and fired. The first ball passed through Noonan's arm and leg. The second hit the finger of a bystander, who was doing his best to get out of the way, and then lodged in the thigh of the bartender in the saloon.

Arrested, Reddy put up $900 bail, waived a hearing, and jumped bail after the Grand Jury found two bills against him. He crossed again to the east side of the Sierra, where his brother could, if necessary, prevent his being jailed.

There young Reddy became a mine foreman, sometimes a professional gambler, and later a saloonkeeper. However, he seemed unlucky in finding himself in situations calling for gunplay.

On Christmas Eve, 1870, Ned was celebrating with some friends in a Cerro Gordo saloon when some Cornish miners came in. Hard feelings between the Cornish and the Irish had been common in Western mines. With considerable libation continuing through the night, the festivities turned ugly about 6:00 am, when a fight started between Reddy and one of the Cornishmen, Tom Dunn. After knocking his opponent under the billiard table, Ned walked across the room to the water barrel. Dunn got up and stepped toward Reddy, cocking his revolver.

"Scatter!" he shouted to the others. "Fair play!"

"Reddy! Look out," cried someone. "He is going to shoot!"

Ned whirled as he drew his revolver and fired one shot. The ball hit Dunn in the right breast and brought him down. Reddy's move was so fast that some of the bystanders thought it was Dunn who had fired. The stricken Cornishman was carried to his cabin, where he died a few hours later. Reddy was cleared on grounds of self-defense.

In October 1873 Ned was in Lone Pine when a notorious character named George "Bulger" Rains, alias George Watson, got drunk and went on a rampage. Peterson's saloon was crowded when Rains leaned on the bar and told the proprietor he intended to "have blood before morning," with particular reference to Ned Reddy, who was in the room playing billiards. He was "Chief", Rains continued, and "afraid of no man." And he upbraided the town constable, Russian Steve, for keeping a shotgun behind the bar for such people as himself.

Losing his patience, Steve offered to put guns aside and fight it out in the street with Rains, who agreed. Steve handed his revolver to Ned Reddy and threw off his heavy coat. But Rains stood in the doorway, his revolver drawn and cocked. He pointed it at the crowd standing there, with Russian Steve and Ned Reddy in the forefront. He could, declared Rains, "shoot any son of a bitch in the house," and he "would do it if any man made a move."

A shot shook the room and Rains fell dead, a ball through his temple. His six-shooter dropped by his side. Of the fourteen men in the room, no one had seen Reddy fire Steve's revolver.

At the hearing, Ned was released on grounds of self-defense. In his accuracy, at least, he had come a long way since Sonora. Two fast shots, two men dead.

After that, the otherwise likeable Ned Reddy was known as a man not to be crossed. Later, when he ran the Capital Saloon in Darwin, a man who had shot another gave himself up to Ned Reddy as the nearest thing to a law man in town, and at popular request Reddy escorted the man to Independence. His reputation spread to the Mother Lode, where, according to the *San Francisco Call*, "he became known...as a man who knew no fear."

As Inyo County had been created in 1866, Pat and Emily soon moved to the new county seat of Independence, where the superior court had plenty of mining litigation and criminal actions in this still-untamed frontier. They settled into a comfortable home on Kearsarge Street, where Emily created a productive garden, enjoyed giving fruit and flowers to her friends, and earned from one of them the compliment:

"In taste and success as an amateur horticulturalist she has no superior."

With her warm heart Emily won the friendship of the town's population. Once, after a trip east to visit relatives in Pennsylvania, she returned to a rousing ball held in her honor at the county courthouse.

From the first, Pat Reddy dominated the Inyo courtroom. In civil suits his wizardry won case after case. It was Reddy who led Belshaw's fight to rule Cerro Gordo with his Union Mine; when the jury ruled otherwise, Reddy continued to keep the case tied in legal maneuvers while Belshaw exploited the prize.

In 1875, Reddy was joined in his law firm by Emily's brother, Judge A.N. Conklin, who came west from Pennsylvania. Through the 1870s, though there were two or three other lawyers in Inyo County, Reddy & Conklin represented one side or the other in nearly every case before the superior court.

Following his Irish instincts, Reddy took the cases of the underdogs and outcasts—often those charged with murder, robbery and assault. No matter how obvious the evidence against the defendant, Reddy was able to gain an acquittal or a light sentence on the slimmest technicalities.

It was clearly not for money that he fought these battles; most criminal defendants had none. It was the challenge of the difficult case and the exhilaration of victory that motivated Reddy. And in American law every accused had a right to legal representation; when others refused to take the impossible case Reddy would—and win. Stage robbers caught red-handed, gunmen who killed their man in a barroom full of witnesses—most all were set free through Reddy's wily maneuvers.

Once when some saloonkeepers were arrested for selling whiskey on election day, Reddy got them off by demonstrating that, since the polls were not open the required number of hours, the election itself was invalid.

Despite his personal charm, Pat could not pursue this course without making enemies. When hard feelings developed between Reddy and Judge Theron Reed, someone claimed that Reed accused Reddy of plotting to kill him, and that Reed had bought a Smith & Wesson to defend himself. Reddy's response in the *Inyo Independent* was devastating:

"If Judge Reed is the author of this infamous lie, I take this method of informing him to come prepared with proof. . .or else stand branded as a willful and cowardly maligner of a well earned and dearly cherished character."

No response came, but before this issue died away Reddy had to meet another. In March 1875 a young Panamint merchant named Bark Ashim shot and killed Nick Perasich in a Darwin saloon. Reddy took Ashim's case and got him released on bail pending trial—highly unusual in a murder case.

At this point the sheriff of Eureka County, Nevada, sent word to the sheriff of Inyo County that friends of Perasich were coming down to lynch Ashim. The Inyo under sheriff then asked local citizens to help guard Ashim, who had sought refuge back in jail. But only one or two citizens stepped forward, the rest being among many in Owens Valley leaning toward a vigilance committee to deliver "justice" to Pat Reddy's clients. The under sheriff then called upon Captain Alexander B. MacGowan of Camp Independence for help. MacGowan headed a detachment of soldiers guarding the prisoner through the night of March 25.

While the Perasich force did not appear, a more serious threat was afoot in Owens Valley itself. Besides the Ashim case, public opinion was inflamed over the escape of another convicted murderer. Reddy was suspected of complicity, though he was many miles distant at the time. Some thirty citizens, half of them from Bishop Creek, decided to revive the vigilante movement that had plagued California since the Gold Rush.

On the night of April 1 the local Independence contingent camped east of Owens River. Next night the full body, reinforced by the Bishop Creek crowd, met in the sagebrush near the Inyo County courthouse and jail, which they decided to attack at 11:00 p.m. But the courthouse windows being strangely dark and quiet, some in the mob suspected the sheriff had prepared a warm reception with his corporal's guard. About half a dozen of the conspirators deserted.

Then someone proposed to fire Pat Reddy's stable and, while the townspeople were diverted to this emergency, descend on the jail to get

Ashim. If "circumstances made it at all necessary," they would hang Reddy and kill his law clerk.

To this foul suggestion one man raised objection. It was hardly right, he said, "to hang a lawyer for doing his best for his clients."

Before long the brave vigilantes melted away in the night. The *Inyo Independent*, always a friend of Reddy's, sent them a parting shot: "it would have been the merest chance. . .if a single individual of that gang would have got away alive."

Fearing another attempt to capture Ashim, the sheriff had spirited him to Camp Independence. As for Reddy, he did not miss the opportunity to raise his fist at his enemies. On April 7 he wrote in the *Independent:*

> I will continue to practice my profession, will defend or prosecute whenever I see fit to do so, and I hereby hurl defiance in the teeth of all midnight assassins and marauders, and especially to those conspiring against me. True, you may murder me, but you cannot frighten or intimidate me.

And his final challenge: "I am ready to meet them all, at any time, night or day."

With that he swore out a complaint on conspiracy charges against those who could be identified. Six days later four of the vigilantes wrote the newspaper that they, at least, had refused to fire Pat Reddy's stable, and that "Mr. Reddy has always conducted himself as a gentleman and a citizen above reproach."

The vigilance movement cooled off, and so did Pat Reddy. Emily, who had borne the crisis in the confidence that her husband would prevail, now sought peace and quiet with a visit to her relatives in Pennsylvania.

In November 1875 Bark Ashim was found not guilty on grounds of self-defense, and later in Carson City he escaped on the stage under a hail of bullets from the Perasich crowd.

But the acquittal prompted new venom against Pat Reddy. On the word of an unknown Inyoite, the *Carson City Appeal* savagely attacked him as "a king of the desperadoes" and his brother Ned as one of "that dangerous class." But three weeks later, on numerous testimonials received from Owens Valley, the *Appeal* retracted all in a glorious eulogy to Pat Reddy.

The gifted lawyer continued to defend accused murderers and get them off free. Finally the *Independent* proposed a solution: increase the salaries of California district attorneys to $5,000 a year, and offer the Inyo job to Pat Reddy. Then the editor put the good lawyer on the spot.

"Will Mr. Reddy tell the public. . .what *he* thinks about it?"

For once Pat kept silent, but the idea of increasing salaries was seconded by newspapers in neighboring Kern and Tulare counties, which were having similar problems. In Inyo, the Darwin *Coso Mining News* endorsed Reddy and upped the ante to $6,000 per year.

Five months later, Reddy quietly succumbed to the pressure and allowed himself to be appointed assistant district attorney, at an unknown salary. For a time, justice was more even-handed in Inyo County.

By the beginning of 1876, Darwin was running full blast around the clock. Four furnaces were turning out more than 400 bars of silver-lead bullion every day, filling the air with smoke and cinders. Nadeau's teams, now increased to sixteen mules each, were hauling them to Owens Lake, where the steamer *Bessie Brady* transferred them across the water to the south-bound Bullion Trail. Pat Reddy's Defiance furnace, running in three shifts, was yielding 150 bars daily, each weighing about 90 pounds and worth between $150 and $200.

On the west slopes of the Coso Range, where piñon pine covered the northern exposures, a small army of woodchoppers and charcoal burners was producing the fuel brought to the furnaces by muleteers and their pack burros. But by 1876 this source was nearly exhausted. Col. Sherman Stevens, who was already supplying the wood for Cerro Gordo from Cottonwood Creek in the Sierra, had the answer. He built two adobe kilns on Owens Lake and brought in a second lake steamer, the *Mollie Stevens*, to haul the charcoal to the nearest point to Darwin.

Deep in Mount Ophir, the earth shook with explosions inside at least eight shafts and tunnels as the miners burrowed into the ledges. Ore in the Lucky Jim and Christmas Gift mines, owned by L.L. Robinson's New Coso Mining Company, was said to run $400 to the ton; in Reddy's Defiance mine, between $150 and $225. The *Independent* declared "it has no superior as a base metal mine in the world."

For the horse population, wagons brought hay from the green fields in the Bishop Creek area of Owens Valley, where a farm community was flourishing on trade with the mining camps of Inyo. Bishop Creek, with its false front buildings, looked like any other frontier town in Owens Valley or elsewhere, but it was the beginning of civilization. With women and children, schools and churches, it was sprouting roots unknown in the mining camps. So respectable and God-fearing was Bishop Creek that the brawling mining towns called it "Gospel Swamp".

By this time Pat Reddy had placed a few hundred shares of Defiance stock on a San Francisco exchange, where they sold at $3.00 a share. By contrast, L.L. Robinson's New Coso Mining Company, having assessed its stockholders 50 cents a share in October 1875, found its stock dropping from an initial $3.25 to $1.50 within two months, then to 50 cents by March

1876. Robinson's famous acumen was failing him while the mining district was booming.

As for Darwin itself, the *Kern County Courier* conceded it was larger than Bakersfield. Some of the frame buildings at Panamint were torn down, hauled to Darwin and rebuilt. Working at $4.00 a day, miners and furnacemen paid 25 cents for a meal, $1.00 per night for a bed, and for boarding a horse, $1.50 a day. A cut above other hotels was Mrs. Henry's Florence Lodging House, offering:

"Single rooms and spring mattresses, clean, sweet bedding, rugs on the floor, curtains at the windows, pictures on the walls, bowl and pitcher in the room."

Darwin was not only up to date, but a cosmopolitan town, with Yankees, Mexicans, native Californians, Cornish and Irish miners, and a Chinatown.

By May 1876 Darwin had a six-piece brass band, headed by editor T.S. Harris of the *Coso Mining News* and including Constable Billy Welch on the horn and Ned Reddy on the tuba. When their instruments arrived they hired an instructor to teach them, and as the *Independent* declared, "we have no doubt some if not all of them will succeed." Sure enough, the Darwin Brass Band was ready to sponsor an Independence Day grand ball, which was said to be "the finest affair of its kind," and which, moreover, passed off with "no accidents or serious disturbances of any kind worth mentioning."

All of this prosperity was not without drawbacks. Road agents found employment on the highways, particularly the stage route from Caliente to Darwin. In little more than two years, from January 1875 to February 1877, robbers stopped the Darwin and Panamint stages five times. One Darwinite, Oliver Roberts, claimed stage holdups were so common that if a man in Darwin said openly to another, "Let us go rob the stage," no one paid any attention.

In the evening of October 12, 1876, a lone bandit stopped the stage about two minutes east of Indian Wells, and got $2,400 from the Wells Fargo treasure box. Ten days later on the street in Darwin, Deputy Sheriff Jim Wales accosted Fred Gillette and accused him of having "something to do" with the robbery. Gillette drew his pistol and shot Wales through the leg. Wales then shot Gillette in the shoulder. Both were arrested, then released, and recovered from their wounds.

Soon Wells Fargo's crack detective, James B. Hume, arrived with a warrant for Gillette and took him for examination to a justice of the peace at Indian Wells. Gillette was released, but Detective Hume was not a man to give up easily. He rearrested Gillette and in a court trial, got him sent to San Quentin.

As the robberies continued the shotgun messenger on the Darwin run, R. H. Paul, made a point of pursuing the culprits and successfully tracked down some of them. Taking offense at this, their confederates laid for him on the "big hill" below Wild Horse Mesa, a few miles south of Darwin..

At daylight on February 14, 1877, while the six-horse team labored up the grade, three gunmen stopped the stage. Paul was not on the box beside the driver, but passenger Jack Lloyd was asleep in the front boot. When the stage halted he raised up to receive a charge of buckshot that killed him instantly. Panic-stricken, the horses lunged onward while a shot from the robbers tore through the driver's sleeve.

As the stage and lathered horses whirled into Darwin, the gathering citizens finally took stage robbery seriously. But by the time a posse raised dust out of Darwin, the trail was too cold. When the villains are caught, warned the *Independent*, "it will be hardly worth while to put them through the inconvenience of a formal arrest."

With everyone carrying weapons and drinking more or less heavily in the saloons at night, shootings were also common. Even respectable citizens could get involved in violence.

In February 1876 the editor of the *Coso Mining News*, T. S. Harris, was knocked down and beaten on the street by a saloonkeeper who disagreed with editorial policy.

In March 1876 the owner of the Centennial Saloon, Col. J.R. Norton, got drunk and brandished his six-shooter with much braggadocio until he finally shot and killed his bartender. During the hearing the deputy sheriff went to the nearby camp of Lookout to get a witness, who came under protest. Later the witness tried to shoot the deputy in a Darwin restaurant, and was stopped by a bystander who grabbed the pistol as the hammer came down on his little finger.

And in May 1877, after the Cerro Gordo madames had taken their gay-gartered flock to Darwin, Lola Travis was attacked by a rejected customer with a foot-long knife. With her little pistol she gave him a fatal dose of lead.

Another Cerro Gordo prostitute, Nancy Williams, popularly known as "Featherlegs", had settled in Darwin to run a boarding house. In September 1877 she was brutally murdered by parties unknown who were probably after money. A large part of the town population walked in her funeral procession, and the *Independent* called her "one of the kindest and most liberal of women, alleviating distress by her means whenever an opportunity offered, and giving from her purse to all public enterprises."

While Darwin was in its prime, property values were a temptation for white-collar shenanigans. In 1875 some roughs were foiled in their attempt to jump some town lots. But the biggest jumper of all appeared in

July 1876. Imitating Pap Kelty in Benton, a former land register named George H. McCallum filed on 320 acres of land. The register in Independence, thinking it was pasture land, forwarded the application to Washington for patenting. But at the Land Office in Bishop, the receiver was Tom May, the same who was involved with the Benton case. He found the filing included the entire town of Darwin. The application was stopped in Washington, and McCallum's little plot to seize the whole camp was frustrated.

Through the summer of 1876, Darwin and Pat Reddy seemed at the height of prosperity. In the Defiance Mine, a new blind lead discovered in June was yielding assays ranging from $234 to $465 per ton.

"It is the finest body of ore to be seen outside of the famous Raymond & Ely mine [at Pioche, Nevada] when it was in all its glory," wrote editor Harris of the *Coso Mining News*.

While all of Darwin was celebrating, the true situation was less than happy. Unlike some other mining camps, Darwin had more furnaces than it needed. In the first excitement, funds raised by selling stock and assessing stockholders were used to develop furnaces, roads and offices more than the mines themselves. And as it turned out, the rich ore bodies were the exception; most ore was lower grade that could scarcely be milled and roasted at a profit.

Thus the owners of the four furnaces had a difficult choice. They could run only on higher grade ores, shutting down when these were exhausted to wait for the next rich strike. Or they could continue running around the clock, using low grade as well as higher grade ores.

Since stopping and starting a furnace are costly processes, to say nothing of the difficulty in holding together a good work force during a shutdown, the operators of all four New Coso and Defiance furnaces chose to keep them running around the clock on low- as well as higher-grade ores.

But other factors also combined to reduce their margins. The hope of lower freight rates with the approach of the railroad was not realized until August 1876, when the Southern Pacific was completed over the Tehachapi Pass, and Mojave became the terminus of Nadeau's mule teams. And while the lead content in the bullion could previously be counted on to cover the shipping and refining in San Francisco, the price of lead began to decline.

The truth was that the two major producers in Darwin—L.L. Robinson's New Coso and Pat Reddy's Defiance—were losing money.

The real situation became apparent in June 1876 when the Defiance stopped paying bills and operated on the faith of unpaid suppliers. Reddy hurried to San Francisco, where he consulted with investors. Through the summer and early fall the Defiance limped along while he made more trips to San Francisco.

On October 31 he met at Darwin with the local Defiance creditors. He had a proposal, he said, but first he wanted to show them the Defiance Mine. Next morning Pat escorted a large crowd through the Defiance tunnels and winzes, showing by torchlight the ore bodies awaiting the attention of the miners. Unlike most other ore, galena shines unmistakeably, even to the untutored eye.

Meeting with the creditors again that evening, Reddy proposed giving them a mortgage on the Defiance to the value of the debts, which would be paid in increments as the ore was mined and the bullion sold. To a man the creditors stepped forward and signed up.

Armed with this Darwin agreement, Reddy took stage and train back to San Francisco to seek the concurrence of creditors there. And he arranged for finances to be handled in the trusted office of Egbert Judson.

By the end of the year both Defiance furnaces were in "full blast". One furnace was running on the higher grade Defiance mine ore, yielding bullion with greater silver content. The other furnace took custom ore from the chloriders, thus opening more mines that had been idling.

In January 1877 the Defiance company was paying creditors out of Judson's office after each bullion shipment. The *Independent* declared in April that "before the summer wanes Darwin will be about the liveliest mining camp on the Pacific Coast."

"There are more mines now being worked and more furnaces in operation than ever before."

Through most of 1877 Darwin was again riding high. By the spring of 1878 the Defiance mine and furnaces were still running, and a new blind lead had been discovered.

But the New Coso Mining Company was in trouble. Robinson's company had assessed stockholders eight times without paying a dividend. The *Coso Mining News* accused it of "criminal mismanagement."

In February 1878 the New Coso master smelterman, J.J. Williams, caught the stage for an urgent meeting in San Francisco with L.L. Robinson and his colleagues. On his return Williams announced that the board of directors had voted to reduce the wages of furnacemen from the usual $4.00 per day down to $3.00.

The men were adamant. They had formed a Workingmen's Club—part of the labor union movement then sweeping the West's hard-rock mining towns. Dedicated to the $4.00 day, the unions were further inspired to confront the big companies by the Workingmen's Party then gaining strength in San Francisco under the fiery leadership of Denis Kearney.

In answer to the New Coso announcement, the men at Darwin held protest meetings and signed a pledge not to work for less than $4.00 a day.

They were joined in the pledge by most other furnacemen in town, and were supported by most of the merchants. Confronting Williams, the furnacemen demanded that he prove the need for reducing wages. When he was unable to answer, the talk became ugly. They threatened his life and ran him out of town.

At this the New Coso shut down its furnaces. For three months Darwin idled along with only the Defiance and one or two smaller furnaces running. The town was desolate; the laughter and the clink of glasses in the saloons died away.

In the second week of May, Williams returned to face his adversaries. Hiring men willing to work for $3.00 per day, he fired up the furnaces on May 20. At this the members of the Workingmen's Club held a new meeting and determined to shut down the New Coso, "peaceably, if possible; by force, if necessary." They threatened harm to anyone willing to work at the New Coso for $3.00 a day; thus intimidated, some men left town. And the union sent messengers to the nearby camp of Lookout, in the Argus Range, for reinforcements.

While the town boiled with reckless talk, some thirty union men resolved to march down to the New Coso furnaces and stop the next shift. They threw a skirmish line across the trail from Darwin to the furnaces—a tactic that had been used in labor troubles in Pioche. One of the skirmishers was C.M. Delehanty, a young engineer and machinist favorably known in Western mining camps from Virginia City to Darwin.

"He was always considered a good-natured, clever fellow," as the *Bodie Standard* put it, "but considerably given to bragging."

The *Inyo Independent* called him "a brave and generous man. . .a good and upright citizen." He was known to be a friend of Deputy Sheriff Billy Welch.

In the saloons, Delehanty swore he would have nothing to do with the resistance. But by evening, influenced as one one observer wrote, "by bad men and worse whisky," he declared:

"The furnaces should shut down or I will leave my body in the trail."

Meanwhile, the union men told Williams that the first shift could finish its work, but a second shift would not be permitted on the ground—unless the company would pay $4.00 a day.

At this, Williams talked with Billy Welch, the constable and deputy sheriff, and Frank Fitzgerald, the town constable at nearby Lookout. Two more—Oliver Roberts and Bill Hagan—were deputized.

Welch had on previous occasions proven his courage in making difficult arrests, and would not hesitate to shoot if he thought necessary. Fitzgerald also had a reputation as a brave officer who brooked no nonsense. At the tender age of nineteen, Roberts had already killed more than one man, at least according to his own account. Though far outnumbered, this was an eagle's brood.

About 10:30 p.m., to protect furnacemen going on the next shift, they started walking in the dark to the furnaces in the wash below town. They were armed with Winchesters and six-shooters, though Welch himself carried a double-barreled scattergun, loaded with buckshot. Behind them marched some of the furnacemen going on the next shift.

At about 100 yards from the furnaces two of the men in the skirmish line ahead suddenly appeared in the darkness, about 10 yards in front of the law men. Welch immediately called out.

"Who's there? Speak quick."

Without replying, one of them continued advancing. One of the law men fired and the man fell, saying:

"I have got it."

Examination showed it was Welch's friend, C.M. Delehanty. The other man, Jim McDonald, disappeared. Immediately shots were fired by other skirmishers. More bullets came from union men stationed at the furnaces. Due to the darkness, and also because the firing was principally from pistols, no one else was hit.

But Delehanty lay in a pool of blood by the trail, his groin and abdomen pierced with eight holes. They removed his boots and carried him to Ned Reddy's saloon, where he died about an hour later.

At the inquest next day the coroner's jury reported that deceased "came to his death by gunshot wounds inflicted by parties to the jury unknown." But Oliver Roberts later wrote that Welch had fired the shotgun; the multiple wounds confirm this. The *Independent* supposed that Welch "did more than simply challenge."

For the rest of that day Darwin seethed with anger. Fist fights between partisans brought several knockdowns, and pistols were drawn but not fired. One of the unionists, John McGinness, was known as a "dangerous man", and had killed another man a few months before. He now berated Frank Fitzgerald for his part in the affair.

That night both men were in Lookout, and McGinness renewed his tongue-lashing of Fitzgerald. Finally Frank, taking no more, drew his six-shooter and told McGinness to go for his. Since McGinness was unarmed, Fitzgerald asked him to "heel yourself". McGinness went out and soon came back shooting. Fitzgerald shot him through the chest and both arms. When he died an hour later someone said, "he got no more than he deserved."

With news of the labor battle at Darwin racing through Inyo County, Pat Reddy drove his buggy down from Independence and joined a public meeting in the Darwin schoolhouse on the night of May 23. Though usually on the side of miners unions, he gave an impassioned speech against "the wrongs committed" by the Workingmen's Club. The citizens assembled approved of resolutions accusing the union members of lawlessness and blaming them for Delehanty's death. They would hold the officers and members of the Workingmen's Club "personally responsible for any violence or violation of the law" by the club or its members.

Delehanty's fate quickly sobered Darwin. The New Coso furnaces continued operating intermittently at $3.00 a day, without interference. But new bodies of rich ore were not to be found. By October 1878, though a half-dozen mines were operating and numerous stores were still open, Darwin's population had dropped from some 1,500 to not more than 300. Times were less than lively in the streets, and the Darwin Hotel soon closed its doors. In the following spring the New Coso furnaces were operating largely on ore from Lookout and other Panamint Valley mines.

At this declining moment Darwin was struck by a terror feared in all frame-built Western camps. At 2:00 in the morning on April 30, 1879, a blaze—probably set by a firebug—started in the empty Darwin Hotel.. Quickly it swept through town in what one observer called "probably the largest fire that ever occurred in Inyo County." Fifteen buildings—all of those on both sides along Main Street—burned to the ground. Due to the late hour the only casualties were two stable horses. Among those who turned out to battle the inferno was Peter Taylor, who was superintendent of the Emigrant Silver Mining Company. He led the force that saved the lower part of Main Street.

For a time the town was shattered by the disaster. Only two stores had carried insurance. In their fury the Darwinites swore that if they caught the culprit "it would be a warm camp for his already damned soul."

Though Darwin's population sank to 70 people by February 1880, the town continued to bounce up and down with the price of lead. One month all would be quiet. The next, signalled by the whistles of two remaining furnaces at the Defiance and the New Coso, Darwin was stirring again with mule teams hauling ore to the furnaces and happy men filling the boarding houses and saloons.

At least Darwin was on a sober, paying basis, as one visitor put it, by practical miners "who have learned to count the profits of mining, not by the fluctuations of stock, but by the strokes of a hammer or the revolutions of a windlass."

During this time, Darwin's calm was shattered with one ironic event: Constable Billy Welch, who had apparently killed Delehanty in the New Coso strike, was stabbed to death in a hand-to-hand fight in June 1880—a victim of his hazardous profession.

Through these months, Pat Reddy leased the Defiance furnace and mines to other operators, and by the fall of 1881 he and his head smelterman, J.S. Gorman, bought out the other Defiance mine owners.

But since the mine had been worked only intermittently, a veteran Darwinite named Martin Mee had claimed it by right of relocation. News of this raced through the Eastern Sierra in mid-November. On the 22nd Pat Reddy, brother Ned Reddy and a friend, all conspicuously armed, darkened the mine entrance. By the time the story reached Independence the three had become "an armed force of some ten or fifteen men." But it required no more than the Reddy brothers to convince Martin Mee of his grievous error.

Neither Darwin silver nor his triumphs in the Independence courtroom could hold Pat for long. In 1878 a convention had been called to write a new constitution for California. Reddy was easily elected by a joint committee of both parties as the delegate from Inyo and Mono counties. There his towering presence and eloquent tongue cut a commanding figure among his colleagues. But the convention became a raucous arena between conservative elements and Denis Kearney's Workingmen's Party, which was trying to pack the document with what Reddy later called "insane and communistic schemes." When the Kearneyites hooted at delegates who voted against them, Reddy railed against such intimidation.

"A principle is involved," he declared, "and I came here to do right."

Reddy returned to Independence for a hero's welcome, telling his constituents to vote against the constitution. But in the end it was adopted over such objections, and many of the Workingmen's proposals later became standard legislation in the rest of the nation.

By 1883 the Defiance had produced three quarters of a million dollars in silver and gold, not counting the prodigious yield of lead. Mine and furnace continued to be operated off and on into the Nineties, with Ned Reddy as superintendent for some of the time. The Defiance had helped to make Pat Reddy comfortably independent, if not wealthy. Today, in the ghost town of Darwin, his name is perpetuated in the sign, "Reddy Street" at the corner of Main Street, where a few frame buildings stand in silence at the foot of Mount Ophir.

10. ENTER GEORGE HEARST

>When George Hearst bought into the Modoc Mine in the Argus Range, he helped create Lookout, the unique "town on top of a mountain." Down below, the Minnietta Belle Mine added to the excitement. Across Panamint Valley, it took ten big kilns to supply the charcoal for the furnaces. With Hearst's furnace and Lookout's saloons roaring round the clock, times were wild and good, but late at night you had to be careful not to fall over the side.

In February 1875 three mining investors from San Francisco stepped off the stage in Darwin. One was George Hearst, a veteran mining man with the reputation as, according to a contemporary, "the most expert prospector and judge of mining property on the Pacific Coast." Familiar with the lead mines of Missouri as a boy, he was graduated from a local mining school at the age of 18. A dozen years later he joined the Gold Rush across the plains, and in 1859 pioneered the stampede to Washoe. He made a fortune in the Comstock and another in the Black Hills with the Homestake Mine, still one of the richest gold mines in the United States. What Hearst lacked in higher education he made up in hard work, good sense, and a "nose for ore"—a miner's mining man.

"I think I was naturally a mineralogist," he later commented. "I was born that way."

Hearst's arrival at Darwin set Inyo mining men agog. But after canvassing the mines for two weeks, he and his friends left without buying—but with the benign pronouncement that Inyo's prospects were "unusually brilliant".

However, not all of Inyo's treasure had been uncovered. Already, Darwin's riches were spurring men to prospect nearby. Somewhere, they thought, must be that mountain of silver called the Gunsight.

East of Darwin the Argus Range rears more than 8,000 feet high on the west wall of Panamint Valley. Ten miles from Darwin, a spur of the Argus called Lookout Mountain thrusts itself into that valley.

Here, on April 22, 1875, the same kind of silver-lead ore that had spawned Cerro Gordo and Darwin was discovered by prospector B. E. Ball. It was not the Gunsight, but it was high-grade argentiferous galena. Ball and his three fellow prospectors called it the Modoc.

Eleven days later one of them, Jerome S. Childs, located the Lookout, and the group claimed the nearby Confidence Mine the same day. As usual the four offered their combined mines for sale in San Francisco. On August 5 they sold for $15,000 to a group of investors headed by none other than George Hearst. Already familiar with the tawny hills around Darwin, he found that his stage trip had not been in vain.

Also with Hearst in the Modoc investment was Captain George W. Kidd, who had won his title as the owner of steamboats on the Sacramento River, but spent much of his time in mining ventures. Later this same year, when the Bank of California would have to close its doors, he would be one of the first to step forward with financial support.

Now he was joining George Hearst in a remote silver mine on Lookout Mountain. With their colleagues they formed the Modoc Consolidated Mining Company. Heartened by a rich blind lead discovered in the Lookout Mine, they made the dirt fly in the Argus Range.

Into the heat and desolation of Panamint Valley came a crack superintendent, C. J. Barber, whom one friend referred to as "able, genial and respected." By the spring of 1876, under his driving energy, crews were grading a road through Darwin Wash into Panamint Valley, then blasting it along the cliffs around the west side of Lookout Mountain and up to the Modoc shaft on top.

By late May, Hearst and his friends incorporated the Modoc, issuing 100,000 shares that sold on the market for $3 apiece. With this capital they moved to buy, deliver and install their furnaces.

But the first question was, where to put the furnaces? Superintendent Barber chose the valley floor on the east side of the mountain—easy to supply with charcoal and other necessities, and easy to haul away the bullion. The ore would have to be packed from the mountaintop, but at least it would be downhill. Besides, the lode could later be tapped by a tunnel from below, greatly aiding the removal of ore.

From its headquarters at 310 Pine Street, San Francisco, the Modoc company sent James Stratton to survey the situation. Believing that men and animals could not work in the 100-degree heat on the floor of Panamint Valley, he decided that the furnaces should be placed on top of the mountain near the mine.

Accordingly, by the summer of 1876, Remi Nadeau's mule teams were plodding past Darwin into Panamint Valley with the furnace machinery for the Modoc. Early in July the teams arrived with the bed-plate for the engine. The iron casting being too heavy to lift, the men hitched it to another team to snake it off the wagon bed. When it hit the ground it smashed to pieces. Hearst's company had to order another in San Francisco, and the Cerro Gordo Freighting Company had to get it off the train at Mojave for another ten-day haul across the desert.

California State Library
A pioneer in the Comstock, George Hearst later bought the Modoc Mine and launched the town of Lookout, Inyo Co.

California State Library
Edward A. "Ned" Reddy, Pat's kid brother, came out best in shooting scrapes in Sonora, Cerro Gordo and Lone Pine.

Eastern California Museum
After George Hearst bought the Modoc Mine in 1875, he built these works to process the ore on top of Lookout Mountain in the Argus Range. Here sprang the town of Lookout, famous for its magnificent view of Panamint Valley and for its wild citizens, who were not much interested in the view.

Nadeau Collection
By 1877 all the nearby pines in the Argus Range had been cut for mine timbers and furnace fuel at Lookout. Across the valley in the Panamint Range, ten large kilns were built in Wildrose Canyon. Their charcoal product was hauled by Nadeau's mule teams to Lookout.

By September, engine, boiler, water jacket and bellows were delivered on top of the mountain. On October 9 the first of the two furnaces was fired, turning out 160 bars of bullion per day. At the end of 1876 the Cerro Gordo Freighting Company had hauled to Mojave 10,000 bars worth some $400,000. On January 5, 1877, the Modoc company declared a 50-cent dividend.

Meanwhile, the magic name of George Hearst had brought others from Darwin and Panamint rushing to Lookout. Among them came the claim-jumpers, with property held by "right of shotgun." More mines peppered the mountain, drawing upon the piñons and junipers of the Argus Range for mine timbers and charcoal. An army of woodchoppers, coal-burners, muleteers and animals supplied the camp along numerous trails. On top of the mountain, a sloping mesa accommodated the town of Lookout—two general stores, three saloons, company offices and as many as thirty other structures built of stone or lumber.

Here in this windswept promontory, with no trees and little natural water, life was down to necessities. The one redeemer was the magnificent view of Panamint Valley. But in their off hours, most residents were too busy looking into the bottom of a whiskey glass to admire the scenery. And with the nearest law courts a two days' journey away in Independence, fights and shootings were frequent.

Chief among the new mines at Lookout Mountain was the Minnietta Belle, located early in 1876 at the south footing of the hill by prospectors Ed Coffey, W.T. McClain and P.J. McDonald. Combining it with the Mountain View and Keystone mines, they sold the three for $30,000 to several Northern California investors late in April. This group promptly raised money by incorporating and selling 100,000 shares at $2 apiece, following this with a stock assessment of 25 cents in October.

To spend this handsome treasure from the pockets of stockholders they hired Richard C. Jacobs, the same who had pioneered Panamint, as superintendent. By the year's end Jacobs was erecting a ten-stamp mill, preferring this simpler process to the furnace method at the Modoc. With the first bullion shipped by Nadeau's teams in May 1877, the *Inyo Independent* declared:

"The Minnietta Belle is perhaps equal to any mine in the county."

By 1877 Lookout was roaring with bullion production. With the Argus denuded of trees, James Honan and O.B. Morrison built ten rock-walled charcoal kilns across Panamint Valley in Wildrose Canyon, where the piñons and junipers grew in profusion. By July they were turning out charcoal, which was hauled across Panamint Valley on a new road built by Nadeau for his teams. Much of the charcoal went to Jacobs' mill at the Minnietta Belle. For the Modoc furnaces, a switchback trail was cut into the cliff, and a continuous train of burros carried the charcoal in sacks up to the Modoc.

To keep up with the output of furnaces and mill, Nadeau needed a shorter route than the old bullion road through Darwin. From the foot of the Slate Range grade, he had a road surveyed northward through Panamint Valley to Lookout. Unique among desert thoroughfares, Nadeau's "Shotgun Road" ran in a straight line, crossing ravines with fills and culverts.

By the spring of 1877 his teams were hauling bullion down Panamint Valley, across the Slate Range crossing, past Searles Lake, swinging westward through Salt Springs and south of China Lake to meet the Bullion Trail at Panamint Station near Indian Wells.

Thus in 1877 Inyo County was a giant beehive of silver output. Cerro Gordo and Darwin were still turning out bullion. Lookout was producing the silver-lead bars from two furnaces and a ten-stamp mill. Even at Panamint the Surprise Valley mill was running intermittently on custom silver ore from the chloriders. As a modest version of the state of Nevada, Inyo was calling itself California's "Silver County".

For Lookout, the boom times had their first stumble in November 1877. The Minnietta Belle, defaulting on payments, was hit with foreclosure by creditors, including Remi Nadeau and Richard C. Jacobs himself. The company survived the $22,000 judgment granted to the creditors in May 1878. But two months later the Cerro Gordo Freighting Company and its owners, M.W. Belshaw, Egbert Judson and Remi Nadeau, got another judgment against the Minnietta Belle of $6,700 and took over the property. Belshaw himself stepped in to manage the mine and mill.

At the same time, change had overtaken the Modoc. Superintendent C.J. Barber, after installing the furnaces in a spot where he did not want them, left the company. Four months later the job fell to A.N. Guptil, who managed the works for a year when costs were exceeding income and Modoc kept assessing its stockholders.

By this time the Modoc directors had decided what Barber had already told them—that the furnaces should be located at the foot of the mountain. Before moving them, however, the Modoc would drive a tunnel from near the valley floor to intercept the main body of ore, make connection with the shaft from the top, and thus ventilate the mine and remove the ore with the benefit of gravity.

In the first 400 feet the tunnel was all dead work, as expected. But at that point the miners struck a blind lead that would yield 350 tons of rich ore. According to Oliver Roberts, at the moment of this discovery Superintendent Guptil hired him to ride the 200 miles to Mojave, where there was a telegraph office, and buy 10,000 shares of Modoc for Guptil's account.

After the purchase was confirmed, the messenger telegraphed news of the strike to Modoc headquarters in San Francisco.

When the price went up Guptil may have made a killing, but he lasted less than a year with Modoc Con. C.J. Barber returned as superintendent in August 1878. He found the company in precarious straits. Assessments were still hitting the stockholders and driving down the price of Modoc shares, and the company had borrowed funds to carry on the work.

Clearly Barber had to speed the connection between shaft and tunnel to get the mine back on a paying basis. In October he brought in a steam-driven hoist for the shaft and an automatic Ingersoll drill for the tunnel. Using an air compressor driven by a steam engine, the new Ingersoll could drill the holes for the blasting powder much faster than the miners could drive them by sledge hammer. It was the first machine drill in Inyo County.

Barber was so proud of his new-fangled equipment, and so anxious to put the best face on Modoc operations, that he threw a grand party on October 25. The affable superintendent proved he could manage a gala event as well as a mine.

From Darwin some 24 guests—husbands and wives, dressed in their best—took horse-drawn carriages over the Argus and up to Lookout. Among them were J.J. Williams, superintendent of the New Coso mines and mill, and Joe Le Cyr, a Darwin businessman who would later migrate to Calico.

Together with another contingent from Lookout itself and from a gold camp in nearby Snow's Canyon, they filled the boarding house for a splendid banquet. At 9 pm they poured into the meeting hall, which was festooned with red, white and blue bunting. When a two-piece band struck up the Grand March, the men rushed for partners and the dancing began. Singing and recitations filled what was called a "variety party" until 4:00 in the morning

"Strange to say that in Inyo not a soul there got boozy," observed the correspondent of the *Inyo Independent*.

After a hearty breakfast in the boarding house, the guests were presented with a collection of horses, mules and burros, saddled and bridled, to take them on the mine tour. First they rode to inspect the hoisting works, then plunged down the east face of Lookout Mountain on the trail worn by the charcoal burros. At the bottom, deep in the earth, they witnessed the Ingersoll drill driving its holes into the tunnel face.

"I confidently believe," wrote the correspondent, "that if the Almighty put ore in Lookout Hill, C.J. Barber will find it."

At two in the afternoon on the 26th the weary Darwin contingent mounted the carriages, which had been brought down from the top, and rolled back home. Impresario Barber had pulled off a triumph that was the high water mark of life in Lookout. And he had convinced those who owned part of the Modoc that, despite the low price, Modoc stock was worth holding.

By March 1879 the miners were working 350 feet deep in the shaft. They had already pushed the tunnel into the mountain 1300 feet—well past the point for connection with the shaft. To speed the connection, Barber started driving a "raise" upward from the tunnel, narrowing the gap to only 200 feet.

But tunnel and shaft, already among the longest and deepest in Inyo County, were harder to work the deeper they went. Most of the machinery was needed to pump air to the miners at the tunnel face. Little more was available to aid the regular stoping work that provided ore to the furnaces. For months the furnaces had produced no bullion and the company had earned no money. What little ore came out of the Modoc was shipped for treatment to the New Coso furnaces at Darwin.

In April the company assessed stockholders for the eighth time. As there had been only one dividend, many stockholders were forfeiting their shares rather than pay up more "mud". Barber had, in fact driven the shaft and the raise to the point where only 84 feet separated the two. But the tunnel, though driven 400 feet beyond the planned connection point, had still not found the expected main ore deposit.

Early in October 1879 the secretary of Modoc Consolidated took train and stage to Lookout. After going over the ground he returned to the city and reported on the mine. There was, he said "nothing in it."

At this, Modoc Con stopped spending money. Checks written by Barber to meet payroll and other expenses were bounced. Writs of attachment were levied by employees, one of whom rode into Darwin for the deputy sheriff. Creditors took over the operation, stoped the mine, fired up one furnace and turned out bullion to pay what was owed them.

In 1881 the Modoc was leased to Frank Fitzgerald, the constable who had figured in Darwin's labor dispute of 1878. Like so many in California's last frontier, he had many talents—peace officer, census enumerator, operator of the stage line from Darwin to Lookout, and now mine superintendent.

Firing one of the furnaces, Frank began producing bullion. Even the Wildrose charcoal kilns, which had closed in 1879, started supplying fuel again to the Modoc. But by the end of 1882 both water jackets were burned out. Thereafter the Modoc ran intermittently until the early 1890s, when the dropping price of silver wrote the Modoc's epitaph.

A stronger survivor was the Minnietta Belle. Starting in 1883 it was owned and operated by Jack J. Gunn, who had run a saloon in Lookout and managed the boarding house at the Modoc Mine. Canadian by birth, Gunn had been a Confederate soldier in the U.S. Civil War and a familiar figure in Bodie before he came to Lookout. Remaining in the Gunn family, the Minnietta Belle was worked intermittently till the end of World War II.

George Hearst, who had helped to launch the Modoc Con, had retreated before the end. His name was not among the stockholders when the last assessment had fallen in May 1880. Before he died in 1891 he had a seat under the Capitol dome, where he prided himself on being "the silent man of the Senate." On the 21st birthday of his son, William Randolph Hearst, he made him a present of the *San Francisco Examiner*. It was George Hearst's greatest investment.

Lookout, the unique "town on top of a mountain", is marked by a few rock walls, numerous foundations, two rusted water jackets and countless broken bottles. A glory hole yawns near the top of the shaft that never reached its destination. The charcoal trail, worn in head-high defiles by the old burro traffic, still finds its way up the east side of Lookout Mountain. And across Panamint Valley in Wildrose Canyon, ten charcoal kilns stand as tombstones for the heyday of Inyo silver.

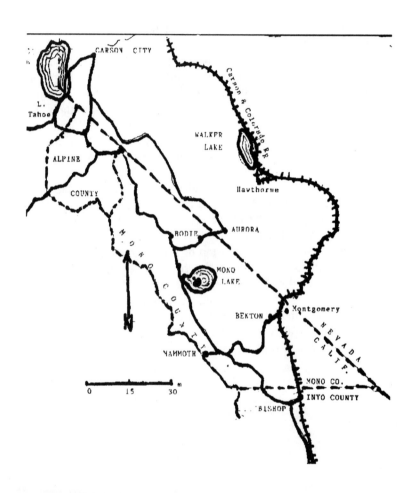

11. THE CONSCIENCE OF GENERAL DODGE

Mammoth it is, and mammoth it was—at least in the minds of General George S. Dodge and his friends who bought the mine and fostered the timberline town near the crest of the High Sierra. For a time Mammoth City and Mammoth common stock were on everybody's lips from Bodie to San Francisco. Within four years of its discovery, Mammoth was dead and so was General Dodge.

During the spring thaw of 1877 four men rode into part of the headwaters of Owens River, on the east face of the Sierra Nevada. They were headed by James A. Parker, a man of many accomplishments—Gold Rush pioneer, experienced miner, and co-founder of the first newspaper in Inyo County. One acquaintance called him "a prospector of the lucky, prodigal type."

Twenty-seven miles southwest of the silver camp of Benton, they rode straight into the Sierra forest in what later came to be known as Mammoth Lakes. The crags about them were granite—hardly a famous host for silver or gold. But they were part of a legion of miners who had, for twenty years, searched for the famous "Lost Cement Mine".

The legend of that vanished gold strike had tantalized prospectors since its reported discovery in 1857 by Argonauts on their way to California at the end of the Gold Rush. According to the tales passed by word of mouth, it had been lost, then found, several times over in the next few years. The story was full of mythical maps and mysterious deaths in the rich tradition of such tales.

Ore samples, shown from time to time, were a reddish cement-like rock spangled with flakes of gold. Mark Twain, living in Aurora when the Lost Cement fever was at its height, wrote how the principal exponent of the legend, Gid Whiteman, showed an ore specimen in town.

"Lumps of virgin gold," wrote Mark, "were as thick in it as raisins in a slice of fruit cake."

Much of the prospecting centered west of Benton around Deadman Creek, named for one of the victims in the grisly tale. But since no color had been found there, the Parker party decided to search the next canyon south, where Mammoth Creek flowed out of a string of lakes that sparkled like emeralds in the granite setting. With visions of golden lumps dancing in their heads, the four men rode through pine and willow underneath what is now known as Red Mountain, east of Lake Mary. Some 800 feet up the escarpment, the reddish outcrop jumped out before their eyes.

Scrambling up the cliff, they found promising indications. On June 20, 1877, they staked their claim, the Alpha. Maybe it wasn't the Lost Cement Mine, but it would do till someone else found rock with those golden raisins.

Running out of provisions, the Parker party trekked the sixty miles to Bodie, carrying some specimens with them. The rock averaged $86 to the ton in gold and silver. This was good enough to stir some interest and to remind two men, A.J. Wren and John Briggs, that they had located some claims two years before on the same mountain. The pair promptly rushed back and located the Mammoth Mine on July 30. Half a dozen others straggled in during the summer of 1877 and joined in creating the Lake Mining District, with Jim Parker as recorder.

By this time other ore samples had been sent to Bodie, where they assayed from $100 to $260 per ton in gold and silver. Parker and his friends, as interested in selling their claims as in working them, naturally declared that Lake District was "the largest bonanza outside of Virginia City."

Through the winter of 1877-78 the Mammoth fever built up steam as the boys waited for the heavy snows to thaw. By late March road builders were surveying a toll road west from Benton to the mines. Passable for light wagons by the end of April, it carried a heavy traffic through Long Valley to Casa Diablo Hot Springs and into the pine-clad Sierra. From the south another toll road was constructed from Bishop Creek to the lake country by J.L.C. Sherwin. The steepest section north of Round Valley is still known as the Sherwin Grade, though the famous old switchbacks have been supplanted by a modern highway. By April 1878 an upper Owens Valley resident described the traffic:

"Men bound for the lake district. . .pass through here daily. Some go afoot, others horseback while a few navigate by vehicles."

On the side of Red Mountain scores of miners were working some 300 claims. At the foot of the mountain, north of the willow thickets lining Mammoth Creek, a town was rising from the sagebrush. For a mile up and down the canyon, lots were staked off and fenced. At least twenty log houses were going up, including a stable and a hotel complete with saloon. At night the steep and primitive street was host to some 125 inhabitants, each of whom believed he was on the ground floor of another Comstock.

One of them wrote of "the *mountains* of gold and silver-bearing quartz rock, in some of which gold can be seen with the naked eye."

Stirring the excitement, the superintendent of the Alpha Mine was happy to show news correspondents into the tunnel and along the outcrops. Soon the Alpha was leased to a Comstock firm, which promptly pushed the tunnel into the side of Red Mountain.

By mid-May a stage line was running from Benton to Mammoth, bringing not only more miners but also the men with capital looking to buy them out. Among the earliest were several gentlemen of Bodie and San Francisco headed by General George S. Dodge.

Though still in his late thirties, Dodge brought a lifetime of management experience. Born in 1839 at Irasburg, Vermont, he early sought his fortune in Boston, where he joined a mercantile firm. With the onset of the Civil War he donned the blue as an officer among Vermont volunteers. For some time he was stationed with the Quartermaster Corps at Norfolk, Virginia, achieving the rank of colonel in September 1864.

As Chief Quartermaster of the Army of the James, he distinguished himself in the capture of Fort Fisher, North Carolina, "disembarking upon an open coast men and material for the siege and assault." For this he was breveted a brigadier general—at twenty-five, certainly among the youngest of that rank in the Union Army.

In character, Dodge was wholly reliable, full of energy, quick to seize opportunities, with a reputation for "shrewdness and tenacity of purpose." Physically he was handsome, with dark hair and moustache, and what one acquaintance called a "magnificent physique". He seemed one of those select few destined for big things.

Mustered out of the army in March 1866, Dodge was appointed U.S. Consul General to the free city of Bremen, then an independent German state. Leaving that post, he sailed for California and threw in his lot with the circle of Comstock financiers. Like so many others, he would typically buy and develop a mine, create a corporation, place its stock on one of the San Francisco exchanges, and when investors ran the price up, he would in the vernacular of the day "make a big turn and a good killing." The huge fortune he amassed came, not so much from the bullion product of his mines, but from the advance in the price of mining stocks.

In 1871 Dodge was attracted by claims of fabulous diamond fields in Wyoming and Arizona. In company with other San Francisco capitalists, including William C. Ralston of the Bank of California and mining magnate William M. Lent, he bought for himself a one-eighth interest, and tried to buy still more. But in 1872 the scheme proved to be what one newspaper called "the greatest fraud ever palmed on the commercial community of

California." For once—perhaps the only time—Dodge lost money in one of his high-flying investments.

By this time he had established himself as a leading citizen of Oakland. His Victorian mansion and grounds covered the entire block from Eighth to Ninth Streets and from Madison to Jackson. Here he and his elegant wife entertained not only the elite of Bay Area society, but also visiting Eastern notables, including Gen. William T. Sherman and Gen. Phil Sheridan.

In 1877 Dodge decided it was time to enter politics. When he ran for mayor of Oakland on the Tax Reform ticket, the San Francisco *Alta California* declared he "will make a good mayor." But Dodge was defeated and thereafter confined himself to mining ventures.

During the same year he was one of the first major investors in Bodie, where he bought and incorporated the Lucky Jack and the McClinton mines, reaping a handsome profit on both. When the Mammoth fever struck in the spring of 1878, Dodge was ready for his next "killing".

In Lake Mining District, the owners of the Mammoth Mine were happy to show Dodge and party their mines, some 800 feet up the side of Red Mountain. They would sell not only the Mammoth, but four other mines, including the nearby Headlight. They shoved ore samples at him and quoted assay figures in gold and silver. Dodge sat down on a pine stump in the main street and looked up at the distant outcroppings.

"I don't want to know how rich they are," he answered. "They'll do for a deal anyhow."

Without climbing the mountain to look into the mine, Dodge and his friends offered $10,000 cash and $20,000 worth of stock in a corporation that yet existed in their imaginations. The owners accepted on the spot and Dodge and party took the next stage for Carson and the railroad, stopping long enough in Bodie to stir some excitement about Mammoth. There was just one large ledge, but it "can be traced for several miles and located the whole length."

On June 3, back in San Francisco, they incorporated the Mammoth Mining Company with 100,000 shares at a par value of $100 apiece—a capitalization on paper of $10 million. The stock sold readily, but at much less than $100 per share. They got a boost from the *San Francisco Stock Report*, with a story headed "A Mono County Bonanza".

"The condition of the property held by the company," it declared, "justifies brilliant hopes."

And Edward Clarke, one of the original Dodge group, wrote with a promoter's flourish to the *Bodie Standard*:

"A grand field for mining enterprise is opened and the future is pregnant with promise."

While selling stock to investors, Dodge drove into the mountain with three tunnels, intending to strike the lode at different levels. He ordered a sawmill for construction of buildings and the machinery for a twenty-stamp, water driven quartz mill. Despite other obstacles, the ore was "free-milling", requiring only pulverizing and chemical amalgamation, without need of a furnace.

Soon Remi Nadeau's sixteen-mule teams were plodding up the Bullion Trail from the railroad at Mojave, loaded with an iron water wheel and other strange-looking castings. To provide a steady flow of water, Dodge's company built a dam and headgate at Twin Lakes, with a ditch to carry the flow eastward down to the mill.

By the end of June the sawmill, its whistle echoing across the crags, began turning out lumber by day and shingles by night, filling the canyon with its whine and clatter around the clock. In late September the stamp mill was finished, awaiting construction of the tramway and chute that would bring Mammoth ore down from the mountain.

Meanwhile, Mammoth City was poised in anticipation—like a beehive about to be disturbed. Seven mines had been incorporated and their shares selling on the San Francisco exchanges. Joining Mammoth City were other fledgling towns—Pine City, on up the road near Lake Mary; Mill City, clustered around the quartz mill half-a-mile below Mammoth; and Mineral Park, another half-mile eastward down the canyon. But so far no bullion had been shipped. That summer men wanting work were sleeping under the pines waiting for action.

In this wilderness of excitement, some voices of caution were raised. Joe Wasson, highly respected mining man, wrote of the Mammoth investors in his guidebook to the Mono County mines:

"Whether they have a bonanza, or only a white elephant, is. . .an open question."

John Hays Hammond, later to win worldwide fame as a mining engineer, looked in on Mammoth and told the company's officials that the mines were "of slight value". Later he wrote:

"They had gone to the extent of erecting a mill in anticipation of ore that existed in imagination only. The worthlessness of the enterprise was apparent to any trained engineer."

But Hammond was too hard on Mammoth, which after all was basking in the glitter of some rich samples. Despite such warnings, the Mammoth company marched bravely on.

The complex system of mines, tramway, chute, ditch, water wheel and stamp mill was cranked into action in October 1878. But the water ditch from Twin Lakes soon filled with snow and had to be replaced by a covered flume. A windstorm blew down part of the tramway, which had been built on stilts. When rebuilt, it was still slow going because the sensible mule laid down on the track at every strong wind. The ore chute was smashed by the first load of ore. And as winter came on, heavy snows broke through one part of the quartz mill, requiring another shutdown.

Still, the mill ran through much of the winter. Despite heavy snows, the company shipped its first two bars of bullion on February 8, 1879.

Up to this time the Mammoth boom had been moderate. One of the heaviest winters on record had blanketed the towns with many feet of snow and stopped all incoming mail, even by pony rider. But with two feet of snow still on the ground by April, the fortunes of Mammoth suddenly took a turn.

On April 1 the miners in the middle tunnel struck a ledge four feet wide and found ore sparkling with free gold. Alongside it was a silver-bearing vein twelve feet wide. Within two days specimens "almost filled wth particles of gold" were shown in Benton. A few days later, letters from Mammoth reached Bodie carrying the news. Declared one:

"The Mammoth mine has struck it rich—the richest body of ore in gold I ever saw."

The superintendent of the Mammoth mine rode into Bodie, the closest telegraph point, and wired the news to San Francisco. Another well-known mine owner came into Bodie with specimens of the blind-lead strike.

"The richness of the ore is simply beyond description or belief," he told his listeners.

At this, Bodie rang with excitement. A horde of men rode or walked out of town on the sixty-mile, snow-heavy trek to Mammoth. There the incoming miners were joined by investors, many of them pure speculators, from the whole Eastern Sierra. From the south, freight wagons were crowding the road. In a single day, 23 teams swung through Round Valley and toiled up the Sherwin Grade.

Fanning the frenzy, Sherwin himself wrote to the *Inyo Independent* of a ledge four feet thick that averaged in the thousands per ton. His toll road, he quickly added, was the best route. Actually, the road up the three-mile grade was so bad that the mountains echoed with the teamsters' profanity, and Sherwin began improvements as fast as melting snow would allow. Meanwhile, other enterprisers were building a toll road direct from Bodie to Mammoth, along the shadow of the Sierra, avoiding the dogleg through Benton.

> **Bodie & Lake Stage Line.**
>
> Stages will leave Bodie for
>
> **Pine City, Casa Diablo, and King's Ranch.**
>
> Every day, commencing on TUESDAY, May 2, 1879, and connecting with the through mails both ways,
>
> MAKING THE TIME THROUGH TO MAMMOTH IN TWELVE HOURS!
>
> **Fare to Mammoth, $15.**
>
> This is the shortest, quickest and best route to Lake District.
>
> Office—With Clugage & Co., on Main street, Bodie.   H. W. LAWTON, Agent.
> CON OGG,   H. C. BLANCHARD,
> Proprietor.   Superintendent.
> ap23tf

As the new army of rainbow hunters poured into Mammoth, town lots jumped to $1500. Many men were old hands lured by every strike along the Eastern Sierra. Ned Reddy and Senator Bill Stewart just looked in and returned to Bodie and Virginia City. Others threw in their fortunes with the new diggings. From Benton came George W. Rowan, the same who had led that town's fight against Pap Kelty. He promptly opened a large general store and built a second sawmill, which turned out more lumber for some fifty buildings going up to the happy song of saw and hammer..

By May 1879 some 2,000 men and perhaps 125 women were filling the towns and mountainsides of Lake Mining District. The rush brought two semi-weekly newspapers: The *Lake Mining Review*, at first printed in Bodie, and the *Mammoth City Herald*, whose editor brought his hand press from Benton, where he had edited the *Mono Messenger*. The first issue of the first paper on May 29 was overflowing with notices from eager advertisers.

One storekeeper offered everything from jewelry through musical instruments to "guns, pistols and ammunition."

Another sold "groceries, liquors, wines, cigars, clothing, mining material, fuse, caps, powder, shovels, wheel-barrows, picks, axes."

The Mammoth Saloon advertised "Club Rooms attached where playfully disposed parties can amuse themselves"—referring to back-room gambling.

The Nevada Saloon: "Club Rooms with an unchained tiger in the back room"—meaning a faro table.

The editor, experienced in frontier towns, gave the boys plenty of salty humor and pungent comments. When one owner ran some jumpers off his lot at Pine City, the editor warned claim jumpers out of Mammoth:

"The chances are that all such will be treated with liberal doses of cold lead."

Through the summer of '79 more fortune-seekers were arriving by stage from three directions—Benton, Bodie and Owens Valley—and by saddle train over the Sierra from the stage connection at Fresno Flats (now Oakhurst). This route was not recommended for those with bad nerves or a short temper.

By this time fourteen mines were incorporated and selling on the San Francisco exchanges. The Lake District recorder was swamped with new locations. In Mammoth City alone the streets were lined with six general merchandise and hardware stores, two clothing shops, three variety stores, two pharmacies, two livery stables, six hotels, six boarding houses, five restaurants, two breweries and 26 saloons. Mammoth City had 125 frame buildings, while Pine City had thirty structures of boards or logs. As for residences, most everybody had moved from tents to log cabins. There were three doctors, three policemen equipped with a jail, a private school with one teacher, and a preacher who yet had no church.

As for other necessities of life, table fare was largely from the can, but farmers in Bishop Creek and Round Valley supplied some vegetables and fruit, which at first were low on quality and high on price. The real luxury peculiar to Mammothites was an ample supply of trout, caught by Paiutes over the divide on the South Fork of the San Joaquin and sold by them in Mammoth for 50 cents a dozen.

These and other commodities were purchased mainly with Mammoth Mining Company paychecks drawn on San Francisco banks. These passed like currency throughout the district, and saved the firm the expense of shipping in gold coins. Small change was made with U.S. silver coins, which were so scarce in this remote camp that merchants accepted most international coins (Mexican, Canadian, British, French) that approximated them. There were no coins less than a quarter, and the smallest item—a pin or button—sold for a quarter or its international equivalent.

At least two women ran boarding houses, and with some professional men bringing their families, the Mammothites enjoyed a rousing social life. Picnics, dances and surprise parties ruled the summer months. At Lake View Park on Lake Mary, night-long dancing was the custom on weekends. Three sail boats took happy excursionists across the water. No occasion—weddings, holidays, lodge installations—was missed for a grand party.

But this remote camp was some 150 miles by road to the nearest rail connection at Carson City. With saloons predominating along Mammoth Avenue and seven prostitutes (five of them Chinese) plying their trade, it was a man's town, rough-hewn and provincial. One editor in search of news for his local column thought it necessary to include: "A dead horse is laying on the trail as you go south." The Mammoth Hotel, claiming to be the leading hostelry in Mammoth City, advertised that "The Fresno Flats saddle train arrives and departs from our door." In the middle of town, two or three Chinese opium dens openly flouted California law, as did others in Bodie and many other western towns. When the miners descended on the saloons on payday, fist fights were frequent, though gunplay was minimal. Claimed the *Mammoth City Herald:*

"Mammoth is the most peaceful and orderly mining camp on the Coast. We have no roughs, no tramps, and no desperadoes."

Through the summer the main street was filled with wagons and mule teams, and prospectors with horse or burro preparing for another foray in the mountains. When a newcomer appeared he was seized by miners showing him rock specimens and quoting assay figures. In the bar of one hotel one of them befriends a stranger.

"You've not been much in mining regions?"

"No."

"No? Then if you want to become rich—immensely rich—mind, I'll let you have fifty feet and try to get my friend Dick. . .to sell you as much in the Real Jink, which is the biggest thing ever discovered on the Pacific slope. . .I wouldn't do this for everyone, but have taken a liking to you, my dear fellow."

The stranger asks how much.

"A mere song, my dear fellow, say five hundred for me and the same to Dick."

At that crucial point the stranger is saved by the breakfast bell.

Despite all its mining activity and the high spirits of its citizens, Mammoth was still living on hope and enthusiasm by the end of summer, 1879. The town shook to the blasts deep in the side of Red Mountain as the Mammoth, Headlight, Monte Cristo, Don Quixote and True Blue were being driven toward the ledges. The *Herald* carried a report each issue on the feet gained in the Mammoth tunnel and the joint Headlight-Monte Cristo tunnel. But little bullion had been shipped.

Anticipation was the same in the offices of the Mammoth Mining Company at 302 Montgomery Street, San Francisco. General Dodge and his colleagues could at least be happy with the price of Mammoth common stock, which reached $20 per share after the big strike in the spring of 1879. But so far the only dividends had been Irish dividends. Subscribing to Mammoth newspapers, the stockholders waited for every fragment of news from the shining Sierra.

"The officers of the mine are looking for it in San Francisco with an interest that surpasses ours," observed the *Mammoth City Times*, a successsor to the *Lake Mining Review*.

As for the mill and mine, the company sent a new superintendent, Col. William H. Hardy, who stepped off the Bodie stage at the end of July. Hardy turned out to be a hard-driving, strong-willed boss. Though he soon won the respect and approval of Mammothites, he brooked no interference with his single purpose of getting the Mammoth enterprise to turn out bullion.

One miner living on the side of the mountain objected to blasting near his home; Hardy fired him and he had to leave camp with his family. Another miner asked for back pay in order to start a boarding house; Hardy fired him and forbade his employees to patronize him. A reporter from the *Los Angeles Express* called Mammoth "a one-man town" and Hardy "the Czar of Mammoth."

But Hardy was making things happen in Lake Mining District. Since the mule-powered tramway was too slow in delivering ore, and the chute had proven a disaster, he started blasting the roadbed for a new tramway at the end of August. Designed to bring the ore cars down the mountain by gravity alone, it included numerous switchback curves and a cover for winter snows that were marvels to local observers.

Before assuring that the novel tramway would work, Hardy started enlarging the quartz mill from twenty to forty stamps with the aid of steam power. In August, Nadeau's eighteen-mule teams were hauling in the huge boiler, the twenty-foot flywheel and other large machinery.

Meanwhile, both newspapers were trying to make up in words what Mammoth lacked in bullion. In early July the *Herald* declared, "there are a hundred ledges rich in gold and silver not to be equalled in this State, and scarcely surpassed by the famous Comstock." It was, he concluded, "but a question of a short time when we will have a camp second to none on the coast." Later that month, the editor joined other visitors on a candlelight tour of the adits, winzes and crosscuts of the Mammoth Mine, led by the company superintendent and the mine foreman .

"Suffice it to say," he wrote, "every nook and corner is filled to overflowing with the richest of gold and silver ore ready for the mill. . . Ore above, ore below, ore on every side."

In early September another visit came from Captain S.D. Prescott of Benton, whom the editor called "a celebrated mining man." Skeptical of claims in the granite Sierra, Prescott had previously downgraded the Lake District. He came out of the Mammoth Mine and announced:

"You may tell the boys that I take it all back. I now believe Lake is going to be one of the most important mining districts on the coast." There was, he said, enough ore to keep the forty-stamp mill going for twenty years.

On September 15, with a blast of the steam whistle, the huge mill rumbled into action, its wheeze and clatter giving little annoyance to Mammothites awaiting prosperity. By early October, Superintendent Hardy was gathering stocks of timber for the mines, more wood for the fireroom of the mill and coal for the ironworking shop—all before winter covered the roads with deep snow.

At this juncture, with snow already on the ground and all hands awaiting completion of the new winter-proof tramway, the time was ripe for another big discovery. On the night of October 7, from out of the middle tunnel, the Mammoth night foreman came tumbling down Red Mountain, his metal boot heels striking sparks on the boulders as he leaped. At the bottom in the timekeeper's room the breathless foreman tossed a sack on the floor.

"Look at that!" he shouted.

Some twenty pounds of quartz glittered with streaks and flakes of gold—perhaps 50 percent gold. The mine foreman was sent for. He burst in, took one look, and scooping up several rocks he hurried down the hill to show Superintendent Bill Hardy in the company's main office at the mill.

Next morning the news was all over Mammoth City. Quickly the boys surged over to the mine office. There, in chunks weighing up to 50 pounds, more ore had been brought from the tunnel.

"All were literally alive with gold," claimed the *Lake Mining Review*. Some gold "splatches" were the size of peach tree leaves. Far richer than that which had caused the excitement in the spring, this ore had been found in the same tunnel, which now came to be called the "Bonanza Tunnel".

Across town the blind lead strike was on everybody's lips. As the *Lake Mining Review* reported, "predictions of Mammoth's future greatness are as plentiful as flies around a sugar barrel." And the *Herald* solemnly added, "the Mammoth mine can show the richest gold-bearing quartz in the world."

The telegraph brought the news to San Francisco and promptly sparked a surge in Mammoth stock. Having previously fallen to 8½, it leaped up, then as second thoughts prevailed, fell back again. By the end of October it was down to 6½.

After running 41 days, the Mammoth mill shut down for the usual "cleanup". It had produced some $49,000 in bullion—hardly spectacular, but excused on the ground that low grade ore had been used while getting the wrinkles out of the new workings.

On November 12 Superintendent Hardy tested the covered tramway. Each holding four tons of ore, the iron cars rolled down the track controlled only by special powered brakes. After emptying at the mill they were hauled up again by a span of mules.

But the Mammoth & Mill City Grand Trunk Line, as the *Mammoth City Times* called it, had some serious flaws. The grade exceeded 10 percent, the curves were too sharp, and the tracks were not banked at the curves. Moreover, the shed housing was so close to the cars that there was no room for a person to crawl on top, and the brakeman was expected to stand on the coupling between two cars—as the *Times* later observed, "about as complete a death trap as it is possible to invent."

Both Superintendent Hardy and the mine foreman had voiced their concern, but the San Francisco office had insisted on this design.

In the first actual operation on the second day, two loaded cars were going down the grade when the brakeman found the brakes were not holding and the cars were out of control. To save his life he somehow crawled back over the second car and dropped unhurt to the track behind. At the first curve the cars jumped the rails, broke through the wooden shed and landed on the mountainside.

Shutting down the mill to await more ore, Bill Hardy made changes to the ore cars and installed stronger brakes. On November 18 they went down without mishap, and the mill started again. Then a storm brought rain and snow which, entering the shed through its windows, iced the rails.

On November 29 the brakeman started two cars rolling down the track. The wheels locked and skidded along the rails. Before he could jump the cars reached a curve, flew off the track and through the shed. The brakeman, knocked senseless, was pulled from the wreck and later recovered.

This was the end of the Mammoth & Mill City Grand Trunk Line. With winter coming on, there would be no escape from snow and ice. Besides, Bill Hardy could find no one willing to ride the brake. After spending some $14,000 on the project he abandoned it, shut down the mill again, and turned to rebuild the old mule-drawn tramway tottering on stilts. Declared the *Mammoth City Times*:

"It was about as amateurish a performance in railway building as we ever heard of."

When this news reached San Francisco the Mammoth stock fell to $4.00 per share—and by December 18, to $2.25.

At this, Bill Hardy had enough of the *Mammoth City Times*. Forbidding his employees to subscribe to the paper, he had friends purchase a delinquent debt against the *Times* and then enforce payment. The editors offered in vain to retract, but in the end they had to sell out to a new management.

Meanwhile, winter was again descending on Mammoth. At first, stages, freight and pony mail could get through, and wagon-boss Austin Lott of Nadeau's teams announced he would keep them coming up from Owens Valley as long as he could "see the tops of the trees."

But beginning December 18 the season's first big storm struck the Eastern Sierra and practically isolated Mammoth. Despite this, Hardy had both the tramway and the mill running before Christmas, and somehow managed to get himself through to the railroad at Carson City and then to San Francisco. There he would explain the situation and hope to bolster the company's officers and stockholders.

On Hardy's return he encountered a worse challenge. With the road from Bodie under many feet of snow, he hired a sleigh and pushed south with several companions. On the way they finally had to abandon the sleigh and plow through on horseback. When this proved impossible they pushed on with skis. Floundering through snowdrifts and losing their way once, they trooped into Mammoth the afternoon of December 29. One thing could be said of Bill Hardy—he was as severe on himself as he was on others.

It was to be one of the worst winters on record. In Mammoth, snow was far too deep for wagon traffic. There were no horses in camp, and as one resident put it, "we couldn't use them if they were here." Most houses were buried in snow, with entry by burrowing down to the front door. At one time the newspapers had to be delivered to some houses down the chimney, with the news carrier presumably warning the occupants beforehand. People spent a large part of their time digging tunnels to their neighbors' houses to see if they were all right. The local liveryman had to sink a shaft to find the ridge pole of his stable. For weeks at a time, stage, freight and mail communication with the outside world was at a standstill.

At the Mammoth works, mining continued and ore piled up at the tunnel mouths. But the mule-drawn tram was shut down for repairs, and the mule took the opportunity to decamp for Benton and points east. Since chemicals and other supplies were cut off, the stamp mill stopped in early January.

At the end of December 1879 the exodus had begun.

"A crowd of ten or twelve men left for Bodie this morning," reported the *Herald*, "and several pairs of snowshoes are missing."

After the mill closed the trickle rose to a flood. As a correspondent of the *Inyo Independent* wrote:

"Men have been and are leaving in twos, in sixes, in dozens; on snow shoes, on barley sacks, on anything, for Bishop, Benton, Bodie, Candelaria—anywhere."

Those who stayed busied themselves shoveling snow, playing poker or bucking the tiger in two or three saloons, the rest being empty. As one observer wrote:

"The storekeepers are sitting reading, idly smoking, or lounging against the door casings, vainly looking for a stray customer." Unfortunately, most potential customers were broke or in debt. "The trouble is, there is no money."

The population of Mammoth City itself was down to some 500 people, but the women, now numbering about 60, did their best to keep spirits high with parties and dances—the Knights of Pythias banquet, a grand leap year ball, and theatrical presentations by the Mammoth City Dramatic Club. And the *Mammoth City Herald* tried to maintain optimism with glowing forecasts of Mammoth's future:

"Before another year shall roll around this will be among the most prominent mining camps on the Pacific Coast."

Though Hardy soon had the mule tramway and the quartz mill running again, the relentless winter stalled the tramway, and hence the delivery of ore to the mill. The company had spent tens of thousands on mine and mill, the dam at the lakes, the water flume to the mill, the broken ore chute, a mule-powered tramway and a gravity-powered tramway, and still the celebrated ore from the Mammoth tunnels could not be delivered fast enough to produce sufficient bullion to pay expenses.

In March 1880, Col. Bill Hardy was recalled to San Francisco. Though he would not say that this was the end for him, the boys gave him what was clearly a send-off party. On Saturday night, the 20th, almost all the men of Mammoth, accompanied by a brass band, surrounded his office at the stamp mill. Wine, speeches and song filled the moment.

Hardy may have made enemies, but he had made a lot more friends. Next day he took stage for Carson and left Mammoth for good.

With the spring breakup, Mammoth came to life once more. Merchants, hotelkeepers and saloonkeepers were repairing, painting and wallpapering. Supplies were arriving by mule team, and Mammoth Avenue once again resounded with the crack of the blacksnake and the oaths of the skinners. Miners returned to work, and the mountain rang to the pneumatic drill and the powder blasts.

"The Mammoth mine will without a doubt be proven to be the bonanza of the Sierras," predicted the *Herald*.

On July 1, with a toot of the whistle and the rattle of dropping stamps, the mill started again. The tramway was delivering ore for the monstrous batteries at Mill City.

But the well from which the company drew its sustenance had now dried up. The fifth stock assessment on June 16, 1880, was not answered. On July 31 the column-and-a-half delinquent sale notice in the newspapers told the story. Mammoth common stock, which in May had rallied to $3.25 with the spring activities, sank to $1.60—a trend that was not helped by a general drop in Western mining stocks. The *Herald* went from a semi-weekly to a weekly, commenting that "Mammoth is dull—very dull."

Up on Red Mountain a fourth tunnel below the others was expected to show richer ore at that level, but it still had some 300 feet to go before reaching the giant ore body.

Mammoth strove mightily to preserve its spirit. Good cheer reigned at the Independence Day celebration at Lake Mary, where the entire population of Lake District turned out for horse races, rowing contests, wrestling matches, sailboat races, fireworks, speeches, music by the Silver Star Band, and a flotilla of boats crossing the lake filled with happy faces. Early in April a grand ball in Mammoth raised funds to help a new public school.

But by September, with the approach of another winter, the same population of celebrants began trekking out of Mammoth, some of them leaving quietly in the night without paying their debts. More than half the stores had closed. By October 16, Mammoth stock dropped to 43 cents a share.

As though bringing the drama to a brilliant climax, an arsonist started a fire in George Rowan's store on the morning of November 15. Within minutes the blaze spread to nearby buildings, leaped across the street, and enveloped the upper end of town in what the *Herald* called "a sheet of roaring, seething flame."

At first people ran about trying to save anything they could. Then a number of men organized to save the lower part of town. One house was hauled out of the way to create a fire break, while across the street an open space already existed west of the *Herald* office The stalwarts stood on the roofs, dousing fires as soon as they started. When it was over, forty buildings lay in smoldering ashes.

Then in early December a savage snowstorm uprooted trees, blew down buildings and toppled part of the mule tramway. Snow caved in the roof and blew in the doors of the *Herald* office until press, type and furniture were buried under tons of white.

"What with fire and the fury of the elements," wrote the editor, "Lake District has surely had her share of misfortunes."

Later in December, Mammoth stock sank to six cents a share. Starting in January 1881 the usual reports of progress in the Mammoth and the joint Headlight-Monte Cristo tunnels failed to appear in the *Herald*. In what was undoubtedly an understatement, the paper noted on January 15:

"During the past week quite a number of miners have left this place for Bodie."

In the same edition the paper shrank from four to two pages, and the last issue appeared on February 12. On July 6 the Bank of California foreclosed on the Mammoth Company, and the entire property was sold at Sheriff's sale on the Mono County courthouse steps in Bridgeport. By the end of October the stamp mill was being dismantled and Mammoth City had only two businesses—a store and a brewery. In an election for recorder of Lake District, only 24 men were on hand to vote. As the *Bishop Creek Times* reported:

"Mammoth City is almost deserted."

By 1885 one visitor reported "not a soul to be seen" in Mammoth.

"Houses overthrown by wind, tables, chairs, beds, etc., in wild confusion. Roofs breaking and broken down under weight of snow; clanging doors and shattered windows showing the deserted interiors."

General George S. Dodge, founder of the Mammoth Mining Company, scarcely outlived the town he had fostered. Two months before he had first visited Mammoth, Dodge had been under the care of a doctor. Later in the same year he had taken a trip East "for a rest", but when he returned to California his wife remained in New York.

The truth was that Dodge had syphilis, and it was reaching the advanced stage. His spinal cord became paralyzed, and he had to take to bed. By the time of the big strike in the Mammoth Mine in the spring of 1879,

Patrick E. Purcell
In 1878 Gen. George Dodge bought the Mammoth Mine, started a 40-stamp mill, and spawned Mammoth City.

Stephen Ginsburg
Dodge's huge mill was built before his manager could solve how to get the ore 800 feet down the steep mountain.

Nadeau Collection

During its heyday in 1879-80, Bodie had 8,000 people and over 600 buildings—the biggest town in Eastern California. This view was taken from Bodie Bluff, which produced more than $20 million in gold and silver. Bodie had "the wickedest men and the worst weather out-of-doors."

Eastern California Museum

With its false fronts and shaded boardwalks, Bodie was the archtype of the Western frontier town. Here a 12-horse team raises dust along the dirt street. Based on Bodie's tough reputation, a hard case visiting other towns would try bullying people by calling himself a "Bad Man from Bodie".

Dodge appears to have been out of the company. A writer in the *San Francisco Daily Exchange* quoted him as saying he "always knew it was a good mine," and hoped his friends "would make some money." But for himself he did "not have any further ambition in this world than to depart from it with a clear conscience."

Suffering severe pain, Dodge progressively lost his sight, his speech and finally his mind. He left his magnificent Oakland home and took quarters in the nearby Galindo Hotel at the corner of Eighth and Franklin Streets. There he died on August 24, 1881, at the age of 42 years. The *Oakland Tribune* carried a routine death notice. The San Francisco *Alta California* referred to him as "a prominent citizen" who "at one time was a candidate for Mayor of Oakland."

Whether a "clear conscience" was an issue in the minds of the other figures in the Mammoth Mining Company is a moot question. At the time there were various charges that the Mammoth was just a stock promotion, or worse, a stock swindle; that the mine and mill stopped because the officers were pocketing some of the stock assessments instead of using them to run the business; that they were using the stock assessments to "freeze out" the small stockholders; that the ore which was so dazzling in two blind lead strikes never went into bullion because it was highgraded by employees or managers of the company.

Some or all of these tricks may indeed have been practiced to a certain degree at one time or another in the Mammoth. None of them was uncommon in Western mines during an era unregulated by public agencies. But serious objection can be raised against each of these as the cause of the debacle.

The stock price reached its peak in the spring of 1879. If stock promotion was the only real objective, this was the time for the promoters to sell and get out, and perhaps some of them did. But the continued expenditure on mine, mill and tramway speak of a company trying to make money. If this was a sham staged by the board of directors, they went to a lot more trouble and expenditure than necessary.

If the officers pocketed assessment funds, this can neither be proved or disproved, but it is not difficult to see where the funds went—into the elaborate rathole on Red Mountain.

If the directors were playing the "freeze-out game", it could only have made sense if a surge in the stock were staged after the freeze-out. But after the fifth assessment went unpaid, the stock continued to slide.

If the top Mammoth men highgraded ore, it could hardly have been done on a scale big enough to ruin the company, especially under the eyes of all of Mammoth's citizens, who were watching their every move.

As diagnosis favors the simplest solution, so the Mammoth fiasco can most easily be laid to poor management—failure to realize the cost of getting the ore down hundreds of feet of steep mountainside; failure to use professional engineering on that problem; failure to locate the mill at the foot of the mountain instead of half-a-mile away; failure to recognize the excessive cost of hauling in heavy machinery to a camp nearly two miles high and 150 miles from the nearest railroad; failure to get the water-driven, twenty-stamp mill to turn out bullion at a profit before enlarging it to a steam-driven, forty-stamp mill; and finally, failure to realize the severity of weather that made most operations impossible or marginal during nearly half the year.

Perhaps the biggest act of mismanagement was committed by Dodge himself, who had cavalierly dismissed an inspection of the mines and their riches with the comment, "They'll do for a deal anyhow." Perhaps it was the onset of his advanced affliction, which attacks one's ability to walk, that discouraged him from climbing the mountain.

For the real problem with Mammoth was that, while there was and still is plenty of ore, and there were some pockets of high-grade, most of it was worth at the time about $20 per ton. And considering the monumental costs of operation which should have been obvious, this was not pay dirt.

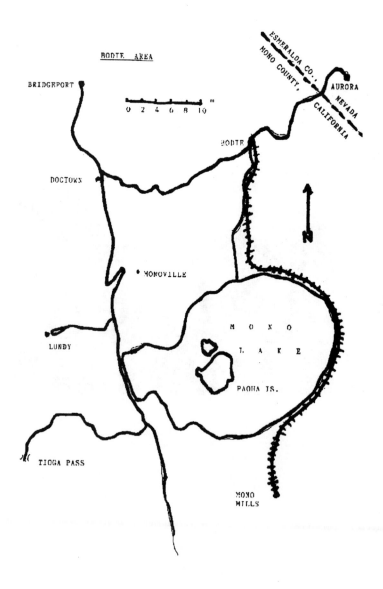

12. BAD MEN OF BODIE

> Though silver was king in Eastern California, the biggest mining district was more gold than silver. Bodie was a paradox in other ways, too. Named for Bill Bodey, it spelled his name wrong. It was discovered in 1860, but its first big rush came 17 years later. Pat Reddy, one of its leading characters, repeatedly championed the rule of law, but when the Miners Union ordered George Daly out of town, Pat advised him to leave. In fact, while Bodie had a squad of lawmen, its name became synonymous with lawlessness.

Gold actually came before silver in Eastern California. As early as 1857, before the Washoe discovery, placer gold was found on Dog Creek, north of Mono Lake and south of Bridgeport Valley. Here, about half a mile below the present turnoff to Bodie, there sprang up a small locality known, naturally enough, as Dogtown.

Then on Independence Day, 1859, a Dogtowner found more placer gold at what became Monoville, north of Mono Lake and east of the present Conway Grade. By the spring of 1860, Californians were rushing over Sonora Pass to these new diggings.

For a time Monoville flared brightly, with 700 people and more than 20 saloons. It was so remote that law was but a crude informality. In the justice court, one witness rode his mule into the building, gave his testimony, and rode out without dismounting.

Hard upon Monoville's rise, gold was discovered further east and Aurora quickly eclipsed the other camps. The influx brought forth Mono County, created in 1861 with Aurora as the county seat. Three years later Aurora had to give up the distinction after a new survey found it to be in Esmeralda County, Nevada. But through the early Sixties—with several thousand people, brick buildings and two newspapers—Aurora was the metropolis of the Mono-Esmeralda region.

Meanwhile, among those rushing to Monoville had been William S. Bodey, a tinsmith from Poughkeepsie, New York, who had said goodbye to his wife and two sons and sailed around the Horn to California in the 1849 rush. For ten years Bill Bodey had placer mined in the Mother Lode, sending money home to his family when he could. He was in Tuolumne County when he heard of the Monoville strike. In the fall of 1859 he tramped over Sonora Pass and prospected to the northeast of Monoville.

According to the story, Bodey was riding past what later became known as Bodie Bluff when he shot and wounded a rabbit. In trying to dig it out of a hole he discovered something else—gold. With his partner, E.S. "Black" Taylor, he made a crude home—half dugout and half cabin—at the north end of what would later become the town of Bodie.

That winter—one of the worst on record—Bodey and Taylor went to Monoville for supplies and on the way back were caught in a withering snowstorm. When Bodey collapsed, Taylor tried to carry him on, but finally had to leave him about three-fourths of a mile from their camp. Taylor soon returned for him, but could not find him in the blanket of snow. His body was not uncovered till the snow thawed in the spring of 1860. But he gave his name to the "Bodey Diggings" that sprang up near his old cabin.

Led by Black Taylor, prospectors were combing the bluff for outcrops that summer. On July 10 they met together and formed the Bodey Mining District. Within a year one of the claims recorded was the Bunker Hill, later renamed the Standard, one of the two great bonanzas on Bodie Bluff.

By this time the community that was sprouting had a slight change of name. Owners of a ranch on the road to Aurora ordered a sign, "Bodey Stable", from an Aurora sign painter. He misspelled it, and the sign "Bodie Stable" established the moniker for the budding town.

At the end of 1862, Bodie was moving from the prospecting phase to the stock promotion phase. Some eleven mines had been combined into the Bodie Bluff Consolidated Mining Company, with none other that Leland Stanford—then the California governor—as president. In mid-January 1863 the superintendent took ten pounds of ore samples over to Aurora, where they assayed between $50 and $400 per ton.

All at once Aurora rang with what one correspondent called "the greatest excitement". Two other groups of mines were consolidated and incorporated. Bodie Diggings became a town, with streets and lots surveyed and frame buildings going up. At the behest of the mining companies, noted mineralogists arrived and produced favorable reports. Inspired by one of these, some New Yorkers combined some of the mines, formed a stock company called The Syndicate, and brought in the first stamp mill.

In September 1864 Bodie was visited by the noted journalist and author, J. Ross Browne, who reported 15 to 20 "frame and adobe" houses, with more "springing up in every direction." People were carrying on a brisk trade in town lots, though Browne cautioned:

"A pair of boots, I suppose, would have secured the right to a tolerably good lot."

In fact, the excitement of 1863-4 was premature. So far the ore was hardly rich enough to yield big profits. Financial support from stockholders languished. So did Bodie for a dozen years.

Meanwhile, the Bunker Hill mine had passed through several owners. In 1874 it was acquired by two Scandinavians, Peter Essington and Lewis Lockberg. Pushing down an incline shaft for 120 feet without success, they were ready to abandon the effort. But due to their poor timbering work, part of the hanging wall caved in and revealed a rich body of ore. After working it for two years they sold the mine in 1876 for $67,500 to a group of San Francisco financiers, including William M. Lent.

Renaming it the Standard, Lent and his friends incorporated in April 1877, erected a twenty-stamp mill and a tramway to supply it with ore from the mine. Superintending it was Will Irwin, one of the Lent investors, who was known in the West as a capable mine operator.

As early as 1876, California and Nevada mining people had learned that Bill Lent had bought into Bodie. Among those hearing this news were Warren and Edwin Loose; as young men they had come West years before to help build the transcontinental railroad. They were big, rough-hewn Westerners, used to hard work, and taking no imposition from anybody. Ed Loose located the Bodie claim, which adjoined Lent's mine on the southwest, and soon took in his brother as a partner.

In extending the Standard Mine below ground, Superintendent Will Irwin continued to expose more rich, gold-bearing veins. And he reckoned that the same veins would continue into the ground under the Bodie claim. In San Francisco, the Standard group moved to get title to the Bodie.

In his book, *Bodie Bonanza*, Warren Loose of the Loose family of Bodie history states that the Lent people sent a hard case named Burkhart from San Francisco to run the Loose brothers out of Bodie. But when he and some confederates arrived and made their threats, the friends of the Looses in the saloons all over town got together and, in turn, ordered Burkhart and his pals out of town.

Shortly afterward, Bill Lent swung off the stage in Bodie. He first inspected his Standard Mine, and it is presumed that if the Standard had indeed entered Bodie ground, he inspected that, too. So when he met with Ed Loose to make a deal, he may have known more about the Bodie than Loose did. When he offered around $40,000 for the Bodie Mine, the Loose brothers sold.

Bill Lent hired more miners for the Bodie, replaced the windlass with a horse-powered whim, and pushed down the shaft with round-the-clock shifts. In the fall of 1877 he paid his Standard stockholders the first of many monthly $1.00-per-share dividends.

By this time, the magic name of Bill Lent, and the fame of the Standard Mine, brought the first rush to Bodie. In the fall of '77 they came from silver camps all over Nevada and from California's gold fields—all the caravan of miners, speculators, merchants, sporting men and sporting women.

One early arrival was George Daly, an Irishman from Australia, son of an Anglican minister. As a youth he had joined the rush to California and taken up the printer's trade in San Francisco. By 1870 he was foreman of job printing at the *Alta California*, where he resisted the typographer's union.

Walking home from work one night in December, he was attacked by half-a-dozen men. Two of them aimed revolvers at him, and a third struck at him with a knife. Bundled up from the cold, Daly could not reach his own pistol, but blew a police whistle. His assailants, hearing the approach of policemen, disappeared. Having recognized one of them, Daly next day made a citizen's arrest. The attackers had jumped the wrong man.

Shortly afterward, Daly sought his fortune in Nevada's Comstock Lode, where he took charge of the job shop of the *Virginia Enterprise*. By the fall of 1877 he had gained some experience managing a small mine. Joining the rush to Bodie, he first became superintendent of a mine across the Nevada line near Aurora.

Another prompt arrival in the '77 rush was Frank Kenyon, veteran Western editor, who launched the *Bodie Standard* in October. High on its front page: "Our Standard, Mono County, has the richest gold mine in the world."

That fall, Main Street was fast forming with more false front buildings, using lumber from nearby saw mills. In December the *Standard* editor described the "Lively times in Bodie":

> The stores, shops and saloons have been doing a big trade. . . .Wagons, long trains and stages arriving daily with freight and passengers; interested crowds eagerly discussing the latest strike, or some new discovery; capitalists and prospectors joining forces, or driving quick-business-like bargains; the rush and stir of superintendents hurrying their winter supplies to safe and convenient shelter while the favorable weather lasts.

Through the winter of 1877-78, Bodie was booming despite the bad local weather that would become famous. For a mile along the side of Bodie Bluff, the ground was pocked with other mines—the Bechtel, the Syndicate, the Noonday, the Red Cloud and many others. Their ledges seemed to angle downward toward some junction—the stem of what the boys called a "true fissure vein". Later, they would find that the deeper they went, the more the gold content would be joined by a silver content, as well.

By the end of March, Bodie had 1500 people; about 600 of these had no jobs, representing the drifters and hopefuls who attended every mining excitement. Businesses numbered more than sixty, of which seventeen were saloons and fifteen were bawdy houses. Town lots were selling at from $100 to $1,000.

In the saloons up and down Main Street, old friends from camps all over California and Nevada were greeting each other with the invitation to "take something"—meaning, "Let's have a drink." Since refusing the offer was a heinous insult, "somethings" were taken many and often.

With everyone carrying revolvers, such tradition soon led to mayhem. As Bodie earned a reputation for violence, Western newspapers spread the legend with the mythical character, "Bad Man from Bodie". Accordingly, hard cases arriving in other camps would demand a free drink at a saloon by claiming to be "a bad man from Bodie!"

Apparently the original such bad man was John Braslin, alias U.P. Jack, alias Rough-and-Tumble Jack, who was described by a Nevada newspaper as "one of the roughest and toughest customers ever known." In January 1878 he was boasting of his physical prowess in a Main Street saloon, claiming to be the undisputed "chief" of Bodie. He went outside with one challenger, and the two cooly drew guns at a distance of two feet and unloaded them at each other. Back into the saloon reeled U.P. Jack. The other, with one arm shattered, reloaded by holding his gun between his knees. A moment later Jack got a dose that proved sufficient to end the career of Bodie's first bad man. His antagonist died some days later.

After that, shootings—usually unpunished—were weekly and sometimes daily events. One arrival noted six fatal shooting scrapes during his first week in town. Even the president of the Bodie Miners Union died in an exchange of gunfire. Another disgruntled miner left a meeting in the Union Hall and returned with his six-shooter, drawing a hail of bullets that sent him to Boot Hill.

Bodie quickly earned a reputation for the "wickedest men in the West." A little girl, whose family was leaving another town for Bodie, ended her nightly prayer with: "Goodbye, God; we're going to Bodie." And a Comstock undertaker complained:

"As soon as the local talent get to thinking they're tough, they go to try it out in Bodie and the Bodie undertakers get the job of burying them."

Through the spring of 1878 came the gold hunters, either to dig it out of the ground or out of people's pockets. One of these was Eleanor Dumont, an ornament in every gold and silver rush from California to

Montana. From a slight fringe above her lips she was known as Madame Moustache. Her business was gambling, and as usual she was dealing faro or twenty-one in a leading saloon. As the *Bodie Standard* noted, no woman on the Coast was better known.

"She appears as young as ever, and those who knew her ever so many years ago would instantly recognize her now."

Madame Moustache would find her luck running out in Bodie. Scarcely a year after her arrival, gambling losses were too much to bear. They found her body, along with an empty poison bottle, in the sagebrush near the road to Bridgeport.

In early May, 1878, the stock of Bill Lent's Bodie Mining Company rose from 50 cents a share to $4.70. Rumors were floating through town of a rich blind lead in the Bodie. By June 4 the stock had sunk to $1.00 and then recovered to $3.50. In the next day's issue of the *Bodie Standard* the editor hinted of developments "which to chronicle would astonish the public, but for good reasons the particulars are a secret as yet." But before the paper went to press it had to make room for a report from Will Irwin, superintendent of both the Standard and the Bodie mines; At the 250-foot level in the Bodie, the main west cross-cut had exposed a vein of ore about 2½ feet wide and averaging about $100 to the ton. The *Standard's* headline:

OUR SECOND WIND
Big Strike in ``Bodie"
A remarkably rich ledge

The "second wind", after the demonstrated richness of the Standard Mine, meant a second bonanza for Bodie. The Syndicate stamp mill agreed to take 1,000 tons of Bodie ore. When the first load of rock was hauled to the mill on July 10, as the foreman observed, it "showed gold in every piece." It was so rich that for the next few days, one guard—armed with a Winchester—was placed at the mill dump and another rode with the wagon, around the clock.

When the twenty-stamp mill started pounding the ore on July 12, Bodie stock rose to $12 a share in San Francisco. Among those selling at this heroic price was George Daly, now superintendent of two Bodie mines, who got rid of his 50 shares and, with two cases of wine, threw what a friend called "a nice little party."

When the results of the first week's run were known, Bill Lent in San Francisco received a telegram from Bodie. With this in his pocket, and snappily dressed from his pink salmon gloves to his well-waxed moustache,

Uncle Billy swaggered into the San Francisco Stock and Exchange Board. Bodie stock jumped to $18 per share.

Three weeks later the Syndicate mill finished working the 1,000 tons of Bodie ore, producing $601,000 in bullion, or roughly $600 per ton of ore. Bodie stock jumped to $57.25 per share. Among those who decided they had sold too soon was George Daly.

Still, this was simply the first excitement. The treasure in the Bodie kept expanding as new drifts and crosscuts were opened, while Will Irwin charged the miners with secrecy. But the story was too good to hold. At the gigantic Independence Day celebration, some miners were passing around high-grade specimens where the gold sparkled to the naked eye. Near the end of July a correspondent from the *Virginia Enterprise* descended into the Bodie and watched while 26 ounces of ore were ground in a mortar to produce 20 ounces of gold. A journalist from the *Bodie Standard* toured the mine and, with a pen dipped in delirium, reported ore from $2,000 to $4,000 per ton at every shift.

"This is one of the grandest bodies of ore which has ever been uncovered, not even excepting the Comstock."

With such tidings, the Bodie Mine was split in two, with 100,000 shares in a new Mono Mining Company selling at $12 each. Erecting its own mill, the Bodie promptly began paying generous dividends. So much bullion was being shipped that many thought it had to be a stock promotion play. Was the Bodie shipping those gold pigs to the railroad at Carson and then hauling them over the Sierra and around by wagon through Sonora Pass to ship them out again?

After the Bodie strike, as one resident wrote, "The town of Bodie ran wild." Every night, Bodieites filled the streets and saloons to get the latest news from the bowels of the Bodie. Most of them, if they did not own Bodie stock, held a little of the other mine shares, which rose in sympathy with the Bodie stock. During the day, up till 3:30 p.m. when the telegraph brought the closing quotes in San Francisco, hundreds and sometimes thousands of shares were sold directly on the streets of Bodie. Many a miner who had bought some Bodie at 40 cents a share was now rolling in riches.

In the other camps, from Sonora to Eureka, the excitement was the same, creating a second and bigger rush. This time those who had hung back lost all restraint. Stages from all directions—Carson City, Sonora, Mojave—were loaded inside and topside, the passengers complaining of the dust, the bad roads, the poor food and high prices at the way stations. At Dexter Mills, last stop before Bodie, the travelers slept in the barn, according to one of them, "on hay over which numerous chickens have ranged for several months."

By the end of July, Bodie had 2,000 people, and by December at least 600 frame buildings. As Main Street was stretched in both directions, houses were being rushed to completion at both ends of town, as well as on several side streets and back streets. Main Street itself was jammed with buggies, stagecoaches, and most of all, wagons with goods for the happy merchants.

"If all signs do not fail," boasted the *Bodie Standard* in October, "we shall soon be called upon to chronicle a population second only to Virginia City."

Through the relatively mild winter of 1878-9 the boom roared on. In December, Bodie had some 5,000 people, 47 saloons, two banks and two newspapers. With February's early thaw, the stampede rose again, well into 1879. In Virginia City the gold hunters had to book a seat on the stage days in advance. On one Bodie-bound stage there were so many passengers that they had to leave behind much of their baggage.

Bodie, in fact, was so obsessed with gold that any discovery created a frenzy. When a woman had Indians dig a well at her property on Green Street, she noticed them carrying away the dirt. Early in 1879 a friend panned out some of the dirt and found it "lousy with gold". When she staked and recorded a claim, the boys came rushing in, despite a heavy snowstorm, and peppered the adjoining ground with claims. The upper part of town, including Main Street, sprouted with stakes. By late afternoon nearly every resident there had located a claim of his own to forestall intruders. Before the excitement faded, the boys had staked some 200 claims, of which about half were recorded. It turned out to be some placer ore washed down from the veins on the bluff, and Bodie relaxed to await the next excitement.

At this time Pat Reddy was visiting Bodie on law cases. He decided the wild and booming camp just suited his ways and his profession. From Independence he brought Emily northward and established a home at Bodie in April 1879.

Pat's brother Ned had already joined the rush in March. While most stampeders were traveling on foot or horseback, Ned and three friends left Inyo in two buggies, smartly drawn by high-stepping horses. Chalfant of the *Independent* could not resist comment:

"Kid gloves and carriages for the next prospecting outfit, if you please."

In Bodie, Pat and Ned found old friends from Aurora, Montgomery, Benton, Cerro Gordo, Darwin and every silver strike along the Eastern Sierra. Fresh from the battles in the California state Constitutional Convention, Pat now settled easily into this hail-fellow life. He loved to make the traditional rounds of saloons in the evening, swapping drinks and tall stories. His friends were legion—from mine superintendents getting his advice to down-and-out prospectors enjoying his generosity. Once when he returned to Bodie after an absence of only two weeks he was greeted by a brass band and big party, complete with speeches, running into the late hours.

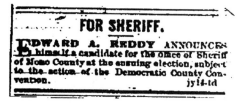

Ned Reddy, too, fell easily into the Bodie scene. With his no-nonsense reputation he was nominated by the Democrats for sheriff of Mono County, and was Grand Marshal in a Democratic parade in the campaign of 1880. Endorsed by the *Bodie Standard-News*, he even got a boost from the neighboring *Inyo Independent*:

"In the probable event of his election, there will be no foolishness among the roughs while Ned is around."

But Ned and other Democrats lost in a Republican year.

In pursuing his law practice, Pat Reddy continued to defend the defenseless. By September 1880 he was as notorious for saving murderers in Bodie as he had been in Inyo County.

"We doubt," wrote a correspondent from Bodie, "if any criminal lawyer in the United States can anywheres near show such an extended record of unvarying successes in defending manslayers."

And the same writer suggested what had already been apparent in Inyo County—that many people resorted to the knife or revolver in the faith that Pat Reddy would get them off. Stung by the rising tide of opinion against his practice, Reddy is said to have declared he would stop taking such cases.

Still, Pat would not be intimidated, in or out of the courtroom. In 1880 an accusation of cattle rustling fell upon Edwin Loose, the man who had discovered the Bodie Mine in 1876. To Bodieites, Loose was favorably known for his exhuberant disposition, his large and powerful frame, and his athletic exploits. His accuser was one Pauly Plane, the notorious head of a gang of horse and cattle thieves, who was trying to divert suspicion from himself.

The trial opened at the Esmeralda County Courthouse in Aurora early in the morning, with Pat Reddy for the defense. Reddy was warned, under pain of death, not to cast suspicion on Pauly Plane by revealing his criminal record. Backing this threat were friends of Plane's, well armed and scattered throughout the courtroom.

Sizing this situation, Reddy spoke aside to a friend, who mounted a fleet horse and pounded the fourteen miles to Bodie. Just before the afternoon session, a dust-caked wagon filled with Reddy's friends whirled up at the courthouse, the four horses sweating and heaving. With six-shooters at their sides, the men filed into the courtroom and took seats in strategic places.

In cross examination, Reddy asked Plane whether he had ever served time in prison. Knowing Reddy had the facts at hand, Plane in labored breath admitted it was true.

"That is all," said Pat, smiling triumphantly.

Pauly Plane stalked from the the courtroom, from Aurora, and from Esmeralda County. Pat's client, big Ed Loose, was so happy he turned cartwheels in the courtroom and, it is said, broke the floorboards each time his heels landed. Long afterward the patches in the floor were known as "Loose's Trail".

In Bodie, the clink of coin, the shuffle of cards and the raucous laughter in the saloons sang a song of high prosperity. To share more fully in the good times, the Mechanics Union went on strike in February 1879, demanding an increase from $4.00 a day to $5.00, and a decrease from 12 hours to eight hours per day. Since its 125 members were mainly employed at the steam hoisting works of a dozen principal mines, this meant a sudden halt in the general high life. On February 12 the members met in the Miners Union Hall, then marched out to make the rounds of the mines.

At the Mono Mine they found their most recalcitrant opponent in Superintendent George Daly, whose anti-union sentiment went back to his days as a printer in San Francisco. He locked the hoisting works, flatly turned down the strikers, and ran Old Glory to the top of a flagpole as a symbol of defiance.

Next day, Bodie was reeling from meeting after meeting, rumor upon rumor. Emerging from breakfast at the Palace Restaurant, George Daly was stopped by a band of men.

"We wish you to go up to the meeting at the Hall," declared the leader.

"What meeting?" asked George.

"Mechanics Union."

"I had nothing to say to the Mechanics," snapped Daly. "What I had to say had been said at the Mono hoisting works."

With that he refused to go. The leader then turned to the others.

"Men, do your duty."

Two of them took hold of Daly and hustled him toward the Miners Union Hall. Just before they reached the hall, George shook himself away, jumped from the board walk to the street, and pulled out his six-shooter.

"I'll kill the first man that attempts to lay a hand on me!" he shouted, aiming at the leader.

While the little group of men faded away, Daly crossed the street and started to enter a store. One of the bunch called after him:

"All right, young man, you're heading for it. You'll get plugged yet."

As for the Mechanics Union, it had made the mistake of failing to consult with the Miners Union, which consisted of nearly 1,000 men, most of whom were idled by the strike. On February 14 a spokesman for the bigger union sided with the mine superintendents and told the Mechanics Union to back off. Next day the mines, hoisting works and all, started up again on the old four-dollar, twelve-hour basis. Many union members rankled, especially against George Daly.

Around this time one of the Bad Men of Bodie threatened Daly's life. George was walking down Main Street with two companions when they encountered the Bad Man and two well-armed confederates. As Daly's friends stepped out of the way, George hit his opponent in the face, drew his revolver, and asked if he still wanted to fight. The Bad Man left Bodie.

With its first labor dispute over, Bodie's bandwagon rolled on, and through the summer of 1879 the camp afforded some new ornaments—a theater, a brass band, a fire company, but not yet a church, though Catholic and Protestant services were sometimes held in the Miners Union Hall. The summer began, in fact, somewhat quietly.

"There has not been a man killed for a month or two," reported an *Alta California* correspondent.

One night in June, two men got into

an argument in a crowded saloon, and both moved their hands toward their guns. As the *Alta* described it:

"The crowd made such a stampede for the street that it carried the large doors, glass and all, with it."

It was a false alarm, but in July the town erupted with a real tragedy. The powder magazine on the Standard Mine property was stored with giant powder and black powder for use in blasting the mine faces underground. On July 10 it suddenly exploded with a force that leveled nearby buildings, killed seven men, rained rocks on the buildings in Bodie, and was felt twenty miles away in Bridgeport. With the Standard works on fire, the miners and millmen struggled to put it out until the fire boys rushed with their new Babcock engine up the hill to douse the flames.

With some 40 people injured, the Miners Union Hall was turned into a hospital, full of wrenching scenes of suffering men and families searching for loved ones. Since no one knew the cause of the blast, the many who fled the town were not quick to return. As the *Alta* correspondent wrote:

"The hills are black with people."

Though mine accidents were not uncommon in Western camps, Bodie would not suffer one worse than this.

Bodie's compassionate spirit was shown in other ways. In 1879 someone remembered that Bill Bodey's remains lay down the canyon in an unmarked grave. "Poor old Bodey!" said the boys. Nothing was too good for the town's noble founder! A search was launched forthwith, and his bones were brought back in triumph. For days they were on public exhibit, subject to handling and close scrutiny by every civic-minded citizen. On November 1, 1879, amid much speech-making (and doubtless elbow-bending), Bill Bodey's bones were laid to rest in Boot Hill.

What, then, of a suitable monument to Bodie's founder? For months the boys waited while an elaborate headstone was ordered and hauled in by mule team. But when it arrived the month was September 1881, and Bodie was stunned by news of President Garfield's death by assassination. Nothing was too good for the nation's martyred President! The monument was erected as a memorial, and Bill Bodey's bones remained in a transplanted, but still forgotten grave. Bodie's heart was big, but it was not always constant.

PIONEER BREWERY,
—AND THE—
MAMMOTH SALOON.
Frankenburger & Davison, Proprietors.

THE MAMMOTH IS MANAGED BY THE Pioneer Brewery firm and is fitted up in the building formerly occupied by P. HOGAN, on Main street, in a style unsurpassed in the town of Bodie.

A Beer Hall is Attached.

Orders for beer may be left at the Mammoth and will receive prompt attention. BEST BRANDS OF LIQUORS AND CIGARS constantly on hand.   apr8tf

In its heyday between 1878 and 1881, Bodie ran full blast around the clock, both above and below ground. On top, fortunes were passing across the bars and gaming tables of the hell-roaring Bonanza, Rifle Club, Champion, and the score of other saloons that graced Main Street. Underneath, in the shafts of nearly 30 active mines on the side of Bodie Bluff, more than $20 million was being recovered in gold and silver ore.

The Standard and the Bodie Mine below it on the hillside were the two bonanzas. Both were controlled by Bill Lent, who was considered the "boss" of Bodie, though he lived in New York and San Francisco. But nothing stayed the same in Bodie for long, and more claims were continually being located, giving rise to hopes of new bonanzas..

One of these was the Savage, located in June 1877. The next October, George Daly became superintendent of the Mono Mine, which had been split from the Bodie. Taking an interest in the nearby Savage, George suggested that a further claim to the East should be located, which was done in November 1877, with stakes driven to show the boundary. George then bought both claims and combined them in the Jupiter Mine. By the summer of 1879 the Jupiter shaft was down 400 feet, with several winzes and crosscuts on the way down, some of them in good ore.

In April 1878, some other miners had staked a claim in the same area and called it the Owyhee, then a popular way of pronouncing and spelling "Hawaii". One of them was John Goff, a shift boss in the Bodie Mine and a popular member of the Miners Union. An Irishman from County Tyrone, Goff had made a reputation as a gunfighter in Idaho.

Actually, the Owyhee encroached on the south boundary of the Jupiter and on four other claims. To confirm their possession, the Owyhee miners sunk two shafts on the lower end of their ground, stopping work in September. George Daly was unaware of this activity, since it was out of sight from his Jupiter shafthead.

When it was time to do their assessment work for the year 1879, the Owyhee men decided to sink a shaft at the north end. This was in sight of the Jupiter works. Near the shaft was a dugout cabin about ten by twelve feet, owned by Peter Burke, an employee at the Standard Mine. Two of the Owyhee men—one of them Patrick Reynolds—told him they were going to use the cabin to protect their property, and that he would have to leave whether he "wanted to or no." On the morning of August 11, 1879, six Owyhee men armed with Winchesters and six-shooters walked up the hill and took their positions.

"We went there to get possession of the ground," one later declared.

Early on the morning of August 12, George Daly rode his horse to the Owyhee cabin. Some of the stakes marking the Jupiter boundary had been pulled up, though a few remained. Eyeing the men standing there, George got off his horse.

"Well, boys, what are you doing?"

"We're sinking a shaft," replied their spokesman, Pat Reynolds.

"What do you call this?" demanded Daly.

"The Owyhee."

"We call this the Jupiter," Daly countered.

"You do, hey," growled Reynolds. "The Jupiter ain't got any ground here."

"Well, I am playing this for the Jupiter, all within its posts," snapped Daly, pointing to his remaining stakes.

He was, he said, superintendent of the Jupiter, and told them to "cease work." Reynolds then revealed their strategy: Daly could buy anything they had. But he could not put them off "so long as shotguns would hold it."

Daly was unmoved.

"Maybe the Jupiter could get some shotguns."

With that he mounted and rode to the Jupiter works. One or two days later he hired the surveyor who had staked out the Jupiter boundaries in 1878, and went over the ground with him again. For over 600 feet the claimed Owyhee line ran inside the Jupiter ground.

Going to the Owyhee shafthead, which was also over the Jupiter line, Daly accused the men of pulling up the Jupiter stakes. He would replace them, he said, "and keep them there if it would take a shotgun on each post."

Hearing that George Goff was one of the Owyhee claimants, Daly went to him, reviewed the conflicting boundaries, and tried to "settle the matter amicably." Goff answered that Daly could buy the Owyhee for $4,500. He and his friends knew it would cost Daly far more to take the issue into court. When Daly refused the offer and pressed his case further, Goff ended the conversation with:

"Go to Hell."

Always the fighter, Daly set about proving the Owyhee people had jumped the wrong man. On Daly's orders his mine foreman, Joe McDonald, took control of the Owyhee shaft on the afternoon of August 21. With his men he pulled out the windlass and shoveled earth back into the shaft. Over it he placed a wooden cabin and stationed four men there to hold it.

Late on the afternoon of August 22, Daly was walking down the hill past the Owyhee shaft. He was intercepted by George Goff.

"Did you send them sons of bitches up there to put up that house to take possession of it?"

When Daly agreed, Goff made a quick move with his hand, resting it on the pistol at his hip.

"If you don't take them sons of bitches off I will kill every one of them before morning, and I have a good mind to kill you now."

Daly stuck his hand in the side pocket of his coat, and without drawing his revolver, pointed it at Goff—in effect, getting the drop on him.

"About this thing of killing men," he drawled, "You mustn't talk that way."

Goff repeated his threat, ignoring the gun muzzle protruding inside Daly's pocket. Daly kept him covered and gave him a tongue-lashing.

"Yes, you are a fighter," he mocked. "You are a fighter from Idaho, a man that has got a record. . .as being a shooter, but you and your crowd can't kill one side of me or scare the other. Now take your hand off your pistol. . ."

That ended the showdown, but not without new threats from Goff. He could "bring more fighters than any other man in Mono County. . .and will kill all the sons of bitches you can bring."

That night the slope of Bodie hill was an armed camp. Four Jupiter men were in the cabin placed over the Owyhee shaft. Six Owyhee men were in Burke's dugout cabin nearby. They carved loopholes in the cabin wall so they could see and shoot out.

About one o'clock at night, two of the Jupiter men stepped out of their cabin for some exercise. From the Burke cabin came a volley of shots. Returning the fire, the Jupiter men retreated inside. Up at the Jupiter hoisting works, foreman Joe McDonald was awakened by the shooting. After dressing, he was standing by the blacksmith shop when a man stepped up and fired at him point blank, missing him in the dark. McDonald emptied his six-shooter at him, without effect.

Early in the morning he and three other Jupiter men reinforced those at the cabin over the Owyhee shaft. While all of them were standing and talking outside the house, those in the Burke cabin began shooting again. The Jupiter men fired back.

"If we stay here we will all be killed!" shouted McDonald. "Let's charge the dugout!"

Running toward the Burke cabin and firing as they went, they came to the roof and began shooting into it. Goff, who had been outside firing with his long Springfield from behind the dugout, came into the cabin. There a bullet through the door hit him in the head. At this, those inside had enough.

"Don't kill us!" one of them cried. "We surrender."

McDonald ordered his men to stop shooting.

"Come out," he called to those inside, "but don't bring your arms with you."

The Owyhee men filed out, but Reynolds had his shotgun to his shoulder, both hammers cocked. With eight muzzles trained on him, he obeyed the order to drop the gun. Stepping inside, some of the Jupiter men found Goff lying by the door groaning from his head wound. While one of them went for a doctor, McDonald and the rest marched the five Owyhee men into the town and turned them over to the night policeman. He soon released them and, in the words of one:

"We knocked around town all day in different saloons."

Their talk across the Bodie bars, together with the death that day of John Goff, stirred the bitter anger of the Miners Union. That afternoon the union members met in their hall amid calls for retribution. John Goff had been killed on orders of George Daly, the man who had defied the Mechanics Union in the previous February. Though the meeting was conducted in secret, rumor said that a resolution was proposed to hang George Daly. This was rejected, but the union men did resolve to take possession of the disputed ground in favor of the Owyhee.

Marching out of the hall, some of them with shotguns and rifles, at least 500 miners filled Bodie's streets and moved up the hill. John Dinan, who was Daly's foreman at the Mono Mine, encountered them and said something that was mistaken for resistance.. He was knocked down. The crowd surged up to the Jupiter cabin. A few shots were fired at it, and some Jupiter men wisely ran out. The crowd burned down the cabin amid general celebration. Then they marched down the hill again. Through it all, Daly was working at his desk in the Jupiter Mine office, a Winchester across his lap.

That night threats were freely voiced against Daly. Some said he would be killed that night. It was rumored that a vigilance committee was formed to hang him. The *Bodie Standard* observed that his death was "highly probable".

Still defiant, Daly continued working in his office while the emotional tide was rising about him. Around 3:00 in the afternoon a policeman arrested Daly on a charge of murder. An hour later he was rushed out of town by a sheriff's posse and lodged in the little wooden jail at Bridgeport, the county seat.

That evening a coroner's jury concluded that Goff had died from a gunshot wound "at the hands of" the Jupiter men, including Daly (who had not been present at the shooting). In the hearing, one or two Owyhee men further inflamed the situation by claiming that the Jupiter men had fired first.

The arrest and removal of Daly probably saved his life. Fearing that the Bodie crowd would follow and take Daly from the jail, the county judge released him on bond. Daly took to horse and rode for Nevada. Next day, from Carson City, he wired Pat Reddy's law office for help. He would, of course, have to return to Bridgeport for trial.

Meanwhile, the frustrated Miners Union was taking its ire out on J.F. Kirgan, deputy sheriff and constable, who had presided over Daly's retreat to Bridgeport. So savage was the recrimination that Kirgan announced he would not, after all, run for sheriff in the next election.

In the second week of September the Superior Court in Bridgeport tried the case of George Daly, Joseph McDonald and three others in the death of John Goff, with Pat Reddy for the defense. When the jury gave a verdict of "Not Guilty", Bodie erupted once more. On the evening of September 20 the miners union met in secret at the hall and resolved that George Daly, Joe McDonald and their employees involved in the Owyhee fight "be ordered outside the limits of Bodie Mining District within the period of 12 hours from the date of this notice." This meant that, on pain of death, the Jupiter men had to leave Bodie by 10 a.m., September 21.

Daly, defiant as ever, published his reply in the same morning paper that printed the Miners' resolution. He called on "all good and true men" to help him and his men to maintain their rights, and he declared the resolution to be "merely the voice of the mob, and will be repudiated by every good man in the Union."

That night Daly led his men to the Jupiter hoisting works and prepared for siege. They threw barricades around the building, cut portholes for sighting and firing, and stocked a supply of guns, ammunition, water and food.

Next morning, a group of leading merchants, bankers and lawyers called a meeting of Bodie citizens for 11:00 a.m. at the fire engine house. A motion to create a vigilance committee—for what purpose was not revealed—was voted down. Pat Reddy called for a citizens committee to meet with the Miners Union "to avert a collision, which would be likely to terminate in serious danger to the entire town." Pat sat down amid a round of applause. After choosing a committee of nine to meet with both sides, the meeting was adjourned until called again by the fire house bell.

Conferring separately with the two sides, the committee found them immovable. Pat Reddy, Will Irwin and others now laid siege to Daly's sense of honor, finally convincing him that keeping up the fight would lead, at the least, "to private assassinations, with reprisals." For its part, the union allowed Daly and his men 48 hours to leave town, and visitation rights for a day each month.

Daly's notice "To the People of Bodie" ran in the *Standard* on September 22. Though he had been willing to assert his rights "even at the expense of my life," he knew that "hostilities would cause great and perhaps *irreparable* injury to the camp." But his departure was due "to the *influence of my friends* instead of the *threats of my foes.*"

The outcome represented the triumph of expediency over justice. As editor T.S. Harris of the *Standard* wrote:

"A power has arisen entirely unknown to the law, which assumes to say who shall and who shall not live in Bodie."

For once Pat Reddy had failed to stand up for the law. He and the other solid citizens of Bodie had abandoned Daly, his men and their families in the name of "peace".

They could have got an injunction against the union, enforced by the corporal's guard of lawmen in Mono County. They could have marched, as the miners had marched, to provide a shield for the Jupiter men at their works. They could have appealed to the governor for the state militia. But none of these could have guaranteed against bloodshed, and it would have taken days for the state troops to arrive.

Bodie was, in fact, a victim of its own isolation. The raw power of half a thousand emotionally charged miners was too much, even for brave men like Pat Reddy and George Daly. Wrote Editor Harris in a triumph of irony:

> When we contemplate how easily peace is obtained—by a simple sacrifice of the honor and manhood of the community—we can but think the Revolutionary Fathers an absurd lot of stubborn old fools. . . .They should have appointed a Citizens' Committee, and then they might have had peace.

Though George Daly knew he could not stay in Bodie, he paid scant attention to the Miners Union order. He visited Bodie on October 10 and shortly after spent several days at a time in town. Then, after visiting Mammoth, Daly left for New York. There he met with an old friend, George D. Roberts, who was strongly invested in the silver-lead mines at Leadville, Colorado. Roberts sent Daly there to manage the Little Chief Mine.

Late in May 1880, Daly was also made acting superintendent of the Chrysopolite Mine in Leadville, at a time when the regular manager had antagonized his shift bosses by claiming the miners were idling on the job. As one of the shifts was changing, several hundred miners descended on the mine and stopped the next shift from going in. When Daly arrived on the spot the strikers surrounded him, demanded an increase in wages, and threatened to seize the mine. Characteristically, Daly told them he would "shoot dead the first man who entered the works."

The crowd then moved to the Little Chief, where Daly repeated his stand. He would relay the strikers' demands to the owners in New York. Meanwhile, he closed both mines. His example was followed by the other mine superintendents as the strike spread across Leadville.

For a month, Leadville was in the grip of anarchy, marked by marching and counter-marching, incendiary fires at the shaftheads, and rampant sharpshooting. Daly reopened and fortified his mines, placed guards armed with Winchesters at strategic points, and as in Bodie, ran up the American flag as a symbol of defiance. His night guards, finding themselves targets for unknown marksmen, returned the fire whenever they could see a muzzle flash. At one point, as the main street surged with hundreds of angry miners, Daly and eighty picked men of his "Little Chief Guards" were kneeling in firing position, their Winchesters at the ready, to hold off any threatening move. And Daly became a leader among the mine superintendents, chairing some of their meetings.

"He was," a reporter for the *Leadville Chronicle* later recalled, "the most commanding figure among the mine managers."

At length the Colorado state militia came in and restored order in the third week of June 1880. In September, after visiting George D. Roberts again in New York, Daly returned to Leadville to manage another mine. Opposition stockholders, including many of its own miners, tried to seize the mine and threatened to lynch Daly. While a mob of several hundred gathered in front of the mine, Daly decided to visit the lower end of the property. Shunning the entreaties of his friends, he sallied out with a revolver in each hand. The sullen crowd, knowing his reputation, made way for him to pass.

Then in February 1881, Roberts sent Daly to run some mines at Lake Valley, forty miles east of Silver City, New Mexico Territory. On August 18, George was in bed with a fever when he learned that Apaches had massacred the people in the nearby town of Hillsboro, and were camped about ten miles away. Hardly able to sit in the saddle, Daly agreed to go after them with a band of his men, in company with twenty cavalrymen. Planning to attack the Apache camp at dawn, they left at 1:00 a.m.

"I will return at nine o'clock," Daly told those he left behind, "but you know some of us must take the bullets."

The Indians had anticipated them and ridden on to Gavilan Canyon, where they lay in ambush. Being in the lead, Daly and the cavalry officer fell in the first fire.

Though his body was buried in New Mexico, Daly's funeral was held in January 1882 at Grace Church in his home town of San Francisco. As in

Bodie, he had been willing to face great odds, but this time there had been no Pat Reddy to talk him out of it.

Back in Bodie, the triumph of mob rule had left the town even more lawless than before. During one two-week period in January 1880, there were one fatal shooting, two knifings, one clubbing, and three pistol whippings. Another newspaper report cited three murders in three days. As one editor wrote:

"Crime runs riot nightly in the criminal quarters of the town, and few arrests are made."

By this time Bodie's bad men were turning their attention to the Concord stages that rocked out of town toward Carson City every day, groaning with gold and silver bullion. During 1880, enterprising road agents stopped the Bodie stage no less than six times.

One of them was Milton A. Sharp, a short, dark-complexioned Missourian, who was exemplary in his personal habits—no drinking, smoking or profanity—and described as "a gentleman in his manner." His one bad habit was robbing stagecoaches.

In company with W.C. Jones, Sharp stopped the Auburn stage in May 1880, then crossed the Sierra and took an interest in the stage line south from Carson City, especially in the narrow canyon between Aurora and Bodie. There Sharp and Jones robbed Wells Fargo's treasure chest twice in June. Then, after revisiting the Auburn stage, they returned to Bodie and planned a double robbery of the downstage and the upstage on two succeeding days.

With no apparent help from Mono County law, Wells Fargo Express had begun putting on one or two guards when a heavy consignment was shipped, transferring stage passengers to a light wagon that followed at a safe distance. And the company brought in its crack shotgun messenger, Mike Tovey—a Canadian who was shy of manner but not of nerve.

On September 4, Sharp and Jones robbed the upstage from Carson, then waited for the next downstage out of Bodie. Next day, while Bodie still buzzed with news of the first holdup, Tovey rode the downstage while the passenger wagon brought up the rear.

Near the scene of the first robbery, Tovey got out and walked ahead. Undaunted by this precaution, Sharp and Jones fired a volley at him. Tovey then retreated to the stage and cut loose at the advancing enemy with a shotgun. In the shooting, Jones killed one of the stage horses, Tovey killed Jones, and Sharp wounded Tovey in the arm. Sharp then fled, and Tovey went to a nearby house to get his wound dressed. Sharp came back, ordered the driver to throw down Wells Fargo's treasure chest, and while his partner lay dead in the road, made off with its contents.

Sharp was soon arrested in San Francisco and brought back to Aurora, where he was tried for the robbery of September 4. Wells Fargo retained Pat Reddy for the prosecution.

Though the evidence was circumstantial, Reddy won a conviction and presented the express company with a bill for $5,000. When Wells Fargo countered by offering $2,500, Reddy adamantly refused anything. But henceforth he turned his legal talents against the company.

A local rancher had found $10,000 in a money belt on the robber killed at the scene. Claiming it was entitled to this money because of its loss in the robbery, Wells Fargo sued the rancher. Pat Reddy took his case and won, charging no fee.

Then when robbers who stopped a Wells Fargo stage in Oregon were apprehended, Reddy got them acquitted—again on a *pro buono* basis.

According to a long-time friend, Reddy won at least twenty such cases against Wells Fargo at much expense to himself but more to the company, which must have wished it had paid without question Reddy's $5,000 bill.

As for Sharp, he escaped from the Aurora jail, was caught and sent to the Nevada State Prison in Carson City, escaped from there in 1889, and was captured again in Red Bluff, California, in 1893. Based on his good behavior in prison and on the recommendation of Wells Fargo's chief detective, James B. Hume, he was pardoned and released in 1894.

In Bodie, the same issue of the *Standard News* that reported the double stage robberies also mentioned three shootings—apparently all in a day's work. As the editor wrote:

"Four men killed in two days is rather too much, even for Bodie."

Many Bodieites were beginning to agree, and as the year 1881 opened, were ready to act.

There was in Bodie a young Irish woman, Johanna Londrigan, from Providence, Rhode Island. While visiting a sister in Chicago around 1867, she had met a French Canadian named Joseph Da Roche, who had a wife and three children. Johanna visited his home often. Leaving his family behind, Da Roche sought his fortune out West. First at Virginia City and then in Bodie, he had a reputation, according to the *Bodie Daily Free Press*, as a "'hard citizen,' a petty larceny thief, and somewhat dangerous." Another observer called him a "notorious desperado".

About 1877, Johanna also came West—first to San Francisco, then to Bodie late in 1878. With a woman friend she opened a laundry. Living at the American Hotel, she met another guest, Thomas Henry Treloar, a Cornish miner. Treloar had fallen down a mine shaft in Virginia City and the accident affected his mind. But he continued to function as a miner and was a member of both the Bodie Miners Union and the Bodie Fire Company, which cited him as "a good and peaceable citizen, an industrious and honest man, one of our most efficient and effective members." One of his assets was a $1,000 insurance policy with nine years of accumulated interest. He was soon wed to Johanna, and when asked why she married "that little half-witted fellow", she replied coldly:

"Oh, I married him for that endowment policy on his life, which will be due in a couple of years, and then I will have money."

Also in Bodie by that time was the French Canadian, Joseph Da Roche, who operated a brick yard in town. He promptly resumed his friendship with Johanna, meeting with her secretly from time to time. Learning of this, Treloar ordered her not to see Da Roche. To a friend he confided that his wife had been unfaithful, and blamed Da Roche. He would, swore Treloar, "kill him with a knife." Warned about Treloar's threat, Da Roche said he would "shoot him if he made any motion."

On the night of Thursday, January 13, 1881, as Bodie was enveloped in a fresh snowfall, most of the regular citizens were enjoying a grand ball in the Miners Union Hall. Tommy Treloar was not with his wife at the dance.

Around 11 p.m., dressed in work clothes, Tommy came into the ballroom and went outside to talk with his wife. Then she came back in and Tommy left. About 12:30 he returned and saw Johanna dancing the quadrille wilth Da Roche. Tommy looked so agitated that the doorkeeper would not let him enter.

"I told my wife not to dance with that man," Treloar confided, "and she said she wouldn't. I want to get her away."

Barred from entering, Treloar left, then returned about 1:00 when the party was ending. Meeting him at the door, Johanna asked if he was going home. Treloar turned away. Leaving at the same time, Da Roche accompanied Tommy down the board sidewalk. In his pocket the Frenchman carried a small, five-shot, double-action revolver. They passed two men, one of whom had heard Treloar's statement about his wife at the ball.

"That's the two gentlemen," he told his companion, "who are likely to have trouble over the wife of the smaller man."

At the corner of Main and Lowe streets, Treloar stepped off the boards and into the street. Hanging back, Da Roche drew his pistol and fired one shot at Treloar through the left ear. Tommy fell onto the frozen street, a great pool of blood quickly spreading on the ice. The two other men rushed up and one of them disarmed Da Roche.

"What did you shoot that man for?" he demanded.

Then they held him until a Bodie deputy sheriff, Jim Monahan, ran up to them.

"That man has killed a man," one of them said.

Monahan arrested Da Roche and took him toward the jail. On the way they encountered T.A. Stephens, a Bodie attorney, who was escorting his wife and Johanna Treloar, who had said she "feared trouble." Da Roche stopped and addressed her.

"Mrs. Treloar, I have killed your husband."

"Good God!" cried Johanna.

Monahan pulled Da Roche along and turned him over to J.F. Kirgan, constable and jailer, at the Bodie jail. Johanna and the Stephens couple hurried on to Treloar's body. Still alive, he was carried to the undertaker's, where he soon died.

About 4:30 in the morning, a deputy sheriff named Joe Farnsworth rushed into the jail and told Kirgan that a vigilance committee had formed to come to the jail and lynch Da Roche. Though Farnsworth was obviously drunk, Kirgan let him take the prisoner, handcuffed and shackled, with him to his quarters for safekeeping. There Da Roche offered the deputy $1,000 to let him escape—an offer that was rejected, according to Farnsworth. The deputy then proceeded to fall asleep. Da Roche took the keys from his pocket, unlocked handcuffs and shackles, and vanished. No vigilance committee made its appearance that night or next day.

The coroner proceeded to hold his inquest Friday afternoon and Saturday. Treloar's funeral was held Saturday afternoon in the Miners Union Hall, attended by a huge crowd whiich followed the hearse to the graveyard on the hill west of town. During the services Johanna was effusive in her grief. But some people talked knowingly about her involvement with Da Roche and called attention to Treloar's insurance.

That evening the coroner's jury found Treloar to be the victim of a "willful and premeditated murder" by Joseph Da Roche, that Kirgan was guilty of gross neglect of duty "in allowing the prisoner to be removed from the jail," and that Farnsworth was "criminally careless in allowing the escape of the prisoner."

On Friday, among the agitated townspeople the suspicion was rising that Farnsworth had accepted Da Roche's bribe. A search was mounted by scores of Bodieites, not only for Da Roche, but Farnsworth. For Da Roche, the *Bodie Daily Free Press* promised "there will be no parley when he is captured." And as for Farnsworth, he would be "roughly handled."

Ned Reddy, a friend of Farnsworth, advised him to get out of town or he "would certainly be a corpse on the following morning." Oliver Roberts later claimed that Reddy hid Farnsworth and helped him escape. In any case, on Saturday night Farnsworth rode to Aurora and caught the stage for Carson City. When he swung off the stage in Carson he was arrested by the sheriff on charges from the Mono County sheriff.

Meantime, the search for Da Roche had narrowed to the community of French Canadians in Bodie. At Webber's blacksmith shop, the searchers made their headquarters, bringing one Frenchman at a time for questioning. As the *Free Press* justified it, "if occasionally they were handled rather roughly and severely threatened it was merely done that justice might come out of it."

Finally they dragged a saloonkeeper named De Gerro from his bed and to the blacksmith shop. Known as a friend of Da Roche, he had just returned from his brother's wood ranch eight miles away. To elicit his information they produced a rope, dangled it from a beam, and told him if he did not talk they would hang him—emphasizing their message with:

"Fix up the noose."

"Pull him up."

At this, De Gerro told them that Da Roche was at the wood ranch. Around midnight ten men, well armed and mounted, pounded out of Bodie, taking De Gerro with them. At the ranch, a low adobe cabin, they surrounded the place, pounded on the door and called for Da Roche. He emerged with his arms spread out.

"Hang me! Hang me!"

The posse put him on a horse and rode for Bodie. About 6:30 a.m. on Sunday, January 16, they plodded into town, the horses' hoofs crunching on the frozen snow. This time Da Roche was shackled in his jail cell.

By 9:00 a.m. word of the recapture was all over town. People began gathering on Main Street, discussing the crime and fanning their indignation.

"At noon," reported the *Free Press*, "the excitement was so great that matters began to look serious."

At a meeting in the Bodie Hotel, Da Roche's fate was heatedly debated. The crowd, now including hundreds, moved down Main Street. At Mill Street, Pat Reddy appeared. They should "do nothing rash," he counseled in a long speech.

"The law would, if properly administered, deal with the prisoner severely and justly."

Reddy was seconded by other speakers. Somewhat calmer, the crowd converged on the jail, where members of the posse and other citizens were on hand to help the sheriff protect the prisoner.

At 2:00 p.m. the sheriff and a deputy brought out the frightened Da Roche and, as the crowd parted for them, marched him up Bonanza Street to the court of the justice of the peace. As the prisoner was seated and the court opened, Pat Reddy volunteered to prosecute.

"You are entitled to secure counsel," the judge told Da Roche.

"I'll take Pat Reddy."

"I am engaged for the prosecution," snapped Reddy.

At this the prisoner chose another attorney. Through that day and evening, the hearing brought out much the same testimony as in the coroner's examination. Court was adjourned until next morning, and the prisoner was escorted through the crowd back to jail.

The hearing was never resumed. A large group of men held a long meeting and constituted themselves a vigilance committee—a Bodie version of the "601"committees that had risen in Virginia City and Aurora.

After 1:30 in the morning on Monday, January 17, about 200 men—some of them wearing masks—marched in an orderly column from a side street into Bonanza Street. Those in front carried shotguns. As they moved toward the jail, others joined them. At the town lockup the leader banged on the door. Inside, Constable Kirgan lighted a lamp and looked out the window.

"Da Roche!" shouted some of the men.

"Bring him out!"

"Open the door!"

"Hurry up!"

Kirgan's answer: "All right, boys, wait a minute. Give me a little time."

Slowly opening the door, he admitted several men, who in turn ordered him to open the cell door. In a few minutes Da Roche appeared in the jail doorway—over his shoulders a canvas overcoat buttoned at the neck, on his face an expression of both submission and defiance.

Guarded by men with shotguns and revolvers, Da Roche was hustled up Bonanza Street, into a back street to Fuller Street, then into Green. There, in front of a blacksmith shop, stood a large wooden frame that served to raise wagon beds for repair.

"Move it to the spot where the murder was committed," ordered the leader.

A dozen men raised and hauled it to the corner of Main and Lowe streets, where it was set upright again. As Da Roche was led under it, his eyes turned upward and his lips moved. The canvas overcoat was removed. One of the ropes hanging from the frame was tied around his neck, the knot resting against his ear. When he moved it to the rear, someone suggested tying his hands and legs. While this was done the iron hooks used for raising wagons dangled against him. Then the leader asked whether he wanted to say anything.

"No, nothing."

Asked again, he responded in French:

"I have nothing to say, only, Oh God."

"Pull him," rasped the leader.

The men at the windlass turned the handle and hauled Da Roche three feet in the air. His legs twitched once. Then he was motionless, his eyes closed. His body swung like a pendulum while the crowd watched in silence. In about two minutes, from behind the crowd, came a loud and commanding voice.

"I will give $100 if twenty men concerned with this affair will publish their names in the paper tomorrow morning."

The men recognized it as the voice of Pat Reddy, a veteran in opposing mob rule. They roared their response.

"Give him the rope!"

"Put him out!"

At this rebuff, Reddy retired gracefully, nursing his disgust. On Da Roche's chest the committee pinned a note:

"All others take warning. Let no one cut him down. Bodie 601."

Next day the *Free Press* condoned the mob action, adding: "De Roche died game." A coroner's jury decided that "deceased came to his death at the hands of persons unknown to the jury." It was rumored that some of the jurymen were among those who had lynched Da Roche.

To the sheriff at Carson City, the sheriff of Mono County canceled his warrant for Farnsworth, since justice had been satisfied in Bodie. But the *Free Press* made it clear that if Farnsworth thought of returning, the climate in Bodie was unfavorable. As for Johanna Treloar, it is uncertain whether or not she collected Tommy's insurance.

To guard against future lawlessness, both by the criminal and by the mob, a Law and Order Association of some 90 men was formed at a meeting in the Music Hall, with Pat Reddy among the speakers. And the *Free Press* declared that, with the stern lesson of Da Roche's fate:

"'The bad man of Bodie' shall only be remembered as a relic of the past."

But Bodie could not end lawlessness with another lawless act. If anything, Bodie fell into a still more callous treatment of human life. Within a week of the lynching, two men fell to fighting in the Divide Saloon. Both were shot, one fatally. This hardly created a stir in Bodie. After all, the dead gunman was a Bodie bad man, and got what he deserved.

In May, one Myers had boasted he would kill Jack Roberts, a deputy constable. When Myers came into the Comstock Saloon, Roberts saw him and shot first.

"Roberts was justified in shooting Myers like a dog," one Bodieite told a San Francisco reporter, "and the people upheld him."

In June, at a Bodie opium den, Dave Hitchell pistol-whipped Jim Stockdale, who went out and got a revolver, returned, and shot Hitchell dead. Stockdale was captured in Aurora and, after a coroner's jury failed to accuse him, was released in Bodie. After all, Hitchell was a bad man, known as the "roughest of the rough". He got, as the *Free Press* declared, "what he deserved".

With the law made a mockery, the rule in Bodie now was: It is all right to shoot someone if, in your opinion, he is a bad man. The general attitude was expressed by the *Standard-News* editor:

> It is a universal custom in Bodie and other camps that when revolvers are drawn for law-abiding citizens to get out of the way as rapidly as possible. . . .After the smoke has cleared away and the victim is waiting for the coroner, the respectable citizens and others crowd into the saloon and discuss the virtues and bad qualities of the man whose light has just been put out. So long as the shooting is confined in certain quarters no great harm is done.

At the beginning of 1881, Bodie had become a rival of Virginia City. In San Francisco, the "Bodies", as the stocks were called, were enjoying a surge, while the "Comstocks" were in the doldrums. With a peak population of 8,000 people, Bodie had easily become the biggest town in Eastern California's silver belt. Gold and silver production reached a record of more than $3,000,000 for the year 1881.

Yet in the same year, Bodie was suffering deep problems. Most of the mines had begun encountering water at the 450-foot level. The bigger mines hauled in the huge pumps that had long been familiar in Virginia City. They were able to push operations down to the 600- and 700-foot levels. Some mines resorted to shareholder assessments to finance the pumps, thus driving down their stock prices. Others could not afford pumps at all, and had to close. A dozen mines produced nothing in 1881.

Still another problem was attacking Bodie. By 1880 the camp's appetite for mine timbers and fuel for the hoisting works had denuded all the hills within easy reach. From the ample timber stands 30 miles away at the south side of Mono Lake, freight charges by ox- and mule-team were cutting deeply into mine profits.

Captain G.L. Porter, who operated sawmills in the timber stands, went to San Francisco and bought the *Rocket*, a 36-foot, propeller-driven steamboat. Shipped by rail to Carson City and by freight team to Bodie, she was put into service in the spring of 1880. Besides towing lumber barges across the lake and shortening the wagon distance to Bodie, she provided transportation for passengers and excursionists. But since she could only carry the lumber part of the distance to Bodie, freight rates were still damping mine profits.

Then in February 1881 several San Francisco investors, including two of the biggest owners of the Standard Mine, incorporated the Bodie Railway and Lumber Company. Its purpose: to tap the Mono Lake timber and supply Bodie's needs at a fraction of the cost.

The road would start along the east side of the Bodie Ridge, where it could serve the mines, then would descend the steep grade southeast of town by sharp switchback curves, and run along the eastern shore of the lake to some 12,000 acres of pine forest at Mono Mills. It would be, boasted the *Standard News,* even crookeder than the Virginia & Truckee Railroad, with "several loops that will nearly rival that of the famous Tehachapi Pass."

Superintending the construction was Col. Thomas H. Holt, who had come across the plains from Kentucky in the Gold Rush and soon became city attorney of San Francisco. On the last day of his term the San Francisco fire of 1851 threatened the city offices. Sacrificing his own library and personal effects, Holt gathered up all of the city's legal records and rushed out in time to save them, and himself. Later he became city auditor, director of schools, and president of the Board of Education.

Known as a man to anticipate problems and take bold action, the good colonel now took on a new kind of challenge in the wilds of Mono County. Advertising for workmen, he planned to launch construction on May 25 with a force of white men grading the roadbed at the Bodie end and a party of fifty Chinese at the other end, east of Mono Lake. When this became known in Bodie, excitement raced through town.

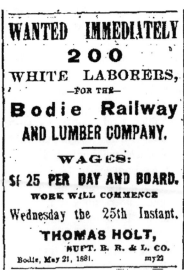

For years the Chinese had been much in evidence in Bodie. They operated laundries and other services. Even their opium dens were patronized by some white men. On one occasion their internal politics erupted in a night-long shooting match. But generally they were an inoffensive element in the community.

The real source of the excitement was the long-standing fear among Western miners and other workingmen that cheap Chinese labor would rob them of their jobs. For years the slogan, "the Chinese must go," rang through California and Nevada. Even Pat Reddy, ordinarily a cham-pion of the underdog, gave speeches urging a halt to Chinese immigration.

So when Col. Holt put Chinese to work on the railroad, the old alarms were sounded. A large group of laboring men packed the Miners Union Hall for a mass meeting the night of May 24. The president of the meeting started by declaring that he wanted "to see the Chinamen driven out of the county" and "the workingmen of Bodie to rise up and prevent the 'cheap labor' scheme." After rousing applause and more speakers chanting the same theme, the secretary of the railroad rose and read a statement. The Chinese, he said, would only work at the south end of the lake.

"We will," he added, "give employment to all the white laborers we can get."

And the railroad would "reach out a helping hand to the many mines now idle because they cannot pay the exorbitant ruling prices for wood." The railroad would "reestablish confidence" in Bodie and attract new investment.

The secretary's words were met with the response that there was "no necessity of employing any Chinamen at all"—amid more vigorous applause. The meeting resolved to get up a list "of all those willing to hold themselves in readiness to take part in removing the Chinese laborers from the local railroad."

That night the group asked the Miners Union to join in the cause. The union members met, and greatly to their credit, decided that hiring Chinese by the railroad was a subject with which they "as a union had no concern." Next morning, May 25, Col. Holt started his grading crews on the eastern slope of Bodie Ridge. The *Free Press* supported the railroad, adding a new argument:

"The road must be finished this fall, or it will not be built at all."

The implication was that if the road had to await the spring of 1882 there would not be enough left of Bodie mining to justify its construction.

Next day one of the anti-Chinese men sent an anonymous letter to the *Free Press* editor from "We the workingmen of Bodie." He warned against siding with the railroad and against the workingmen.

"We will send you with the Chinese if you are not careful."

At this, the editor leaped joyfully into the fray, printed the letter and called its author "a sneak, a lowdown contemptible mangy dog's whelp, a human coyote." And he challenged him to "come out on Main Street. . . and talk to the editor of this paper about driving him out of town."

Meanwhile, the anti-Chinese men had met again on the night of May 25 and named a committee to meet with Col. Holt and "demand that he discharge all the Chinese laborers." Holt refused until he could talk with his company officers. This was reported back to the meeting, still in session, and was branded "unsatisfactory". The assembled men were then called on to "go to the Chinese camp and drive the inmates from Mono County." Some 60 or 70 men signed up, and next morning, May 26, some 40 or 50 actually started the 26-mile trek—in wagons, on horseback and afoot.

Characteristically, Col. Holt was equal to the moment. From Bodie, one of his men pounded ahead on horseback and reported the advance of the horde.

Standing ready for the emergency was the stout little steamer, *Rocket*. Holt loaded the Chinese and a month's provisions onto the boat and one or more of the barges. Then it chuffed across the brine to Paoha Island, six miles from shore. There the Chinese settled in, hoisting the American flag in the manner of George Daly, and boiling gull's eggs in the island's hot springs. When the mob arrived at the deserted Chinese camp, as the *Alta California* correspondent wrote, "they found the birds had flown."

With no other craft on the lake except a couple of rowboats, the angry men could only holler and grimace at the Chinese across the water. They hoped to capture the steamer *Rocket*, which continued to make calls around the lake, but the stout little craft eluded them.

Further south, a force of 21 white men was grading a wagon road for hauling lumber to the proposed railroad terminus. The would-be Bodie raiders tried to enlist their help, but were refused. Then the Mono County sheriff arrived to read the Riot Act. He planned, if need be, to recruit the lower force of white men as a posse to arrest the interlopers.

This soon proved unnecessary. Among the attackers, anger turned to frustration, then to extreme discomfort as they found themselves with little provision and nothing to drink but the alkaline waves of Mono Lake. That night, camping without blankets in the cold wind, they sent messengers back to Bodie for funds to buy supplies, but the relief wagons took the wrong road west of the lake, vainly looking for the sufferers.

Next day most of the valiant chargers stumbled back to Bodie, worn and bedraggled. On the way they received some supplies, including whiskey. As the last of them straggled in, the debacle became a non-subject in Bodie. Wrote the *Alta* correspondent:

"The war may be considered over."

And the *Bridgeport Union* declared, "The Railroad Company is master of the field."

By May 29, Col. Holt had brought his Chinese back to the mainland by steamer and barge, and put them to work. In one stroke Bodie had won back at least some of its good name.

"The Bodie Miners' Union is behaving exceedingly well," acknowledged the *San Francisco Stock Report*, which was not always complimentary to Bodie. And the *Tombstone Epitaph* noted that "the Miners' Union did not countenance or abet the outrage."

Neither did the Miners Union support a strike started by the white workers on the railroad on June 1. Once again a dissident group had failed to check with the policy of the union before acting. Like the strike of the Mechanics Union in 1879, this one lasted two days.

Meanwhile, Col. Holt was driving the roadbed of the Bodie Railway and Lumber Company south toward Mono Lake. Down the mountainside the grade dropped 200 feet to the mile. While there were no "loops" like the one at Tehachapi, two of the switchbacks were so sharp that the trains could not turn. Rather, they would run out on track beyond the turn, then back up to the next turn.

While the narrow-gauge rails were being shipped from Nevada, the white and Chinese workers graded the roadbed far ahead of the tracks. Meanwhile, the sawmill for cutting the railroad ties was being finished in the big timber stand south of the lake. Finally, with two work engines hauling rails to the end-of-track, the 31½-mile line was finished on November 14, 1881.

By lowering the cost of mine timbers and steam engine fuel, the little railroad charged Bodie with renewed life. Over its rugged grades, sometimes as steep as six percent, chugged a trio of quaint little engines with quaint little names—*Mono, Inyo,* and *Tybo.* In May 1882 came a fourth and larger engine, the *Bodie.* Passengers could ride the cars free of charge; the local Paiutes, delighted with the Iron Horse, were the most frequent customers.

Soon the road, renamed the Bodie and Benton Railroad, was projected toward a meeting with the Carson & Colorado Railroad, which had been building southward from Nevada. A junction at Benton Station, which the C & C would reach in little more than a year, would give Bodie rail connection with the rest of the nation.

But the staunch little railroad could not solve Bodie's other problems. Most mines were beset with water, and some were showing poor ore at lower depths. By the last half of 1881, the signs of a faltering camp began to appear. By December, many of Bodie's buildings, both business and residential, were vacant. Prices of even the most famous Bodie stocks were sagging. As if to bolster the camp's morale, Bill Lent dropped in during August 1881.

"Bodie has not yet seen her best days," he declared, adding that "the stock quotations may at any time go higher than ever before."

But Uncle Billy's optimism could not stem the slide. Through Bodie's career the mining fraternity had believed that her "true fissure veins" would be richer the deeper they went. This proved a chimera. In the Bodie Mine, one ledge was widely expected to show improved value at deeper levels. But a crosscut striking it at the 700-foot level found it to be almost barren.

Through 1882 the exodus from Bodie continued, prices of town lots plummeted, and one of Bodie's two banks suspended. In July the Bodie and Benton Railroad stopped further extension of its line, with the roadbed only 15 miles from Benton.

Early in December the Noonday and Red Cloud group of mines was attached by Wells Fargo. Since its great pumps had been holding down the water level in the whole south end of Bodie Bluff, all the shafts in that area operating below the 450-foot horizon had to close.

Bodie stock plunged still further. By early 1883 only six or seven mines were operating, and the retreat of population had left Bodie with some 2,500 inhabitants—still a respectable size for a mining town. At least it was down to the solid citizens with substantial jobs and businesses. The restless crowd of speculators and drifters had gone. In that same year Bodie finally got two churches—the Catholic and the Methodist.

Leading the exodus from Bodie was Pat Reddy. By 1881, enriched by his law practice and such mining ventures as the Defiance in Darwin, he was outgrowing the rough life of Bodie—a view heartily shared by Emily. He already owned, besides the mine and furnace in Darwin, important shares of other mines, a house and four lots in Bodie, and other lots in Bridgeport and the gold camp of Lundy, in the Sierra near Mono Lake. He now invaded the larger arena of San Francisco, opening a law office with two junior partners at 330 Pine Street. He and Emily lived, first, in comfortable quarters in the Lick House on Montgomery Street, and later in a handsome home at 2717 Pacific Avenue.

In San Francisco, Reddy crossed legal swords with the most experienced attorneys on the Coast, and usually won. Among his talents was an acute insight into human nature, which he used in helping fledgling attorneys in their maiden courtroom battles. In 1885 he was prosecuting a murder trial with a young associate, William H. Jordan, who in later years became Speaker of the State Assembly. As Jordan had never seen a trial, he expected Reddy to take the lead. But Pat, recognizing the young man's potential, made him try the case.

"Reddy, you are a good fellow," protested Jordan, "but you have no soul."

"You might as well begin now—never a better opportunity," insisted Reddy.

And Jordan went on to win a conviction.

Pat always kept one foot in the silver country, and starting in 1883, represented five counties for two terms in the state senate. There he championed various causes—the right of women to vote, the rights of miners against the big mining companies, and the movement to limit Chinese labor in California. His active role in the Anti-Chinese League seemed out of character for a sympathizer with underdogs, but his main sentiment lay with his mining constituents, who felt threatened by cheap Chinese labor.

Though Pat loved politics, the talk of nominating him for congressman and then for governor never spread beyond Mono County. His main arena for verbal combat remained the courtroom, where he won a reputation as one of the most effective and feared lawyers in the West.

Soon, in contrast to his opposition to the Workingmen's Club in Darwin, he was championing miners involved in some of the most violent strikes in the Western states.

In 1884 he went to Idaho's battle-wracked Coeur d'Alene silver mines to defend union workers charged by the Mine Owners Association with dynamiting a quartz mill. He got nearly all the defendants off free or with light sentences.

Next year he was in Eureka, Nevada, defending union men accused of forcing non-union miners off the job. There he proved that the mine managers were trying to rig the jury with picked men. In the end the defendants were acquitted.

As late as 1899, when new labor strife brought martial law to the Coeur d'Alenes, Reddy was on the spot writing to the *San Francisco Call* of appalling denials of human rights.

Around 1895, Pat had been afflicted with Bright's Disease, an inflammation of the kidneys now known as nephritis. It was the same affliction that had felled Remi Nadeau. But Reddy fought his way through his declining years with both moral and physical courage.

In 1896 the editor and publisher of the *San Francisco Star*, James H. Barry, took up an old verbal feud with David Neagle, the same who had run the Oriental Saloon in Panamint. Having killed a man in Pioche, Dave had a violent reputation before he appeared in Panamint. After that, as a law officer, he killed a man in Tombstone and shot another in Montana.

As a U.S. deputy marshal, Neagle was retained as a bodyguard in 1889 by Justice Stephen J. Field of the U.S. Supreme Court. Field's life had been threatened by David S. Terry, a former California supreme court justice, who in 1859 had killed Sen. David Broderick in a duel. Encountering Field at the restaurant in a railway station in Lathrop, Terry struck him in the head. Neagle drew his pistol and killed Terry on the spot.

The celebrated trial resulted in the acquittal of Neagle, who was free to pursue his career in California, including San Francisco. Editor James Barry had railed against the verdict at the time, and revived the attack when Neagle assaulted a man in August 1896, calling Neagle a murderer in a banner headline. Neagle promptly announced that he would kill Barry.

At this point Pat Reddy, a friend of Barry's, sent for him and they talked in Pat's office. The lawyer had known Neagle for thirty years and was aware of his mode of operation.

"Neagle will endeavor to provoke you into a quarrel. . . ," said Reddy. "Then he will claim that he acted in self-defense. . . .Even if he should come up and slap you in the face do not strike him back."

Next morning Barry was walking up Pine Street near Montgomery when a man stepped up and spat in his face. The outraged Barry was restrained from hitting the man by the remembered words of Pat Reddy.

"You are Neagle."

"Yes, I am."

Dave then drew a pistol and twirled it in his forefinger. While they stared at each other, Neagle pulled out a copy of the *Star's* offending issue.

"Did you publish that?"

"Yes, I not only published it, but I wrote it."

A moment later, disarmed by Barry's refusal to move, Neagle muttered, "Well, see that it does not occur again." With that he abruptly walked away. Later Barry declared that Reddy had saved his life.

On January 5, 1897, Reddy left the Supreme Court chambers in Sacramento and was walking past the post office at Seventh and K streets when a crowd of people came rushing out of the building.

"What is the matter?" he asked one of them.

"A man is in there trying to murder a woman."

Reddy ran up the steps and into the post office. There an enraged man was waving a cocked revolver and threatening the woman, who was cringing on the floor. Reddy sprang at the man and with his one hand grabbed the revolver. Struggling to turn the gun on Reddy, the man pulled the trigger, but Reddy had put his thumb under the hammer, which fell on his nail. Jamming the man against the wall, Reddy held him with his own might and his one arm until a policeman arrived and made the arrest. It turned out that the man had been quarreling with his sister over a family issue.

Reddy then proceeded calmly down the street, stopping at a drug store to have his injured thumb dressed. To those who asked him about the risk of being shot, the smiling lawyer answered simply that he "objected to any such method of procedure."

To the end Reddy was active in mining and politics. In 1895 he hurried to the gold strike at Randsburg and tried to buy half the famous Yellow Aster Mine from its discoverers. But the wife of one of them, Dr. Rose Burcham, arrived from San Bernardino in time to stop the sale. Pat tried every angle to consummate the deal, but met his nemesis in the strong-willed woman. Over the years the Yellow Aster would yield millions.

In 1889 Pat Reddy had been appointed to the state prison board, and probably due to his influence his brother Ned became for a time the chief of guards at San Quentin. Later, Ned became superintendent of the San Francisco Alms House, where his second wife was the matron.

In June 1900 Bright's Disease caught up with Pat Reddy. For a month, further beset with pneumonia, he suffered in bed at his Pacific Avenue home. Surrounded by Emily, Ned and other relatives, he died early in the morning on June 26 at the age of sixty-one. He was said to have had a smile on his face.

The bond between the brothers was unshakeable, even in death. Three weeks later, Ned suffered a stroke; he died in April 1901 at age fifty-six. Emily, who had devoted her life to Pat and his career, began failing in health and died in May 1904 at the age of seventy-nine. While San Francisco newspapers made scant notice, it was the *Inyo Independent* that gave a full obituary. Crediting her with Pat's success, editor Willie Chalfant declared:

"She was ever watchful over his career and was at once wife, mother, counsellor and guardian angel."

In Bodie, the great Standard Mine continued to sustain the town. But in 1884 it ceased its famous regular dividends. In the 1890s it still paid occasional dividends, but also levied some assessments. With the onset of electric power, the Standard built a power station 12½ miles away on Green Creek and proved the feasibility of transmitting power by electric wire. The Standard continued under various ownerships, with some interruptions, until the mill was burned in 1947.

As for Uncle Billy Lent, who had been a major owner in both the Standard and the Bodie mines in the camp's heyday, he had lived to the age of 86 at his home at the southwest corner of Eddy and Polk streets in San

Francisco, where he died in 1904. He was hailed as the last survivor of those on the ship *Oregon*, one of the first to reach San Francisco in 1849.

One man who would not believe the camp was dead was Jim Cain, an original arrival of the late '70s. Living in Bodie with his family, he gradually bought up the mines and buildings until he virtually owned the whole town. For years he opened the Bank of Bodie every day at 10 a.m.—when the only possible customers were the ground squirrels scampering across the grass on Main Street.

When surfaced highways brought Bodie within the reach of motorists in the 1920s, old Jim Cain showed off his ghost town to wondering tourists, always assuring them that Bodie would come back. I remember visiting Bodie as a boy just before the fire of 1932, marveling at the rows of false front buildings "right out of a Western movie", and meeting Jim Cain—tall and nattily dressed, with a gold watch chain across his vest—as he showed us glittering ore samples in his bank.

Since 1964, Bodie has been part of the California state park system, and is preserved from further decay. Today, from her warped board walks to her sagging false fronts, Bodie keeps her reckless and truculent air. Despite the 1932 fire, Bodie remains the most exciting ghost town in Eastern California. Among its many buildings are the home of Pat and Emily Reddy, one of the churches, and the Miners Union Hall—scene of so many turbulent episodes. Through these streets, as in so many other boom camps, thousands of men and women lived for a few years in that euphoric land between reality and great expectation.

CAN CAN RESTAURANT.

Main Street, (Second door south of Ellis' News Depot)

BODIE, CAL.

This House is conducted upon the

European Plan.

And is

"The" Restaurant of Bodie.

The BILL OF FARE comprises all the delicacies to be found in the market.

EVERYTHING COOKED TO ORDER.

   Game, Poultry, Fish and Oysters constantly on hand and served in the most approved style.

n30-tf   GEORGE CALLAHAN.

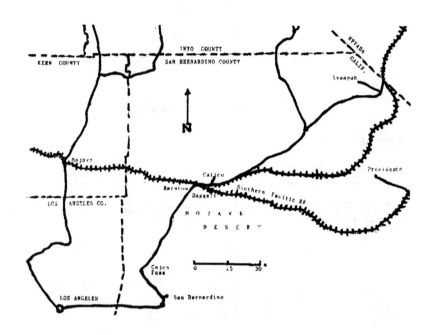

13. THE LAST BONANZA

With Calico located reasonably close to San Bernardino, and with its ores rich and easily worked, it was bound to be a profitable camp despite the dropping price of silver in the early 1880s. What was also good was the extent of its riches, making it the most productive silver district in California, and the arrival of the railroad at its front door. What was bad was that the same riches fostered more greed than usual in the Western camps, and for a time some of the boys stood each other off with Winchesters.

In 1881 San Bernardino, nestled at the foot of Cajon Pass, was still a frontier town. The Mormons who had founded it thirty years before had been recalled to Salt Lake, and been supplanted by gentiles from throughout the Southwest. Its dirt streets were lined with false front buildings, many of them raucous saloons that shocked any Mormons remaining. As new silver strikes were made at Panamint, Ivanpah, Resting Springs or Providence, the town's streets were jammed with the horse- or mule-drawn wagons of the silver trade.

One of the early arrivals in San Bernardino was Lafayette Mecham, a New Yorker by birth and a a storekeeper by trade. In 1849, at the age of twenty, he had ridden the California trail as far as Utah, then on to San Bernardino three years later. In the late 1860s he built a way station at the Fish Ponds, east of the present Barstow. Since the desert Paiutes were less than friendly, Mecham kept his wife and family in San Bernardino while he operated the store in the otherwise-empty Mojave Desert.

During this time an Indian stole a horse from Mecham, who followed in pursuit. The tracks led eastward to a low range later referred to as the "calico covered hills". There he lost the Indian trail but did make mental note of a large outcrop of reddish quartz. Though he described the promising lead to his two sons, no one tried to find it again for nearly a decade.

In 1880 silver was discovered near the Fish Ponds by John L. Porter and Robert W. Waterman, a Southern California enterpriser who later went to Sacramento as lieutenant governor, then governor. From the Minnietta Mine at Lookout they hauled in R.C. Jacobs' ten-stamp mill and started turning out bullion early in 1881.

When the silver bars rolled through San Bernardino, the businessmen who had been supplying silver towns as far away as Ivanpah now took new interest in their own back yard. One of them was John C. King, sheriff of the biggest county in the United States, who had ridden from Texas across the Southwest desert to San Bernardino in 1868. Operating a harness shop with his brother, he was elected sheriff in 1879.

Now, at his office in the basement of the old Victorian courthouse, King talked with G. Frank Mecham, Lafayette's oldest son. Did the Mechams know of any likely indications near the station at Fish Ponds? Frank then spilled the story of the stolen horse and the reddish outcrop seen by his father. King pounced on the revelation.

"You go up there and stake a claim and I will grubstake you."

Two of King's deputies, overhearing the conversation, volunteered to join the party. By the time Frank Mecham rode through Cajon Pass there were four men with him. In the calico hills they found his father's outcrop and on April 6 staked their claims, one of which they called the Silver King. Unknown to them, four miles northward a solitary prospector named Lowry Silver was already working a claim—the first in the district.

Back to San Bernardino they hurried, armed with ore specimens. But their hearts sank when the highest assay showed $8.00 to the ton.

Yet Sheriff John King would not let go. In June he persuaded his deputy, a hefty lawman named Edward Hughes Thomas, to go back. Thomas recruited Frank's younger brother, Charles, who reluctantly agreed on condition that they get back in time for the town's Independence Day Ball, for which he had a heavy date.

"All the mines on the desert could not have kept me away," Charlie later recalled.

Arriving at the claims around June 25, they camped in the shadow of the later Calico town. They found a lone prospector, Johnny Peterson, already on the ground, working a claim. Next morning they began prospecting, young Charlie climbing to the top of the outcrop and Thomas, with his weighty frame, going over the lower ground.

At the highest point of the outcrop Charlie found rock with some curious small "blisters". With his knife he carved into one of them. It was like, he later wrote, "cutting into a lead bullet." With his pick he broke off pieces of rock—each of them showing the rich mineral—chloride of silver! In rising excitement he gathered as many as he could carry. Running down the hill, he started hollering to Thomas, who was back in camp. His friend shouted back, asking if he had "been bitten by a snake."

"I certainly had struck it rich," Charlie answered, stumbling into camp. "Look here, pure horn silver!"

Thomas, true to his name, was the doubter.

"Horn the devil!" he retorted. "It can't be silver; no such good luck."

"Well, that is just what it is, and nothing else."

Next morning they broke camp and started back in high spirits, carryng a sack of ore. This time the samples assayed in the hundreds per ton. San Bernardinans chattered with excitement at the discovery only 80 miles away. A total of six, including two more of King's deputies, joined as the first owners of the Silver King Mine. And Charlie Mecham kept his date for the Independence Day Ball.

Soon the Silver King claimants trekked back over Cajon Pass to the calico mountain, financed by Sheriff King and armed with mining equipment. Accompanying them was a crowd of San Bernardinans with picks and shovels.

While Charlie Mecham and his crew of eight or ten men were pulling down and sacking the surface rock, the others were combing over the nearby ground and staking claims The first ore from the Silver King, valued at $400 to $500 per ton, was hauled by mule team and railroad to San Francisco for reduction at the Selby Works.

At first the miners put up tents and shanties in what became known as Wall Street Canyon. But a summer flash flood sent them up to the small mesa on the east, below the Silver King Mine. By the fall of 1881 a rough townsite, innocent of any surveys or plats, was laid out. With a ravine on each side, the flat offered room for only one street. But tent houses were going up, and town lots were selling at $300 apiece, title resting in the owner's rifle.

"It is as much as a man's life is worth to dispute these claims, " reported one visitor, "as a regular gang of fighters are employed."

Mining claims, more than town lots, were held by "right of shotgun" as greed ruled the Calico hills. One of the most dazzling strikes was the Burning Moscow, where one observer reported "the silver stood out like warts on the digits of a small boy . . ." When five tons of ore yielded two bars of bullion worth $2,200, the four partners in the Burning Moscow fell to quarreling. More than once, at Calico, hard words led to blows.

In San Bernardino on the morning of June 7, 1882, three of them had an argument in Brinkmeyer's saloon. Next morning they had breakfast in Starke's Hotel and reopened the quarrel while crossing C Street at the corner of Third. Johnny Peterson, the pioneer prospector whom the Mechams had first encountered at the Calico site, claimed he was not treated fairly. Enraged, John Taylor drew his Colt 42-caliber five-shooter.

"Don't shoot, John," cried Peterson.

Taylor fired three shots, killing his partner. The third man tried to grab the gun, only to be fired upon himself by the distraught Taylor. The bullet was stopped by a purse in his pocket. Taylor lurched away and shot himself in the chest. Peterson died in minutes. In jail, Taylor completed his suicide with a knife two days later.

First known simply as "the town in the Calico Mountains," the place soon won the name, Calico. By the spring of 1882 it claimed some 100 inhabitants, including eight families, and 300 mining locations. Among the fifteen frame houses were three saloons, two general stores and two hotels and restaurants.

One institution passing for a hotel was Bill Harpold's Hyena House, which consisted of barrel staves on the outside and holes-in-the-rocks on the inside. When the Calico stage rolled in, Bill was on hand with a wheelbarrow and the commanding yell:

"Here y'are, gentlemen! Right this way for Hyena House, best hotel in all Calico!"

Bill's breakfasts were simple and straightforward: chili beans and whiskey. When Hyena House caved in after a storm, Bill was unconcerned as he dragged out his two guests. He was sorry, he said, for the leaky roof.

With no grass and no water, necessities were hauled seven miles from the Mojave River—wood for $9.00 per cord, water packed on burros and sold door to door at five cents a gallon. Lumber for more permanent buildings, hay for horses and mules, and provisions of all types were hauled in by mule team from San Bernardino at 1¼ cents a pound. Yet restaurants charged only 50 cents a meal, and hotels $9.00 a week for room and board.

Through 1882 Calico boomed on hope and expectation, since little bullion had yet been shipped. Certainly, except for the colors of the hills, the place was forbidding. One early arrival wrote:

"There is not a semblance of vegetation for miles, and so far as looks are concerned, it is the worst country in the world."

Yet the Calico air was filled with exuberance. New mines, destined for later fame, were opened in the "suburbs" of East Calico, Bismarck and West Calico. Among these were the Bismarck, Burning Moscow, and the Oriental and Occidental groups of mines—some sacking and shipping high grade ore for miles to other mills in the area.. One visitor in June found every inch of ground above camp taken up in claims, about 400 people in town, and more coming in daily. An arrival taking dinner at a restaurant in the fall of 1882 noted the table talk of men "to whom a few thousand dollars were a mere bagatelle," but who scarcely had enough capital to live on for a few weeks.

"While we ate," he later wrote, "mills and roads were planned, railroads laid out and new camps started as though such things were mere incidents in the day's work."

The excitement swept to San Bernardino, the bustling gateway to the mines, where one observer noted that "everybody in town was carrying specimens in his pockets."

"The road from this town to the mine is alive with men and teams," he added. "There is now more travel through the Cajon than ever before."

Spurred by such fever, experienced mining men were riding in from other California and Nevada camps.

From Mono County came Captain Jim Powning, the mine superintendent who had been wounded in a shooting scrape in Benton, and now settled in a house on the main street.

From Bodie came Ned Reddy, who surveyed the outlook and returned without investing.

From Pioche came "Uncle Billy" Raymond, a founder of the famed Raymond and Ely Mine. Born in New York in 1826, Raymond was short and, according to one chronicler, a hunchback. But he was, as the same writer added, "courageous beyond the average." In Pioche the first claim owned by Raymond and his partner, John H. Ely, was jumped by a party of gunmen who threatened to kill the owners if they approached the mine. Raymond and Ely then retained four gunmen headed by the notorious Morgan Courtney, who drove off the jumpers, killed one, and regained the mine for Raymond and his partner.

Now, in 1882, Uncle Billy Raymond arrived in Calico and, for a start, bought the Garfield and Runover mines, and later the Occidental group. As a Calico employee wrote:

"He wore a full white beard, was very soft spoken and patriarchal in appearance, but stubborn as a mule."

Arriving from Placerville in California's Mother Lode was young James L. Patterson, a Missourian with the energy and resolve to superintend Bill Raymond's mines.

And from Nevada came Judge James Walsh, who had helped to start the great stampede to silver in 1859; he now opened an assay office just above town and soon became superintendent of the Oriental group of mines. Naturally he predicted that the Calico District would become "another Comstock."

Most remarkable of all was Annie Kline Townsend, a widow from Mississippi, who arrived in 1882 at about age forty-two. In a group photo taken at some distance, she has a buxomy, hour-glass figure. Her character was boldly etched—headstrong, audacious, self-confident, and quick to cast blame for her own mistakes.

In partnership with Cinderella Cook, one of the first women in the new camp, she ran a boarding house in East Calico, living nearby with her father and little daughter in Deep Canyon. In her spare time she prospected the canyons and ridges of East Calico—first on foot, later riding a burro. As the *San Bernardino Times* observed:

"She is said to have all the skill and pluck of a typical prospector, shouldering her pick and pan and starting into the wilderness alone with more than the courage of an average man."

In Odessa Canyon, Annie broke off some samples which, when given to the assayer down in Calico, showed $200 to the ton. Back rode Annie to the canyon, where she staked her claim to what she called, flamboyantly, the Golconda. Soon she acquired an equally valuable mine, the Alhambra. She sold a part interest to two other miners—one of them A.M. Rikert, said to be a nephew of Commodore Cornelius Vanderbilt. An experienced mining man, Rikert married Annie and joined in managing the properties.

By the summer of 1882 Calico was bursting from a canvas camp to a board-and-shingle town, with the ornaments of civilization arriving fast. The first school classes opened in September, and the first church service (held in the general store) brought the Gospel to town on October 15.

From San Diego came veteran journalist John Overshiner with his hand press and bank of type. On July 12 his first issue trumpeted the glories of Calico and began carrying ads for its business houses. By October they included notices of Jordan's Hotel—not just a frame building but a "new adobe", located where it is "orderly and quiet", with board and lodging at $2.00 per day; the James brothers' General Merchandise store: "mill, mine and prospecting supplies a specialty"; and the Globe Chop House: "everything conducted in Metropolitan style."

By the fall of 1882, new money was bringing in substance. In October the great Silver King Mine was leased to a group that owned the Oro Grande quartz mill forty miles westward near the present Victorville. Heading the group was Henry Harrison Markham, a leading citizen of Pasadena.

Born in upstate New York in 1840, Markham was a Union soldier in the Civil War and among other actions, was in Sherman's march through Georgia. He practiced law in Milwaukee and came to Pasadena for his health in 1879. Active in Southern California business and politics, he later went to Washington as a Congressman and to Sacramento as Governor.

Mojave River Valley Museum & San Bernardino Co. Parks Collection
James Patterson, mine manager, came out best in disputes with both the Silver King Mine and a payroll bandit.

Nadeau Collection
H.H Markham and friends built the Oro Grande mill and bought Calico's leading mine. He later became governor.

Mojave River Valley Museum

In the 1880s Calico, near the Mojave River east of today's Barstow, was California's leading silver town. The clatter of 100 stamps in the mills processing the ore, together with the revelry in the saloons, made Calico a roaring camp around the clock.

Laws Railroad Museum, Bishop

After the Southern Pacific RR passed Calico late in 1882, silver bullion was hauled by mule team to be loaded onto freight cars at Daggett, said to be "the worst place between Mojave and New York." Here an 18-mule team gives up its cargo to the Iron Horse.

In 1880, Markham had bought a gold mine just west of Oro Grande, and by early 1881 he had joined several friends in building a stamp mill and launching the Oro Grande Mining Company, with headquarters in Los Angeles and Markham as vice president and manager.

Now, with an ample supply of ore for the Oro Grande mill, Markham's company engaged Remi Nadeau to haul it by mule team the forty miles from mine to mill.

But the mine itself was on the mountainside a thousand feet above the town. The mine superintendent, George Barber, had to decide how to get the ore down to the wagons. An enterpriser in his own right, Barber was both cost-conscious and courageous. He could choose to build a switchback road, at considerable cost in time and money. Or he could construct an ore chute down the mountainside, in much less time and money but with the risk of technical failure in a project of this size.

By October, Barber had decided on the chute. To build it he hired a Los Angeles contractor named Mellen, who had already worked at Oro Grande. Mellen arrived with his men and equipment in the fall of 1882.

Nobody in camp had any experience in ore chutes, but Mellen received plenty of advice from onlookers. To determine the pitch required to slide ore down the chute, Mellen ran an experiment and settled on 32 degrees. To get this steep an incline, he started building an ore-car track on a trestle from the dump to the proposed top of the chute. The trestle was nearly finished when Calico suffered one of its periodic windstorms that for three days almost blew the workmen off the track.

By the last week of December the whole structure was finished. At its furthest point the trestle was ninety feet above the ground. The chute was ready to deliver half a ton of ore at a time 1500 feet down to a huge bin on the road just above town..

On January 1, 1883, Mellen ran the first ore car out the trestle and dumped the load into the chute. Below, the unsuspecting townspeople were going about their work when they were shaken by a roar—Mellen's teenage son called it "a freight train in full motion."

In an instant the entire population of Calico ran into the street, eyes up the mountain to the King Mine. Then they let out an equally loud cheer. The chute meant a regular supply of ore to the mill, production of silver bullion, and new coin for the business of Calico.

However, the men pushing the ore car were dubious about the frail trestle, especially in one of Calico's high winds. They demanded a hand rail along the trestle. Mellen told George Barber, who said it wasn't necessary. He had already spent too much company funds on this new-fangled system. But he agreed to look the situation over.

Next morning Barber went up to the mine and, while his colleagues watched, walked out to the end of the trestle. After admiring the structure with great satisfaction, he started back. One of those sudden Calico winds struck him broadside. George dropped to the track, clinging to the rails. With great agility but little dignity, he crawled back to the safety of the ore dump.

"Well, Mr. Barber," ventured Mellen, "shall we put up the hand rail?"

"Hand rail, good Gad! Board it up solid and batten the cracks, Mellen, before I ever go out there again."

The hand rail went up, the ore cars came out of the mine, the rock crashed down the chute, and the mule wagons rolled through town toward Oro Grande. There the ten-stamp mill began turning out bars of silver bullion, each weighing 99 pounds and worth about $1,800.

By this time Calico had been visited by another triumph of technology, the Iron Horse. Early in 1882 the Atlantic & Pacific Railroad (later the Atchison, Topeka and Santa Fe) had reached Arizona Territory and was threatening the domain of California's Southern Pacific. In February the S.P. started laying track eastward from Mojave to block the A & P at the Colorado River.

On November 13 the end-of-track, running on the south side of the Mojave River, reached Calico Station, only seven miles from Calico itself. While the S.P. track moved on to Needles on the Colorado River in 1883, Calico Station's name was changed to Daggett, in honor of John Daggett, a Calico mine owner and at the time, California's Lieutenant Governor.

Daggett now flourished as the transshipping point for Calico. Though the town never exceeded Calico in population, it was tougher and wilder. Saloons and brothels were filled with gamblers and confidence men preying on both new arrivals and through passengers as the trains puffed into Daggett. The S.P. posted warnings for all passengers to beware of the bunco steerers trying to fleece them. Shootings were common, with the culprits usually acquitted on grounds of self-defense.

By December 1884 even Daggett's citizens had enough. When a swamper killed his teamster with a wagon spoke and the inquest jurymen could only blame "parties unknown," the townspeople seized the culprit and hanged him. One correspondent wrote of Daggett:

"It is considered to be the worst place between Mojave and New York."

CALICO HOTEL.

Calico Street, Calico.

Mrs. E. L. Hazen......Proprietress.

This Popular Hotel furnishes first class meals. Everything neat and orderly.

The traveling public will always meet with a generous welcome and will find our table supplied with the best the market affords.

Fresh Bread,

PIES AND CAKES ALWAYS ON HAND AND WILL BE FURNISHED TO HOME PATRONS AT

REASONABLE RATES

Bringing cheaper transportation, the railroad boosted the value of Calico ore and thus boosted Calico's prosperity. Near Daggett the Oriental company built a ten-stamp mill, which it sold to Oro Grande in December 1883. Henceforth Nadeau's sixteen- and eighteen-mule teams cut their ore hauling distance from forty down to six miles. This further slashed freight costs and increased Silver King profits. H.H. Markham and his Oro Grande partners upgraded their stake in the mine from leasing to outright ownership, and increased dividends to happy investors as bullion shipments reached $1 million per year.

Through the early Eighties, Calico was riding high. At some of the mines, hand-cranked windlasses were being replaced by horse-powered whims, while the King installed a steam-driven hoist despite Calico's chronic water shortage. Replacing the old horse trails, wagon roads were graded up Wall Street, Bismarck and Odessa canyons. Tramways and chutes were built to carry ore from mines to roads.

Meanwhile, several stamp mills were sending their wheeze and clatter echoing in the Calico mountains. George Barber, for a time superintendent of the King Mine, broke ground below town and by December 1884 was running his ten-stamp mill—soon increased to fifteen stamps. At its height, Calico had 100 stamps clanking all day—a symphony that spelled prosperity to its people.

The mills in turn opened a way for several hundred chloriders to turn their ore into bullion. By 1884 Calico and its suburbs boasted some 2,000 people. Even when a fire destroyed most of the business district in 1883, Calico was undaunted. With their stores in ashes, the merchants erected new buildings, many of them in the rose-colored adobe of the Calico hills. By October the *Calico Print* was able to report:

"The town is now looking substantial, lively and prosperous . . . . Calico's colors are not the kind that easily fade."

Good times, of course, increased the flow of coin across Calico's fancy bars and gaming tables. While the town was not considered lawless compared to other camps, every man carried a revolver in belt or pocket, and Calico had its share of tough customers. One of these was John Williams, known as a "Bad Man from Bodie." In April 1883 he got into an argument with another and, when they stepped outside, killed him with two shots. Tried in Superior Court in San Bernardino, he was found not guilty on self defense.

On another occasion Williams went into a saloon and, putting his pistol on the bar, demanded free drinks. Only a little intimidated, the bartender established a limit of two such drinks.

Willliams slapped him in the face and, scooping up his gun, stalked across the street to the saloon that had been the scene of his earlier performance. Forewarned, the bartender met him at the door with a leveled six-shooter. When Williams tried to push inside, he was shot dead.

But Calico would not be short of rough characters. When things got especially dull, the town rowdies marched over to the east end and raided Chinatown. With little law enforcement to help them, Calico's Oriental population suffered until they decided to fight back. At the next invasion they counterattacked with laundry paddles and hot irons, forcing the raiders into ignominious retreat. Thereafter, Chinatown was unmolested.

Calico's sports still had plenty of diversion. Their idea of an April Fool's joke was to stuff a dummy and then shoot it full of holes in a mock battle on Main Street. In fact, a favorite outdoor sport of the Calico fraternity was shooting pistols in the air at any time, day or night. The pastime had free rein until the spring of 1885, when the Justice of the Peace cracked down. The first culprit, saving the court expenses by pleading guilty, was fined $10. The second, demanding a jury trial, got a hung jury in the first trial and a guilty verdict in the second. The justice fined him $125, which the *Calico Print* thought "a pretty severe punishment."

Though brought to heel, Calico still had a "shooting spirit". In 1885 its residents seized on a sudden fad for fencing vacant lots. One stranger came upon some young boys driving stakes and stretching wire near the edge of town. He asked what they were doing.

"Fencing in a lot," was the stout answer.

"But you boys are not of age and you can't hold any real estate."

"You bet we can," retorted one. "If anybody tries to jump my lot, I'll shoot him!"

In fact, the growing number of families in town was civilizing Calico. The wives and children of merchants and mine superintendents meant school, church, and socializing. Early in 1885, Calico voters passed a $3,000 bond issue to build a combination schoolhouse, church and town hall at the head of Main Street. Besides classes and church services, the hall became the thriving center of Calico's social calendar. Dances, cake socials, lectures, debates, theatricals, song-fests—all livened the Calico agenda. The hall even hosted a literary society and a dancing class, as well as much-needed temperance lectures.

Fanning the social joys was the *Calico Print*, which was doing its best to prove that Calico had a cultivated society. At one wedding in June 1885, editor Overshiner published a detailed list of the wedding presents and their donors—from "Elegant perfume bottle on silver stand with gold ornaments" to "A very handsome tinted cut glass writing set and thermometer and gold pen and holder."

All this time Calico was roaring with prosperity. With a dozen mines thundering with giant powder and 100 stamps clanking all day, the town roared with the happy cacophony that meant jobs, bullion and money. Every foot of ground above town and in the suburbs of East and West Calico was claimed by working miners. And the riches underground spurred men to roughshod competition.

Nobody felt this more than James Lewis Patterson, the ex-Missourian who superintended the several mines of John S. Doe, who had acquired them from his father-in-law, Billy Raymond. Doe had left Maine for California in 1852, established a construction business in San Francisco, and invested widely in California mines, including some at Calico.

Though an effective mine superintendent, Jim Patterson had been slow to perfect title to the Sweepstake Mine, which he had bought for the Doe interests in April 1884. The sellers had not yet provided the promised deed to the property, and Patterson had not extracted more than thirty sacks of ore by the year's end.

Waiting to pounce on the Sweepstake were two tough-minded Calico men—Ben Tiley and William H. Foster, who was the blacksmith at the Silver King Mine. They well knew the Calico District rules requiring $100 worth of development work on a mine in a given calendar year.

Promptly on January 1, 1885, they moved onto the ground and "relocated" the Sweepstake under another name. Jim Patterson realized what had happened when Foster and Tiley seized the thirty sacks of ore at the mouth of the mine.

On the night of February 3, Patterson arrived at the tunnel mouth with a formidable party of his miners, well armed and ready to take back the Sweepstake. They found, facing them, an equally fierce array of Foster and Tiley men. In their hands were a battery of Winchesters, shotguns and revolvers. In the early hours of February 4 the two forces stood off each other, waiting for the sun to come up over the San Bernardino Mountains so they could see their targets.

Deputy Sheriff Joseph Le Cyr was at his home in Daggett. A State-of-Mainer, he had joined the Union Army at the age of sixteen, fought in several battles with the Army of the Potomac, and turned up at mining excitements out West, including Darwin in Inyo County. Now he was operating a blacksmith and teaming business, doubling as a peace officer. When someone rode down and roused him, Joe LeCyr rushed up to Calico and the Sweepstake.

There, just at daybreak, he talked with Patterson on the one hand and the Foster-Tiley people on the other. If they launched the battle, he told them, there would surely be dead men, and the title to the ground would still be undecided. Instead, they should seek out the lawyers.

Accordingly, guns were laid aside, but miners on both sides remained, armed with picks and shovels. The same day Jim Patterson, acting for John S. Doe, clinched the purchase of the Sweepstake and this time, got a deed. But two days later, Foster and Tiley, with their miners, forcibly ejected the Sweepstake crew from the ground and retook the ore sacks.

At this, Patterson filed a lawsuit which wound up in the Superior Court in San Bernardino. There Judge James A. Gibson, after hearing Patterson's complaint, issued a preliminary injunction against Foster and Tiley. Patterson's miners again took possession of the mine and the ore sacks.

But when Judge Gibson heard the case (both sides waiving a jury trial), Foster and Tiley showed that John S. Doe had not perfected his title before they had relocated the mine on January 1, 1885—a month before Patterson got the deed. So on July 17, Judge Gibson turned about, withdrew the injunction and awarded Foster and Tiley $322 in costs, to be paid by John S. Doe.

This was only the beginning of Jim Patterson's troubles. Another mine owned by John S. Doe and managed by Patterson was the Oriental No. 2, adjoining on the south the great Silver King Mine.

Superintendent of the King at that time was D.F. Edwards. Outspoken, aggressive, Edwards had his own ambitions for mine ownership. On January 1, 1885—the same day that Foster and Tiley had relocated the Sweepstake—Edwards relocated the Oriental No. 2 and renamed it the Raymond Mining Claim. Patterson discovered it on March 2 and made his own relocation on that day under the title of Patterson Mining Claim. Down upon the property came Edwards and some of his men to take forcible possession.

So once again Jim Patterson swore out a complaint to the local justice of the peace. Edwards responded by denying everything, including his possession of the property. On March 5 these papers reached Judge Gibson of the Superior Court in San Bernardino; he promptly issued an injuction against Edwards. Then Edwards filed a cross complaint revealing his relocation of January 1, 1885.

The case was still pending when the good ladies of Calico held an elegant dance at the Town Hall to raise funds for paying off that building's debt. Entitled "The May Day Ball and Ice Cream and Strawberry Festival", the social triumph drew almost everybody in town, dressed in their finest. Opening the celebration was the May Pole Dance, performed by six girls of Calico families. Ida Miller was crowned Queen of the May. The crowd danced and ate its way through the ice cream and strawberries until the small hours of the morning. Among the celebrants were Jim Patterson and two of his friends and employees, W. Stoughton and Jim Marlow.

Around 2 am Patterson and his friends left the ball, stepping out the main door, which was lighted by a lamp. Standing in the shadow of the building were five men, including D. F. Edwards and W. H. Foster, both of them being sued by Patterson They pelted Patterson and friends with a hail of rotten eggs. Patterson was hit in the face, momentarily blinding him. Next moment he staggered as a harder object struck the side of his head.

When Stoughton was hit on the chin with an egg, he drew his revolver and, with commendable discretion, shot at the ground in front of his assailants. They scattered, and he shot again between them. At this point another ran toward the door, throwing an egg. The lamp light revealed him as W. H. Foster. He raced into the sanctuary of the crowd in the hall. But Jim Marlow, revolver in hand, chased him and, with no discretion, fired three shots before Foster escaped out the back door. Foster was unscathed, but bullets had whistled close to several bystanders. Amid shouts and screams, several women fainted, according to the *Calico Print* account.

Out in front, D. F. Edwards emerged from the shadows and asked whom Stoughton was shooting at. He was not afraid of Stoughton, he said, and intended to arrest him. Stoughton warned him "not to put his hands" on him. Edwards went off and found the constable, who deputized him. Edwards returned, took Stoughton's arm and marched him to the justice of the peace, scolding him so violently that a passerby said to him, "Stop abusing a prisoner."

The shootings marked an end to the May Day Ball and Ice Cream and Strawberry Festival.

As for Patterson, he had not fired at his assailants, possibly because, out of an excess of discretion, he had not worn a revolver to the ball. Afterward, Stoughton and Marlow were charged with "exhibiting a deadly weapon in rude and angry and threatening manner to the danger and disturbance of the peace." Stoughton demanded a jury trial and was acquitted. Marlow pled guilty and was fined $50. W. H. Foster, charged with assault, was fined only $20, apparently because he had not drawn a gun. The day after the shooting, the *Calico Print* called it "A Disgraceful Affair" and a week later commented:

> The smoke of battle . . . has entirely disappeared and dove-eyed Peace roosts placidly over the scene that erst was of war. The pop-guns that were so numerous have crawled into pockets of the combatants, there to hide until occasion offers for them to again come forth and bark and perchance bite.

The fight between Jim Patterson and D.F. Edwards soon went underground. Working in the lower levels of the Silver King, Edwards had run a drift into Oriental ground. Patterson realized that Edwards was taking Oriental ore out through a Silver King opening. A surveyor for the King mine accompanied him and confirmed that the stope they were standing in was across the line on Oriental ground. Around town, some of the King miners were making no secret that they were working in Oriental territory.

On September 28 Judge Gibson issued his final ruling—that the Oriental No. 2 belonged to John S. Doe, but that Edwards was only liable for trial costs. At this, Edwards announced he would appeal the case to the U.S. Supreme Court, and he stepped up his attack on Oriental ore. For the next three days after the decision, Stoughton and Marlow, working in the shaft of the Oriental No. 2, heard the rumble of blasts beneath their feet.

Back to the court went Jim Patterson. Back from San Bernardino came a new injunction on October 5 against Edwards extracting Oriental ore through Silver King tunnels. And back from Edwards came notice that he would move for a new trial before the state supreme court.

The case dragged in Sacramento until April 1887, when the supreme court dismissed it. Jim Patterson had lost some battles against Edwards and the Silver King Mine, but now he had won the war.

Early in 1885 Calico was the jewel of the silver camps in the Southwest. The Comstock, Tombstone, the mines of Mono and Inyo counties, were all in decline—partly due to the shrinking of rich ore bodies and partly to the sinking price of silver. But Calico was humming. The King mine, its depths honeycombed with adits and drifts down to the eighth level, was paying up to $30,000 in dividends every month. Other mines were fully developed and shipping bullion. The ores were easily worked, stamp mills large and small were turning out bullion for big investors and chloriders alike, and the railroad running by the front door had reduced shipping costs. In January both the correspondent of San Francisco's *Mining and Scientific Press* and the annual report of the state mineralogist praised Calico in extravagant terms.

Thus a new rush to Calico set in from all directions. By early February every train from east and west was bringing hopeful newcomers. Others, unable to raise railroad fare, pressed through Cajon Pass following the rails. Wrote a correspondent of the *Los Angeles Times*:

> Some were in wagons with their scant household goods, and others on foot unencumbered by any earthly possession beyond blankets and canteens. They were all expecting to 'find a show' in Calico; and for a ticket to that show, many of them had blown in their last dollar.

But the hopes turned to ashes in the realities of Calico. For the chloriders the promising ground had already been claimed. For the job hunters, the mines had matured and were hardly expanding. As early as March, mine owners were talking about reducing wages to $3 per day, especially in view of the plentiful labor market. Calico's main street was thronged with men out of work and out of funds. Their dreams shattered, many who rode in by the trains were walking out on foot. By the end of April the hundreds who had poured into camp had poured out again. Admitted the *Print* editor:

"Calico is not externally quite as lively as it was a few months ago." But with his usual exuberance, Overshiner was pumping the need for investors;

"All the capitalist has to do," he wrote, "is to take a palace car and soon he will be in the midst of mining properties that will pay him a handsome dividend when properly developed. A glorious future is in store for this region."

By summer, always a slow time at Calico because of the heat, the *Print* was protesting too much. At the end of June:

"Mining at no time in Calico district has been in a more prosperous condition than at present."

But underground, the picture was not so assuring. From an output of $500,000 in 1884, the King mine producd only $300,000 in 1885, and $120,000 in 1886. Other mines continued to flourish, but Calico's population began its slow decline.

Heading the exodus was Judge James Walsh, who had led the rush to Washoe in '59; by July 1885 he was operating a big silver mine in Idaho's Coeur d'Alene district.

Captain Jim Powning, the mine superintendent who had been in a shooting scrape in Benton before coming to Calico, left for California's Northern Mines. The fortune he had won in the Comstock he lost in what an observer called "speculations and free-handed generosity." He lived by himself in a cabin at Grass Valley until he died in 1904.

John S. Doe sold his Calico mines to the Silver King in 1891 and died three years later in San Francisco, leaving his wife and daughter a $2 million estate.

Uncle Billy Raymond moved to the bay city, where he died in 1901 at the age of 74.

Still, life in irrepressible Calico went on its cheery path, as decency still vied with the sordid. In 1885 the school teacher reported 54 students in class. But a correspondent reported:

"Saloons are more than numerous. . . .and the number of black-legs and tin-horn gamblers that infest the place is remarked by the newcomer."

Another observer noted that "heavy drinking" was prevalent but resulted in little drunkenness. Aware of temperance sentiment in some quarters, the Calico Pharmacy advertised, "Fine Wines and Liquors for Medical Purposes."

As for prostitution, Calico was one of the few mining camps in California where the brothels were in the middle of the business district, rather than on the outskirts. Among them was a hurdy-gurdy dance house that boasted sixteen girls. It flourished as an accepted Calico business until some of the patrons were robbed. The early victims were too embarrassed to complain to the town constable, but one had enough nerve to blow the whistle, bringing the arrest of the owners and the sixteen employees, and an end to the practice. Prostitution in the middle of Calico was one thing, but robbery!

In March 1885 prostitution itself was challenged when the Justice of the Peace again had the girls arrested. But what the *Print* called "Calico's best citizens" got up a petition to dismiss the case, as the prisoners "were receiving more punishment than is due them," and their behavior was no worse than the same activity flourishing openly in San Bernardino.

But everyone could agree on Calico's best-loved citizen, a black-and-white shepherd named Jack, who earned fame as the only four-legged mail carrier in the United States. For three years Jack delivered Uncle Sam's cargo from Calico to the nearby Bismarck mines. He was usually a playful dog, but once the mail sacks were on his back, Jack was strictly business. Touching him was considered tampering with the U.S. mails. His master, a Bismarck miner, once refused a $500 offer with the reply, "I'd sooner sell a grandson."

Through the mid-Eighties, Calico's noted female mine operator—Annie Kline Townsend Rikert—was proving herself an accomplished businesswoman. Early in 1883, when borax was discovered east of Calico, she staked a borax claim—the Lady Blanch—and hired two men to sack the product. In the summer of 1883 she shipped a ton of first-class silver ore from the Golconda mine to San Francisco for assay; it yielded $900 to the ton on average. A year later she bought out her partners and spent several months in San Francisco seeking investors. She got one mining company interested, and conducted an expert through the Golconda. Annie was present at the board meeting when the directors considered another, unfavorable, report. Then they accepted the report of her expert.

So in August 1884, Annie transferred a half interest in the Golconda and Alhambra mines to the new Golconda Mining Company on its promise to develop them and erect a stamp mill. As the *Calico Print* commented, she "succeeded after overcoming obstacles that would have baffled many less resolute and enterprising operators in mining matters."

Holding a number of shares in the stock company as partial payment, Annie and her husband moved to a comfortable cabin on the spot to look out for their interests and guard against what one observer called a "freeze out game."

But the stock company failed to do the yearly assessment work, as provided in its purchase of the half-interest. The Rikerts took legal steps to regain possession and leased the Golconda mine to another operator.

In November 1886 a Golconda company crew headed by H.G. Tobler arrived on the ground to do the assessment work at last.. Rikert sent word to them not to mine the claim, since it belonged to him and his wife. But Tobler was there to carry out the orders of his company, and he meant to ignore Rikert's order. He reckoned without Annie Kline Rikert.

On Monday evening, November 8, Tobler and his two men were making preparations for firing the first blast next morning at the face of the tunnel. Annie came upon them and demanded:

"Are you the men who are going to do the work on the claim?"

"We are," answered Tobler.

"The ground belongs to me," retorted Annie. "I'm not willing any one should go to work without my consent. The first man that will break ground in the morning will do so at his own risk."

Tobler was not going to be intimidated by a woman. He was starting work next morning, he said. He had brought the other two men to help do the assessment work. He represented the company and if there was any trouble she should come to him and not disturb the others. Annie stalked off.

Next morning the men put their tools in a nearby cabin. Annie arrived with the man to whom she had leased the mine, took out the tools and dropped them onto the dump at the tunnel mouth. Tobler, who was prospecting nearby, came over. He and Annie took a walk up the canyon to talk it out.

Annie proceeded to shower Tobler with threats. She had, Annie confided, many friends. She took a little silver-plated, five-shot revolver from her bag, saying it was one of her friends. If Tobler started work on the ground she would shoot him or get men that would shoot him. She could raise, she said, "a half-a-dozen shot guns and Henry rifles." If she could find the giant powder she would attach it with cap and fuse and blow up him and his men.

While Tobler was receiving this verbal bombardment, Annie's lawyer arrived on the scene. Thus reinforced, Annie again ordered Tobler not to work the ground. Tobler asked for her authority. The lawyer repeated the order. Tobler demanded that he "put it in writing." The lawyer said that he had no papers. Tobler asked what would happen if he worked the ground. The attorney snapped that Tobler "would have to take the chances." Tobler told him to "write me an order to quit and Mrs. Rikert would bring it over and serve it on me." Annie then went to her cabin to get pen and ink, but did not return until evening, when she told Tobler they "would do more about it the next morning." Tobler's answer:

"When papers are served upon me in a legitimate manner by an officer I will leave the ground."

Obviously Tobler would not vacate the field without being able to show his company written authority. Around 11 am next morning, November 10, Tobler and his men were about to fire the first holes they had drilled. None of them was armed. Tobler was up the slope from the other two, looking for a stick to tamp the powder charges. Annie arrived, carrying her little canvas bag. First she addressed one of the men, who referred her to Tobler. She climbed up the hill.

"You have gone to work, have you?" she asked.

"We are at work."

Annie demanded their names, but had no pencil to write them down. Tobler lent her his pencil. She and Tobler sat down while she took the names. Returning the pencil with her thanks, she got up and took a few steps up the slope—ten feet or more above Tobler. She shouted to the men:

"You are all three laying yourselves liable to take the consequences."

Down at the tunnel mouth, one of the men was splitting the fuse for the blast. Tobler was getting up when Annie pulled the revolver from her bag . Seeing this, one of the men hollered a warning:

"Tobie! Look out!"

Tobler, now on his feet, turned to face Annie.

"Leave the ground," she ordered, pointing the pistol at him.

Still Tobler underestimated Annie.

"I will when papers are served on me."

"Leave the ground!" repeated Annie.

"Not until them papers are served." And he called an order down to the men: "Shoot the holes!"

"Don't shoot the holes!" shouted Annie.

The men hesitated.

"Shoot those holes or I will," hollered Tobler.

Annie pulled the trigger. The bullet ripped through Tobler's coat and slightly bruised the skin. Tobler rushed up the hill to stop her. Annie fired a second shot, which tore the side of Tobler's chin. Then another, ripping through the fleshy part of his upper arm. Before she could shoot another, Tobler was upon her, grabbing the gun and twisting it out of her hand. While the wounded Tobler still gripped her arm, Annie tried to clear herself of something:

"I did not shoot you in the back."

"No, you did not."

Tobler then called for one of the men to "take charge of her." Quickly the two men ran up the hill. To the first one Annie cried out:

"See that I am not injured. I did not mean to shoot."

"You will not be injured by me."

Tobler then let go her arm and walked into Calico to the doctor's office. Annie disappeared over the hill and hiked half a mile to the Alhambra mine, where her husband was at work. Rikert and his assistant came down the hill to meet her.

"Mr. Tobler has been shot," she announced.

Back to the Golconda walked Rikert and Annie, who this time was carrying a shotgun. Several men were standing there with Tobler, who had returned. Rikert ordered everyone off the property. Two men escorted

Tobler to the Silver King boarding house, where the town doctor soon arrived and took him into Calico. His wounds were not serious and he soon recovered.

The same day, armed with a warrant, the constable and his deputy hurried to the Golconda Mine and arrested Annie. She was still bristling.

"If those parties had gone according to law, there would have been no difficulty."

As for the shooting, she said that Tobler had taken her arm and twisted her hand, so that "he might have shot himself."

Annie was charged with "assault with a deadly weapon with intent to commit murder" and tried in Superior Court, San Bernardino. The testimony of record is all for the prosecution, with Annie's lawyer simply cross examining the witnesses. As fairly common in Western courts, the chivalrous jury found Annie "not guilty."

According to custom on the frontier, Annie was simply protecting her property. Yet the machinery of law had been available to her in San Bernardino County. Tobler had been hired by the Golconda Mining Company to do a job. He was right in demanding a written authority telling him to stop. In those days he could hardly have gone back to the Golconda company and said he didn't do his job because a woman ran him off at gun point. His mistake was to discount Annie's warnings.

For her part, Annie could have accommodated Tobler by providing lawful notice. But she and her lawyer both seemed to be unsure of her legal ground. She chose instead to shoot an unarmed man.

Some time afterward, Annie sold the Golconda and the Alhambra to the Silver Odessa Mining Company. With her husband she moved to Mexico for another mining venture. In the 1890s they settled in California's Tuolumne County. There they took an interest in the Pino Blanco Mine, built a stamp mill on the Stanislaus River, and subdivided the town of Rosslyn about 1897.

At this time, railroading had captured the California mind, and Annie resolved to become a railroad tycoon. Already the standard-gauge Sierra Railway was being built from Oakdale, northeast of Modesto, to the mining and lumbering markets in the heart of Tuolumne County, with plans for crossing the Sierra to Mono County. Among its backers was William H. Crocker, son of the Charles Crocker of "Big Four" railroad fame. The Sierra Railway ran its first train into Jamestown in November 1897.

Annie was not backward in tackling this formidable adversary. By March 1898 she had launched the Stockton and Tuolumne Railroad, to run from Stockton some sixty-five miles to Sonora and vicinity, with intent to cross the Sierra. She would open offices at Sonora, she announced, in a year's time.

As Annie was president of the company, and as rumor said that Phoebe Hearst, George Hearst's wife, was an investor, the line quickly became known locally as the "Woman's Railroad". By May, Annie had graded 12 miles of roadbed, with rails and ties delivered to Stockton. Speaking in Sonora, she pointed out that while the Sierra Railway was charging $20 freight per ton for ore:

"We could haul this ore out for nine dollars a ton."

In response, the Sierra Railway started pushing beyond Jamestown, and the chief promoter of the line called Annie "a high-toned thief and mesmerizer of large-figured checks."

But Annie had plunged ahead without weighing the obstacles. Rights of way were only partly secured, and the graders had to stop and wait while the land agent bargained with property owners. The proposed 500-foot-high bridge across the Stanislaus River was declared impractical by engineers and damaged public confidence. Running short of cash, Annie fell several months behind in paying her grading company. By August she found her rails in Stockton and her right-of-way property under attachment. When her graders quit and were hired by the Sierra Railway, Annie blamed that company for part of her troubles.

For all her courage and audacity, Annie saw her "Woman's Railroad" founder and sink. She died in Sacramento in December 1906 at the age of about sixty-six.

By the late Eighties, Calico was suffering large troubles of its own. In 1887 the Oro Grande Company spent $250,000 on a great 60-stamp mill near its old one at the Mojave River. But in August it went up in flames. While it was being rebuilt, a fire on the roof of one of Calico's restaurants spread through town, abetted by Calico's famous winds and the lack of fire-fighting water. Just as quickly the townspeople rebuilt their ravaged town, as they had done after the 1883 fire.

But the pluck of the Calico people could not improve the low quality of ore reached in the lower depths of the mines. A reprieve in profit margins came in 1888, when the Oro Grande Company finished rebuilding its 60-stamp mill and laid a narrow-gauge railroad to its mines. Not only the Oro Grande group, but the John S. Doe mines as well, continued to produce, though with ore values cut nearly in half.

Through this period the John S. Doe mines had an average monthly payroll of $4,500, which Doe sent from San Francisco in gold coin. It routinely arrived by train at the Wells Fargo office in Daggett around the 10th of each month. Just as routinely, Superintendent James L. Patterson rode his bay mare the six miles down from Calico, got the sack of gold from the Wells Fargo agent, tied it back of his saddle, and rode back to Calico to pay the miners.

In 1889 a young man named Harry Dodson arrived in Calico from Walla Walla, Washington Territory. He was heavy set, dark complexioned and above average in height. He worked for a time in the Runover stamp mill, one of the Doe properties, where he was a good employee. Unknown to those about him, he had a devious streak and an ample supply of nerve.

On the morning of September 10, Jim Patterson stomped into the Wells Fargo office at Daggett, got the $4,500 in gold coin from the agent, changed $220 of it into silver coins, put the two sacks into a larger gunny sack, rolled it up and tied it behind his saddle. After conducting some other routine business in Daggett, he swung to horse and was riding north to Calico by 9:30 am. About halfway to Calico he saw, less than a mile away, a man on a white horse coming toward him at a gallop. About a quarter mile away the horse slowed to a walk. Patterson recognized it as one from a livery stable in Calico. Then, when less than 100 feet away, he identified the man as a former employee but did not know his name.

With reins in one hand, Dodson held a Smith & Wesson .44 caliber revolver in the other, covered by a red handkerchief. Drawing abreast of Patterson on the left side, Dodson dropped the handkerchief, cocked the gun and pointed it at Patterson's head. For Patterson, the routine vanished.

"I am not armed," he said, as coolly as possible.

"You ain't?"

"No."

"Throw up your hands."

Patterson threw up his hands. Dodson was apparently unable to believe that Patterson was carrying the payroll without being armed. He carefully surveyed Patterson and his bay mare.

"Get off with your hands up."

Patterson dismounted with as much grace as possible while keeping his hands in the air. Then Dodson, still holding his pistol on his victim, threw a leg over the saddle horn and slid to the ground.

"Turn and let me see what you have in your pockets."

From behind Patterson, Dodson went through his pockets. Taking out a pair of kid gloves, he threw them to the ground with the comment that they were "not worth a God damn." Then he pulled out Patterson's purse, containing $40 in greenbacks, two $5 gold pieces and a gold nugget. He looked at Patterson for a moment, apparently determining what to do with him, since Patterson had undoubtedly recognized him. Then, still covering Patterson with his gun:

"Now, you git. Go up that way." He pointed toward Calico. "And not look back."

Patterson started walking, then looked back. Dodson was riding eastward, leading the bay mare with the payroll sack. In about 300 yards he abandoned the white horse, mounted Patterson's bay, and rode eastward as fast as he could through the rocks.

At this point it looked as though Dodson was triumphant, but in fact he had robbed the wrong man. Patterson trudged back to Daggett and posted a reward of $500 for the bandit's capture. Four men rode out on Dodson's trail, which led north on the road to Resting Springs. At Garlic Springs the tracks left the road, and they lost the trail in the darkness.

Between 3:00 and 4:00 am they rode back into Daggett and reported this to Patterson. At daylight he telegraphed to a friend in Calico, John Ackerman, one of his miners who had run a saloon in Calico and was known to be a good man in a pinch. Riding down to Daggett, Ackerman procured a buckboard and provisions for the trail. Before noon on September 11 they were on the road with a two-mule team and a saddle horse.

This time Patterson had his Colt .44 six-shooter in a holster, while Ackerman carried his Winchester .44 repeater. Near Hawley's Station they looked up an Indian, Tecopa John, an expert tracker, who added another Winchester to the arsenal. With Tecopa riding the horse, the trio filled their water barrels and canteen at Hawley's and pressed up the Resting Springs road, Patterson holding the reins. Ten miles above Hawley's, with the Indian tracker riding 100 yards in the lead, they approached Coyote Holes.

Unknown to them, Harry Dodson was at the Holes. He had worn out the mare traveling through the night, and was now on foot with his sacks of gold and silver. He could see Patterson's party approaching from the south. They would surely catch him unless he acted fast. In this treeless oasis, the only hiding place was the spring itself. It was a small pond of murkey water, perhaps two feet deep, resting on another foot of uninviting mud and surrounded with grassy mire. He put the gunny sack and Patterson's purse in the water and laid down in it, face up, with his Smith & Wesson in hand.

With Patterson driving, the wagon caught up with Tecopa John on the rise of ground east of the spring. Tecopa tied his horse to a wagon wheel and, with Patterson, climbed down to the spring, easily following the mare's tracks in the mire. John Ackerman stayed at the wagon, untying the barrels to water the mules. Tecopa walked about 80 yards to the north side of the spring, Patterson to the south, both trying to follow the mare's tracks.

As Patterson passed the spring, Tecopa saw Dodson's form in the water and let out a war whoop. The robber rose up like a ghoul from the grave, blowing water out of his nose and firing his revolver at Patterson. To the alarmed superintendent, the spectre covered with mud and water was "malignant and ugly".

"He looked more like the Devil than a human being," Patterson said later.

But Patterson's own Colt .44 was in its holster tucked in his belt. He ran westward fumbling to get it out, while Dodson rushed after, still firing. Tecopa John ran back toward the wagon, where he had left his rifle. Patterson got his gun free and returned the fire, still running, as he later said, "in mortal danger of my life." He called to Ackerman for help. Ackerman did not hear him, but he could see and hear Dodson shooting. Reaching for his Winchester, he hollered to Dodson:

"Hold up!"

But Dodson kept after Patterson. Ackerman raised his rifle and fired. For a moment all three men were shooting. The last time Dodson pulled the trigger the gun snapped, possibly saving Patterson's life. Then Dodson, taking another jump, fell over, face down.

Patterson had fired three shots, Ackerman four. Dodson had been hit three times, including one on the chest and another back of the head—both wounds fatal. Ackerman turned the body over and they identified Dodson. Two feet away lay his revolver, with two rounds left in the cylinder.

"I am sorry Dodson is dead," said Ackerman. "We will never find the money."

"Yes," agreed Patterson, "we will never find it."

With Tecopa John, Patterson walked back to the waterhole to look for the treasure. The Indian went into the water, put his hand in the mud and pulled out the gunny sack. Patterson threw up his hands and shouted:

"We have got the money!"

Ackerman joined them, dumped the two money bags out of the gunny sack, and washed off the mud. Then they heaved the sacks and Dodson's body into the wagon and rode south, arriving at Calico about 9:30 in the evening.

That night the Justice of the Peace held a coroner's inquest, taking testimony from all concerned. A nine-man coroner's jury found that "the killing was justifiable." When a question arose whether or not it had been necessary to shoot Dodson, Patterson and Tecopa John took two men back to Coyote Holes on September 13 to reconstruct the event by the tracks in the mud and sand. While there, Tecopa started bailing out the pond and one of the men found Patterson's purse, with its greenbacks and its gold nugget.

In Calico, payday was delayed for two days. Dodson's body was buried in the Calico graveyard. Patterson said he regretted that the man had been killed.

After that, John S. Doe honored payday by sending a check from San Francisco, which Patterson then cashed for gold coin in Wells Fargo's office at Calico.

As Calico entered the 1890s its fortunes were tied to the volatile price of silver. From an average of $1.11 per ounce in 1881, Calico's first year, silver fell below $1.00 in 1886. When Congress rescued the metal with the Sherman Silver Purchase Act in 1890, the price briefly climbed to $1.04. But the Sherman Act was repealed three years later. Other world events dealt further blows, and in 1894 the price plunged to an average of 63 cents.

William Jennings Bryan, the great champion of free coinage of silver, headed the national Democratic ticket in 1896 on a platform of gold and silver valued at a ratio of "Sixteen to One". Along with other silver camps throughout the West, Calico went heavily for Bryan against the Republican candidate, William McKinley. But when the telegraph brought news of McKinley's victory, Calico's spirit broke. The mines shut down, the people rode down the mountain with their belongings, and the town was all but abandoned.

Twelve years later a visitor along Main Street found a dozen empty buildings, most of them of the rose-colored adobe, "in possession of the bats and the owls." In the old hurdy-gurdy house, "The sun streamed into the dirty room through great holes in the roof. The lining and gaudy wall paper were hanging in folds like a tapestry long neglected." In one corner sat the piano, disintegrating under a layer of dust.

During World War I the rising price of silver brought a little mining and a flicker of life to Calico. But when this author first visited the town as a youth in 1938, the hills were silent and the town was marked only by the remains of the Silver King works and those wonderful rose-hued adobe walls.

In the early 1950s, the entire townsite was acquired by Walter Knott of Knott's Berry Farm. A relative of Sheriff John King, who had grubstaked the discoverers, Knott rebuilt the town as a labor of love and made it a tourist site. Later he turned it over to the County of San Bernardino, which continues to operate it as a family attraction. But the air no longer rings with the clatter of 100 stamps on the mountainside.

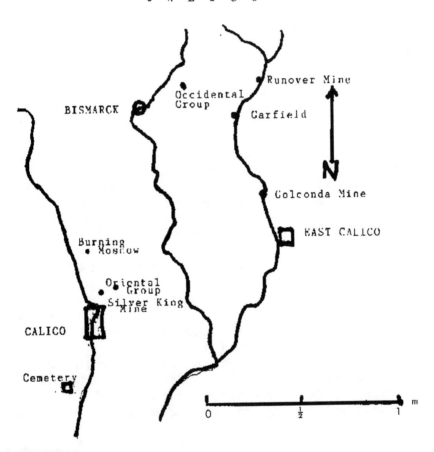

14. SILVER WAS KING

The tumult and the shouting dies; The Captains and the Kings depart...
— Kipling

It was the silver seekers who explored and settled California's last frontier—the Eastern Sierra and the Mojave Desert. Theirs was one of the great testimonies to the triumph of hope over adversity. The men and women who ventured into this wilderness were driven by ambition. They could not know how the venture would turn out. But they saw in this vacant empire, which was theirs for the taking, opportunities bigger than those at home. So they took the chance.

Most of them were jacks-of-all-trades. Storekeepers became mine owners. Miners became lawyers. Sheriffs became mine superintendents. Many pursued two careers at once, and several in a lifetime. Each one saw no reason why he couldn't handle most any job. It was a whole generation of doers and improvisers.

Frontier conditions lasted through the whole era. The trappings of law came early—new counties, with their judges, sheriffs and lawyers. But while this political machinery was in motion, it was flouted openly, as though it were a make-believe system. For protection, most men carried a gun—if not in a holster, then in the belt or pocket. Unlike the Texas cowboys invading the Kansas cattle towns, the Far Westerners were allowed to wear guns in camp while drinking heavily with their companions.

In every town, saloons were more numerous, and filled with more customers, than all other establishments combined. Saloon-keeping was as respectable as any other business, offering what was considered an essential service. Men drank on a scale that would qualify them as alcoholics today. Nobody seemed to see the folly of mixing alcohol with deadly weapons. Thus shootings were frequent, right in a crowded room. The shooters were regularly freed by judges or juries on grounds of self-defense—many of the jurymen had been through the same experience.

Even Judge Lynch was rarely summoned. Newspaper editors condoned and some even encouraged vigilantism, but over an entire generation in a dozen lawless camps where shootings were common, public indignation took summary vengeance only three times—in Alpine County, Bodie and Daggett. Several other times when hotheads were ready with a rope, sterner heads prevailed. It was not so much a respect for the law as it was the Western attitude of minding one's own business. In summary justice, the Eastern California frontier had a better record than the Mother Lode camps in the 1850s.

Women, like the men, risked being hit by a stray bullet, but this was the only hazard for them. They were as safe on this frontier as in the settlements. Besides the hurdy-gurdy girls, there were wives of storekeepers and mine superintendents, and their children. They brought schools and churches as some of the towns matured, though these refinements usually arrived about the time production was declining. Most of the camps never had a church at all.

So any civilizing influence was mainly the presence of good women themselves. Here they were respected and prized even more than in the settlements. Without question the towns had a masculine stamp, but a few women established their own businesses. In most towns, one or more hotels and boarding houses, and sometimes restaurants, were owned and managed by women. In a day when gambling was a widely accepted profession, Eleanor Dumont operated her green table with dignity and respect. Annie Kline Rikert bought and sold mines and sat down to bargain with boards of directors.

And what about the hopes that triumphed over adversity in this hard-bitten environment? Few of the local settlers and practically none of the actual discoverers were crowned with wealth. The few exceptions were M.W. Belshaw at Cerro Gordo, Albert Mack at Benton, Pat Reddy at Darwin and one or two of the McFarlane brothers at Ivanpah.

Much of the control fell into the hands of investors in San Francisco or Virginia City—for Benton and Bodie, Bill Lent; for Mammoth, George Dodge; for Panamint, the Silver Senators and Trenor Park; for Darwin, Lester L. Robinson; for Lookout, George Hearst; for Calico, Henry H. Markham, Billy Raymond and John S. Doe.

And the mines over which the cast of characters fought—did they yield great profit? For the most part, yes: at Cerro Gordo, Benton, Darwin, Bodie, Calico, and to a lesser degree, Ivanpah, Providence and Lookout For some others, no: Panamint, Resting Springs, Mammoth, and most of the Alpine County mines. And in those cases it was mainly the outsiders who lost.

When the curtain fell on the silver drama, what happened to the players—the original silver seekers? Some took their exits and entered other silver or gold dramas in Colorado and Idaho. Some returned to California's coast or central valley and settled in less exciting and risky occupations. But some did stay—not in the minng towns, which decayed in ghostly silence—but in the rich farm lands of Owens Valley and San Bernardino. They found, not quick riches, but permanent wealth of another kind. With their families and their schools and churches they embraced a life just as robust but far more complete.

Even so, their descendants look back to a near-legendary moment when nothing was routine and big things were happening daily—a moment when the American character was painted in bold strokes on a canvas of unsurpassed natural beauty.

APPENDIX

1. THE STORY OF SILVER

In the scale of elements listed by atomic weight, silver (symbol Ag) is 47th out of more than 100 elements. It is heavier than copper and tin, lighter than gold and lead. In the earth's crust it is 20 times more abundant than gold. Its resistance to corrosion, its shiny surface (compared with the moon's color by the ancients), and its malleability that fosters many uses, have traditionally given it a desirability second only to gold.

Creation. Like other metals, native silver is created in the volcanic process by tremendous heat, which can separate and concentrate the elements from one another. This can reach the surface as magma forces its way up through cracks in the country rock.

Silver is commonly associated in its ores with other metals, such as gold, copper, lead and zinc. Usually none of these metals is visible to the naked eye, and their presence must be determined by the art of the assayer. In such case the metal may have been carried in solution into a vein by rising hot water or vapor, by settling of mineral-bearing water from above, or by chemical action involving other minerals, such as sulfides or chlorides.

History. Silver is one of ten elements known in ancient times. As early as 4,000 B.C., silver was worked by the Egyptians, whose tombs have yielded silver ornaments. Both silver and gold were used as media of exchange from about 800 B.C. throughout the ancient world, from Egypt to India.

Mining of silver, along with other metals, began in Western Europe under the Romans. By the Middle Ages silver was mined in many countries, including Britain, Germany, France, Spain, Italy, Hungary, Greece, Bohemia and Russia.

In the New World, the Aztecs in Mexico and the Incas in Peru were mining silver before the conquistadores came in the 16th Century. Since then the Potosí Mine in Bolivia and the Real del Monte Mine in Mexico have been among the most productive in the world.

As for the United States, a fabulous treasure of silver-gold ore was discovered in Nevada's Comstock Lode in 1858, followed by others elsewhere in Nevada, Colorado, New Mexico, Arizona, Utah, Idaho and California. In 1904, discovery of the enormous Cobalt deposit in Ontario made Canada a major silver producer.

Silver and Gold. Silver's value relative to gold was established as early as the second millenium B.C. From about 1600 B.C. in Egypt to around 400 B.C. in Greece, the ratio of value between the two metals was 13.33 parts of silver to one part of gold. From the 4th Century B.C. in Greece to the 16th Century A.D. in Western Europe, the ratio varied more or less from 11.5. Beginning in the 17th Century, silver's worldwide value in relation to gold gradually declined, due in part to more relative scarcity for gold, to 15.3 in 1860, at the beginning of the great silver boom in Nevada and Eastern California.

At this point the price of silver on the New York market averaged $1.35 per ounce. But the tremendous increase in supply of silver, especially from the Comstock Lode, together with adoption of gold as the sole standard of currency exchange by Britain and Germany, forced the price of silver steadily downward. The trend was spurred when silver was abandoned for domestic coinage by the U.S. Congress in 1873.

By the late 1870s the drop in silver prices was seriously affecting the profitability of silver mining. With outcries from the West, Congress tried to stem the tide from time to time, but the remedies were withdrawn as often as they were applied. By 1886 the price of silver had fallen to 99 cents an ounce, and the West's silver era was effectively over.

Silver hit 57 cents an ounce in 1896, and a clamor for relief became a key plank in the 1896 Democratic platform, with candidate William Jennings Bryan crusading for remonitization of silver at a fixed ratio with gold of 16 to 1. But Bryan's defeat spelled the end.

At this writing, after a century of inflation, silver averages less than $6.00 per ounce, and gold under $300 an ounce—a ratio of about 50.

Extracting silver from ore. One of the earliest ways to extract silver and other minerals from ore, such as sulfide of lead (galena), was cupellation, which has traditionally been used in assaying. When the rock is heated to extremely high temperatures, buttons of silver appear. On the same principle, smelting has been used on a large scale to reduce the ore to a molten mass, which then facilitates separation of the metals due to their dlifference in weight.

The ancients added another process known as amalgamation, in which the ore was finely crushed and subjected to quicksilver (mercury), which has a high affinity for silver. This is the process that, in the Western silver era, made use of the famous quartz mills, employing a battery of from five to 100 stamps in crushing the ore, followed by amalgamation. This was employed to good effect at Calico, where the ores were easily worked.

However, much of the ores in the West refused to release their silver simply by crushing and amalgamating. High temperature roasting was first needed to melt the ore. Through the 1860s the Nevada and Eastern California newspapers contained much discussion on reduction processes. The model for all was the Freiburg method of roasting and amalgalmating, perfected at the celebrated mines of that name in Saxony, Germany. While this was adopted successfully on the Comstock Lode and elsewhere, the ores of Alpine County, California, were so rebellious that a satisfactory method for treating most of them was not found before the price of silver sank so low in the 1880s that the quest became academic. Since the turn of the century, an improved process using cyanide has generally replaced these earlier methods.

Prospecting. In most of the West's silver-bearing areas, the barrenness of the mountains make the prospector's job easier. The surface is weathered, helping to expose outcroppings of mineral-bearing veins.

The silver and gold prospectors, familiar with the look of various rock specimens, would first look for "float"—loose rocks different in character from the earth or sand where they lay. The float had been washed down a canyon or hillside from an outcrop of hard rock by the flash rains that can occur in the desert.

The prospector would follow the pieces of float up the canyon or mountainside to the point where they stopped occurring. The outcrop that should appear at that point might be obvious, or it might be difficult to spot because of accumulated overburden. This might require some pick work or even shovel work to find the vein. The outcropping might or might not have commercial quantities of mineral.

Generally, the likeliest place for good samples is on the edge of the vein, where it contacts the country rock, since this is where the most friction and hence the highest heat occurred. The intersection of two veins is especially promising due to the greater violence that had transpired. As a further clue, the country rock next to a vein may be stained with mineral from the vein—reddish brown may mean iron, greenish could mean copper.

The experienced eye of the prospector could identify by color, texture and other properties whether the rock might be valuable. If so, he would explore along the outcrop into other ravines, taking more samples, to see how big the lode might be. If it did not merit further attention he would move on. If it did, he might stake his claim and collect more specimens to get assayed in the nearest settlement.

2. MINERS' JARGON

Adit. A nearly horizontal opening of a mine. Synonymous with **tunnel**, although strictly speaking a tunnel is open at both ends.

Amalgamation. A chemical process of separating silver or gold from pulverized ore by combining with quicksilver (the amalgam). The resulting slurry is then treated again to separate the metal from the quicksilver (mercury). Amalgamation was the main process used with "free milling" silver or gold ore that was pulverized in a stamp mill, without need of roasting in a furnace. However, amalgamation was also used after the roasting process.

Arrastra. A means of crushing ore (Spanish and Mexican in origin) in which the rocks are placed within a circular enclosure and heavy stones or weights are dragged over them by mules or horses moving in a circle outside the enclosure. Amalgamation can be combined with the crushing.

Assessment. 1. Annual work on a mine required by state law and mining district by-laws. If the assessment work was not done, the mine could be relocated by someone else. 2. A levy on stockholders for funds to develop a mine, reduction works or other improvements. If a stockholder did not pay the assessment (generally between 25¢ and $1.00 per share) within a prescribed time, the shares were forfeited and sold on the market.

Bedrock. The solid rock, usually slate, underlying an ore-bearing ledge.

Blind lead or lode. One that has no outcrop on the surface, and is discovered inside the earth after a mine has been opened.

Bonanza. A large body of rich ore in a mine capable of yielding big profits.

Bonding. Putting up a bond to secure an option on a mine until the option expires.

Borrasca. Lack of paying ore in a mine (opposite of bonanza).

Bullion. Bars of metal, usually weighing less than 100 pounds, by which the silver or gold is shipped to be coined or treated further. Much of the bullion from Eastern California was lead with some silver—called **base bullion.**

Camp. A mining town.

Cap-rock. Surface formation that may overlay a body of ore.

Captain. The title given to a mine foreman, who might keep it long after he managed a mine.

Chloriders. Small mine operators who bring their ore to a custom mill for reduction. The term is derived from a common ore, chloride of silver. The mill owner received a fraction of the bullion as a fee.

Claim, or Location. A prospect or mine owned by an individual or partners by virtue of a claim notice on the site (on a post or in a pile of rocks) and a listing by the recorder of the mining district, and documented in the county courthouse records. A claim could be worked or sold.

Cleanup. Collecting the accumulated metal after a run of many days by a stamp mill or furnace.

Cross-cut. A passage at right angles to the course of the lode, usually to determine its width.

Dead work. Driving an adit or shaft through barren rock to reach the lode. This is unproductive work with no return of income until the lode is reached.

Double jack. In the process of blasting loose the ore at the breast of a mine, a means of driving the hole for placing the giant powder. In a double jack operation, one miner holds the long rod with a bit at the end while another miner strikes it with a sledge hammer. In a **single jack** operation, a miner holds the drill in one hand and, using the other hand, hits it with a short-handled sledge. The bit is twisted a quarter turn for each blow.

Drift. A horizontal passage following the vein in order to excavate the ore.

Face. The end of an adit or drift where the work is being advanced. If the face is in ore, it is also called the **breast** of the mine.

Fault. A displacement, vertical or horizonal, of the strata, which marks the end of an ore body unless its continuation can be found.

Feet. The means of measuring the length of a claim. The by-laws of a mining district established how many hundred feet of a lode could be claimed by a locator, who in turn could sell some of the feet. In mines that were incorporated to sell stock, "shares" supplanted "feet".

Float. Loose pieces of ore on the surface separated from the outcrop, which in turn may be found above the float.

Fissure vein. A vein or lode, usually vertical, created by a fissure in the earth's crust, in which the mineral was introduced from below. A "true fissure vein" was believed to be very deep, getting wider and richer the deeper it went—hence a very valuable discovery.

Free milling ore. One in which the precious metal can be separated by crushing and amalgamation without requiring other chemical process or roasting in a furnace.

Furnace or smelter. Reduction works in which the ore is brought to the melting point so that the metal can more easily be separated and then formed in molds, creating bars of bullion.

Galena. A smooth and shiny ore with a high percentage of lead. Often it contains some silver, when it is called **argentiferous galena.**

Giant powder. A type of blasting powder, manufactured in the United States under license from Alfred Nobel, that was much more powerful than ordinary black powder.

Grubstake. Financing the provisions needed for a prospector or prospecting party.

Hanging wall. The wall of solid rock overlaying the mineral ledge.

Hoist. A power-driven (usually steam engine) means of bringing ore to the top of a shaft, as distinct from a **windlass,** cranked by hand, and a **whim,** powered by animals turning in a circle.

Horn silver. Chloride of silver, varying in color from white to gray, and darkening when exposed to light—an ore with a high percent of silver.

Incline. A passage run into a mine at an angle, rather than horizontally or vertically.

Irish dividend. Jocular term for a stock assessment.

Jump. Unlawfully seizing another's claim.

Mill or stamp mill. A means of crushing ore with heavy iron stamps raised and dropped by an eccentrically formed axle, which was powered by a water wheel or steam engine. The number of stamps ordinarily ranged from five to 100, depending on the volume of ore expected to be crushed.

Noble or precious metals. Metals (gold, silver, platinum, mercury) having so little affinity for oxygen that they do not rust. These are the most valuable metals.

Ore. Naturally occurring rock or gravel containing one or more marketable metals.

Outcrop. The exposure of an ore body on the surface. This is the normal means of discovery.

Petered out. The ore body has given out in quantity or quality.

Pocket. A relatively small but rich body of ore.

Prospecting. Searching for new mineral regions or bodies of ore. A **prospect** is a claim in anticipation of finding ore. A **prospector** is one who prospects for ore. If he stakes a claim he may work it himself or sell it to others.

Quartz. A hard silver or gold ore, as distinct from earth or gravel.

Raise or upraise. A vertical passage excavated from below.

Reduction. Extracting the metal from ore by stamp mill, furnace or other means.

Relocation. Staking a claim after the previous owner has abandoned it or failed to do the required assessment work.

Refractory ore. Ore in which the metal resists being separated, even when roasted—hence difficult and expensive to reduce.

Salting. Impregnating a mine face or a rock sample with silver or gold particles to fool mine buyers.

Shaft. A vertical passage sunk from the surface.

Stope. To excavate ore from a vein in a series of horizontal workings, so that each stope is further advanced than the one above or below, usually looking like a set of stairs.

Timbering. Securing the roof and sides of a mine with timbers to prevent a cave-in.

Vaso. A small Mexican oven or furnace to reduce ore by roasting and produce bullion.

Winze. An interior vertical passage connecting one adit or drift with another.

3. WEAPONS

Since firearms are mentioned frequently in the text, a description of those most popular in the 1860s to '80s is offered below:

Pistols. In the early 1860s the most popular were, first, the Colt 1860 Model Army six-chambered revolver, .44 caliber, with separate powder and ball, and a percussion cap. Also still popular at that time was the Colt 1851 Model Navy six-chambered revolver, .36 caliber, also cap-and-ball. Caliber is a measure, in hundredths of an inch, of the diameter of the bore.

After the Civil War, many of the 1860 Army models were converted for cartridge ammunition, in which cap, powder and bullet are combined in a brass shell. This was more reliable and much easier to load.

In 1873 Colt introduced the .45 caliber Frontier Model six-chambered revolver. Known also as "The Peacemaker", it remained with slight modifications as the standard holster gun for the rest of the period covered in this book (through the 1880s). Colt offered it in 4¾, 5½ and 7½ inch barrels. Colt also produced this model in .44 caliber, using the same 44-40 ammunition as the Winchester Model 1873 rifle (44-40 means .44 caliber and 40 grains of powder).

Smith & Wesson began manufacturing breech-loading cartridge revolvers as early as 1857, providing many models in various calibers throughout the period of this book. The .44 caliber single-action revolver was perhaps the most popular model.

Beginning in the mid-1870s, several manufacturers produced smaller revolvers of five or six chambers, self-cocking or "double action", as opposed to the previous "single action" that had to be cocked with the thumb. These guns were tucked in the belt (as the larger Colt six-shooters often were) or in a pocket of trousers or coat. Most men in Eastern California and other Western localities armed themselves in this way, though some wore a holster. The "two-gun man" was, at most, a peace officer in a very tough environment, and was otherwise unknown outside Western novels and movies.

All these pistols were effective only at close range, say up to 25 feet, mainly because of the short barrel (eight inches at most), compared to that of a rifle or carbine. This meant that when fired, the explosion (expanding gas) was only partly used up before the ball or bullet left the muzzle. Hence the bullet had a low velocity and poor accuracy.

In addition, most people firing a pistol tend to flinch as they pull the trigger, which throws off the aim. Moreover, the double action requires such a pull on the trigger that the aim is deflected still further. This is why many gunfights resulted in no casualties, and why bystanders ducked for cover when antagonists drew guns.

Rifles. In the 1850s the traditional muzzle-loading, cap-and-ball rifle was being outmoded by the breech-loader, especially the Sharps, using a paper cartridge for the powder and ball, and still requiring a separate percussion cap (later models had a complete metal cartridge). In the 1860s such single-shot rifles were supplanted by the repeater, using brass cartridges.

The most popular was the Henry, introduced in 1860 by the New Haven Arms Co., which by then was owned by Oliver Winchester. Carried by some Union cavalrymen in the Civil War, the .44 caliber Henry had a lever action to eject the empty cartridge, replace a loaded cartridge in the chamber, and cock the hammer. The magazine under the barrel carried 15 rim-fire cartridges in the rifle version and 13 in the carbine version. It was a reliable, accurate and altogether effective weapon for big game hunting and self-defense at ranges up to 100 yards.

In 1866 Winchester produced a new version with several improvements, including a wooden forestock to solve the problem of holding a barrel heating up with repeated shots. Retaining the Henry's brass frame, this was the first Winchester, but in popular speech it was usually called a Henry.

The Winchester Model 1873 sported several further changes, including a steel frame instead of brass, and used a center-fire cartridge.

With minor improvements it continued as the rifle of choice in the West until supplanted in 1894 by the 30-30 Winchester, using the more powerful smokeless powder that provided better accuracy and longer range.

Shotguns. The shotgun was transformed from a muzzle loader to a breech loader in the 1870s, using a combination brass and paper cartridge. The double-barreled 10-gauge, with hammers on the outside, was manufactured for a generation of small-game hunters by such companies as Colt, Parker and L.C. Smith. Of course, it kicked like the proverbial mule, and many customers preferred the 12-gauge.

With the barrels sawed short to from 18 to 20 inches in length, the shot were scattered more widely and could hardly miss at short range. Using buckshot (about the size of peas) instead of game shot, this was a deadly and greatly feared weapon in the hands of peace officers and Wells Fargo shotgun messengers.

The gauge of a shotgun is established, first, by taking a lead ball that is exactly the size of the bore, then determining the number of such balls that make a pound in weight. Thus a 10-gauge bore is bigger than a 12-gauge.

4 HOW TO GET THERE

Alpine County towns. From South Lake Tahoe, take Highway 89 through Markleeville 34 miles and turn east up Loope Creek on Monitor Pass Road (still Highway 89). In little more than a mile the site of **Monitor** is along the south side of the creek. From this point a poor dirt road goes north two miles, curving eastward, to the site of **Mogul**, identified by a huge slag dump south of the road on the opposite mountainside. For **Silver Mountain,** go south on Highway 4 from the mouth of Loope Creek, swinging right up Silver Creek, toward Ebbetts Pass, for 11 miles. The Chalmers "mansion" is on the right, and a short distance beyond, on the left side of the road, is the site of Silver Mountain, identified by the walls of the old jail.

Bodie. On Highway 395, go 6.3 miles south of Bridgeport and turn east for 13 miles to Bodie.

Benton. Taking Highway 395 to 1½ miles north of Bishop, turn right on Highway 6 and go north 32 miles to Benton Station. Go west four miles to Benton Hot Springs (Benton).

Mammoth City. From Highway 395, take Highway 203 west to the community of Mammoth Lakes. Take Old Mammoth Road south for approximatelly 1½ miles to the site, marked by a plaque on the north side of the road and the large tailings on the side of Red Mountain to the south.

Cerro Gordo. Two miles south of Lone Pine on Highway 395, turn east on highway 136 to Keeler, on the east side of Owens dry lake. Just south of town, take the dirt road eastward up into the Inyo Range to Cerro Gordo. The last two-thirds of this eight-mile road is steep and winding.

Darwin. From Keeler continue south on Highway 136, which becomes Highway 190, for 17.6 miles and turn right (south) on the road to Darwin for 5.3 miles.

Lookout. On Highway 190, from the turnoff to Darwin, continue for 20 miles to Panamint Springs. After 2.5 more miles, turn south on the road down through Panamint Valley for 7.2 miles. Turn west on a road for 4.4 miles. At the foot of Lookout Mountain, a poor road around the north side and then up the back side is sometimes passable. If not, go back to the east side of Lookout Mountain and hike the old burro trail up to the top.

Panamint. On Highway 190, go 2.5 miles east of Panamint Springs, then turn south on the road through Panamint Valley for 23.2 miles. Turn east on a road 3.6 miles to the adobe ruins of Ballarat. Go north, swinging east, up the alluvial fan at the mouth of Surprise Canyon for 10.7 miles to the site. The last six miles are usually impassable except by foot or horseback. Panamint can also be approached from the south through Trona and Ballarat.

Calico. From Barstow, take Highway 15 for 7.6 miles. Turn north on a road for approximately 3.5 miles to Calico. **Daggett** is due south of Calico, crossing Highway 15 down to Highway 40.

Providence. From Barstow, take Highway 40 for approximately 102 miles and turn north on the road to Mitchell Caverns. After 10 miles, take a road branching north and northwest for approximately 6.7 miles to the Bonanza King Mine and, shortly beyond, the remains of Providence.

Ivanpah. From Barstow, go 107 miles east and northeast on Highway 15 to Cal Neva. Take the off-ramp and cross the highway on the overpass, going west a short distance on a surfaced road. Turn north on a dirt road for two miles, then west on a road leading up the alluvial slope toward the north side of Clark Mountain. After 11 miles, park and walk west across a ravine and you find yourself in a shallow canyon that was the site of Ivanpah. At last visit, a number of years ago, the remains of two mills, a smelter, and several rock and adobe walls and dugouts were still there.

Resting Springs. From Baker on Highway 15, take Highway 127 north for 47 miles and turn east for 3.6 miles to Tecopa. Continue for 1.1 miles, and turn left on a dirt road for three miles to Resting Springs.

THOSE WHO HELPED

In acknowledging the assistance of many people, I first want to thank my wife, Margaret Nadeau, who conceived the basic idea of this book. She helped in the research at libraries and ghost towns, made valuable suggestions, offered continued support and encouragement, reviewed the typescript, designed the book cover and title page, and is the publisher.

My son, Remi Robert Nadeau, provided much-needed technical support in getting the most out of computer, scanner and printer. He also provided useful suggestions in the text and illustrations. Without him the book could not have been completed.

Additional technical help was given by my son-in-law, Donald DeVitt.

David Myrick, of Montecito, noted Western historian, provided continuing support with research material, reading of the typescript, suggestions as to content, corrections of fact, and enthusiastic encouragement.

Essential help was given by many librarians, archivists and curators—people dedicated to helping others. They are identified here in their roles when they assisted me, although some have moved on to other situations during the four years that this book was in the making. These and others who contributed to this book are listed here:

Kathryn Bruno and Geoffrey Baykal-Rollins, in charge of Interlibrary Loan at the Santa Barbara Public Library.

Margot Collin and the staff at the reference desk and periodicals room of the Santa Barbara Public Library.

Sibylle Zemitis, Librarian, and the staff of the California History Section, California State Library, Sacramento.

Rebecca Jabbour, Interlibrary Loan, and the staff of the Bancroft Library, University of California, Berkeley.

Jennifer L. Martinez, Curator, Western Historical Manuscripts, Henry E. Huntington Library, San Marino

George A. Dey, historian, of Fraserburgh, Aberdeenshire, Scotland, who generously contributed time, correspondence and considerable material on Lewis Chalmers and his family in Fraserburgh.

Mrs. Lorraine Noble of the Fraserburgh Library.

Alan R. Fulton, Head of Library Services, Central Library, City of Aberdeen, Scotland..

George McGowan Brown, General Register Office for Scotland, Edinburgh.

Nancy Thornburg, Director of the Alpine County Historical Museum, Markleeville, who gave valuable assistance and suggestions, copied many historical items, and contributed historical research.

Frederick Thornburg, who provided personal knowledge on mines and mining in Alpine County, and who took Margaret and me in his pickup truck to Mogul on a road inadvisable, to say the least, for our automobile.

Kim Summerhill, who did much research on Lewis and Antoinette Chalmers in the Alpine County Museum and Alpine County Recorder's Office, and provided copies and interpretation of historical material.

All three of the above kindly reviewed the Alpine Chapter and made corrections and suggestions.

Richard Race, Yerington, Nevada, who provided copies of certain Lewis Chalmers letters.

Demila Jenner, Reno, the historian of Benton, California, who generously lent important parts of her notes gathered over many years on the history of Benton, and read the Benton chapter to catch any errors or misinterpretations.

Richard Fusick and Michael T. Meier, Archives I Reference Branch, Textual Reference Division, U.S. National Archives, Washington, D.C.

Claude Hopkins, U.S. National Archives - Pacific Sierra Region, San Bruno, California.

Susan Haas, Registrar, The Society of California Pioneers, San Francisco.

Patricia Keats and Beth Graham, California Historical Society, San Francisco

Bill Michael, Museum Administrator, Kathy Barnes, Beth Porter and the staff at the Eastern California Museum of Inyo County, Independence.

The staff of the Inyo County Library, Independence.

Beverly J. Harry, Inyo County Clerk-Recorder, and the staff of the Recorder's Office.

Alice Booth and Barbara Moss, Laws Railroad Museum, Bishop.

Jodie Stewart and Mike Patterson of Cerro Gordo, who provided generous hospitality and considerable historical information at the old ghost town.

Ken and Elaine Miller, Ridgecrest, who kindly allowed me to study their Cerro Gordo Freighting Company ledger book and who made extensive copies of pages.

My first cousin, Marilyn Nadeau, Glendale, California, who gave considerable information from her files on Remi Nadeau and his family, joined Margaret and me in further research at the Los Angeles County Archives, and kindly reviewed the chapter on Nadeau for necessary changes

Rose Byrne, Archivist, Arizona Historical Society - Southern Arizona Division, Tucson.

Arlene H. Reveal, County Librarian, Mono County Free Library, Bridgeport.

Phillip I. Earl, Curator of History, Nevada Historical Society, Reno.

Allison Cowgill, Librarian, Reference Room, Nevada State Library and Archives, Carson City.

Robert P. Palazzo, who discovered at the Huntington Library the 99-year leases made between the Earp brothers and Remi Nadeau, and sent me copies.

Mary Paquette and Carlo De Ferrari, historians, Sonora, California.

Lou Juana C. Souza, Tuolumne County Museum, Sonora

Patrick E. Purcell, Wayne, Pennsylvania, who provided information on Gen. George S. Dodge.

Terrence J. Gough, Chief, Staff Support Branch, The Center of Military History, Department of the Army, Washington, D.C.

William W. Sturm, Librarian, Oakland History Room, Oakland Public Library.

Chris Shovey, California History Room Librarian, San Bernardino Public Library.

James D. Hofer, Archivist/Records Manager, and the staff of the San Bernardino County Archives, San Bernardino.

Germaine Moon, Historian, Mojave River Valley Historical Museum, Barstow, California. Mrs. Moon provided considerable information as well as photographs on Calico and Ivanpah.

Robert J. Chandler, Historical Services, Wells Fargo Bank, San Francisco

Betty Ellison and the staff of the History and Genealogy Department, Los Angeles Public Library

Donald W. McNamee, Chief Librarian, Museum Research Library, Natural History Museum of Los Angeles County, Los Angeles.

Robert E. Stewart, Reno, Nev.

Special Collections, Davidson Library, U.C. Santa Barbara

Special Collections, Research Library, U.C.L.A.

For photos and other illustrations I wish to thank the following:
Alpine County Historical Museum, Markleeville
Eastern California Museum of Inyo County, Independence
California History Section, California State Library, Sacramento
Ellen Harding, Photo Section, California State Library

Bancroft Library, University of California, Berkeley
Colorado Historical Society, Denver
Henry E. Huntington Library
Los Angeles Corral, The Westerners
Laws Railroad Museum, Bishop
Mojave River Valley Museum Association and Germaine Moon
Thelma Prater Collection
San Bernardino County Parks and Recreation Association
Grace P. Crocker Collection and Mrs. Patricia Crocker Denton, Bridgeport
Evelyn Cawelti, Camarillo, CA
Patrick E. Purcell, Wayne, PA
Stanley Paher and Nevada Publications
Stephen Ginsburg
Demila Jenner, Reno
George A. Dey, Frazerburgh, Scotland

SOURCES

1. SILVER IS KING!
This chapter is adapted from the article, "Go It, Washoe!", by Remi Nadeau, *American Heritage*, April 1959.

2. BELSHAW'S MILLIONS
Darwin French and the Coso Mines
Primary Sources
Butte Democrat, Oroville, March 24 and 31, April 28, June 2 and 30, July 14 and 21, 1860
Visalia Delta, April 7 and 21, June 23, July 7, 14, 21, and 28, Aug. 4, 11 and 25, Sept. 1, 15 and 29, Dec. 15, 1860
Alta California, San Francisco, July 24 and 25, 1860; Dec. 30, 1863
Mining and Scientific Press, Nov. 30, 1860
Inyo Independent, Aug. 22, 1870
French, E.D. *The Power of Destiny Revealed in our War with Spain and the Philippines*. Los Angeles, 1899.
Secondary Sources
Report of the State Mineralogist. Sacramento, 1880, p.33
History of San Bernardino County (San Diego County). Wallace W. Elliott & Co., 1883, p. 200
Bancroft, Hubert Howe. *History of California*, Vol. 3, San Francisco, 1886, p. 749
Chalfant, W.A. *The Story of Inyo,*. Revised Edition. Bishop, 1933, p. 129
Palazzo, Robert P. *Darwin, California*. Lake Grove, Oregon, 1996, p. 7

Pablo Flores and the discovery of Cerro Gordo
The silver discovery at what became Cerro Gordo has rested largely on the account in Chalfant's *The Story of Inyo*, which is based on oral tradition among pioneers at Cerro Gordo and in Owens Valley, where Flores lived the rest of his life. This is reinforced in two near-contemporary accounts by visitors to Cerro Gordo, where they heard the Flores story. Both of them emphasize Flores' capture by the Paiutes and his escape:
Alta California, Nov. 30, 1868
Inyo Independent, Aug. 22, 1874

Victor Beaudry
Primary Sources
Los Angeles Star, March 30, 1861
Semi-Weekly Southern News, Los Angeles, July 12, Nov. 6, 1861
Los Angeles News, Oct. 2, 1866
Alta California, San Francisco, Nov. 30, 1872 (quotes letter from Victor Beaudry, Oct. 29, 1872, to friend in Los Angeles, as printed in *Los Angeles Star*).
Inyo Independent, July 10, 1875, May 13, 1876, Feb. 10, March 31, 1877, Jan. 4, 1879
Los Angeles Times, April 4,1882, May 18, 1887
Bancroft Scraps, California Mining, Vol. 1, Set W51:1, p. 192, 1868, unidentified source (Beaudry furnace)
Judgment Book A, p. 20, Nov. 6, 1867, Inyo County Recorder's Office, Judgment for Victor Beaudry against Joaquin Almada and José A. Ochoa; Execution Book A, p. 43, April 2, 1868
Judgment Book A, p. 29, May 6, 1868. Judgment for Victor Beaudry against Blas Mendez.
Judgment Book A, p. 65, May 9, 1870. Decree of foreclosure and sale. Victor Beaudry, plaintiff, v. Pablo Flores and five others, defendants.
Agreements Book A, p. 30, Jan. 24, 1868. Contract, Pierre Desormeaux and Victor Beaudry, Inyo County Recorder's Office. Desormeaux to build furnace; Agreement, same parties, p. 50, Oct. 10, 1868. Beaudry acquires all interest in furnace.
Gorman, Mrs. J.S. "The Story of Cerro Gordo", Clipping from *Inyo Independent,* ca 1930. In Eastern California Museum, Independence.
Newmark, Harris. *Sixty Years in Southern California, 1853-1913.* Third Edition, Boston and New York, 1930.

Mortimer W. Belshaw
Primary Sources
Weekly Antioch Ledger, April 19, 1897
San Francisco Chronicle, June 27, 1900, p.5
Deeds, Book C, p. 78, June 1, 1869, Inyo County Recorder's Office. A.B.Elder to M.W. Belshaw
Deeds, Book C, p. 631, Oct. 4, 1870. Joaquine Almada to M.W. Belshaw
Goodyear, W.A. "On the Situation and Altitude of Mount Whitney", *American Journal of Sciences & Arts*, 3rd Series, Vol. 6., July-Dec. 1873, p. 308

Secondary Sources
Guinn, J.M. *History of the State of California and Biographical Record of Coast Counties.* Chicago, 1904, p. 253
Hulaniski, F.J. *The History of Contra Costa County.* Berkeley, 1917, p. 433
Phelps, Alonzo. *Contemporary Biographies of California's Representative Men,* Vol. 1, San Francisco, 1881-2, p. 159
Thornburg, Nancy (niece of M..W..Belshaw). Collection of biographical and genealogical information on the Belshaw family.

Egbert Judson
Primary Sources
San Francisco Chronicle, July 10, 1892, Jan. 10, 1893
Alta California, San Francisco, July 10, 1892
San Francisco Call, Jan. 10 and 12, 1893
Norton, L.A. *Life and Adventures of.* Oakland, 1887, p. 336
Secondary Sources
Dictionary of American Biography, New York, 1928-36.
The National Cyclopædia of American Biography, New York, 1891

Cerro Gordo
Part of this section is drawn from two chapters in the author's *City-Makers,* New York, 1948, and Fourth Edition, Corona del Mar, 1965. Major sources for that work are among the larger body of additional sources, based on subsequent research, shown below.
Primary Sources
Inyo Independent, July 9, 1870 through Oct. 30, 1880.
Almost every issue of this weekly published in Independence contains an item about Cerro Gordo. Other newspapers with frequent stories about Cerro Gordo are:
Los Angeles News, 1868 through 1872
Los Angeles Star, 1868 through 1875
Los Angeles Express, 1871 through 1875
Kern County Courier, Bakersfield, 1871 through 1873
Visalia Delta, 1868 through 1872
Mining and Scientific Press, San Francisco, 1868 through 1883
Alta California, April 3, 1872
Gorman, Mrs. J.S. "The Story of Cerro Gordo", clipping from *Inyo Independent,* ca. 1930. In Eastern California Museum
McClelland, Dr. Hugh K. "A Lively Journey. Sample of Night Life in Cerro Gordo in 1872" In *Inyo Register,* Feb. 10, 1915
Leuba, Edmond. "Bandits, Borax and Bears", *California Historical Society Quarterly,* Vol. 17, June 1938, p. 99

Raymond, Rossiter W. *Statistics of Mines and Mining in the States and Territories West of the Rocky Mountains.* Washington, D.C., 1873 and 1875

Secondary Sources

Chalfant, W.A. "Cerro Gordo", *Southern California Historical Quarterly*, June 1940, p. 55

Chalfant, W.A. *The Story of Inyo,* Revised Edition, Bishop, 1933

Likes, Robert C., and Day, Glenn R. *From this Mountain —Cerro Gordo.* Bishop, 1975

Nadeau, Remi. *Ghost Towns & Mining Camps of California.* Los Angeles, 1954; Fourth Edition, Santa Barbara, 1992

Vredenburgh, Larry M. and Shumway, Gary L. *Desert Fever: An Overview of Mining in the California Desert.* Canoga Park, Calif., 1981

The San Felipe, Galen Fisher and "the Big Suit"

Primary Sources

Prospectus, Owens Lake Silver Lead Mining Co., San Francisco, 1869

Judgment Book A., Oct. 11, 1870 and May 13, 1871. E. A. Reddy *vs.* Gustav Wiss *et als.* and San Felipe Mining Co. Inyo County Recorder's Office

Judgment Book A, Feb.17, 1871. Victor Beaudry *vs.* E.A. de la Meis, Dr. Gustav Wiss and San Felipe Mining Co. Inyo County Recorder's Office

San Felipe Mining Co. *vs.* M.W. Belshaw *et als.* Records of Cases before the California Supreme Court, Vol. 289

Agreement Book C, p. 72, Jan. 13, 1876, Inyo County Recorder's Office. *Union Consolidated Mining Co. of Cerro Gordo*

Inyo Independent, June 13, 1874; Jan. 29, 1876; March 24, 1877; Feb. 23, March 6, Aug. 24, Nov. 23 and 30, 1878

Alta California, Nov. 25, 1889 (Galen Fisher obituary)

San Francisco Call, Dec. 17, 1889

Secondary Source

Nadeau, Remi. *City-Makers.* New York, 1948; Fourth Edition, Santa Barbara, 1992

3. THE RISE AND FALL OF LEWIS CHALMERS

The main sources for this Alpine County chapter are the letter books of Lewis Chalmers, 1867 to 1882 (with some gaps), located in the Alpine County Museum, Markleeville, and the files of the Alpine County newspapers, 1864 to 1879, in that museum and at the Bancroft Library, University of California at Berkeley, and the California History Room, State Li rary, Sacramento.

Alpine beginnings—Silver Mountain, Monitor and Mogul
Primary Sources

Sacramento Union, Aug. 4, 11 and 28, 1863; Jan. 14 and Fe .10, 1864
Alpine Chronicle, April 23, 1864 to Octo er 19, 1878.
Monitor Gazette, June 11, 1864 to June 10, 1865.
Alpine Miner, Jan. 12, 1867; Dec. 16, 1871; Jan. 20 and Nov. 2, 1872.
Alpine Argus, April 23, 1885
Virginia City Daily Union, April 14, 1864
Gold Hill Daily News, Octo er 20, 1864.
Calaveras Prospect, Sept. 23, 1933
Bancroft Scraps, California Mining, Vols. 1 and 2. *Calaveras Chronicle,* May 10 and 28, 1863; *Sacramento Union,* Aug. 24, 1863; *Alta California,* Sept. 30, 1863; *San Francisco Bulletin,* Octo er 6, 1863; untitled clipping, July 26, 1864; *Alpine Miner,* July 7, 1866.
Twenty Years on the Pacific Slope: Letters of Henry Eno from California and Nevada, 1848-1871. Ed. y W. Turrentine Jackson. New Haven and London, 1965.
Brewer, William H. *Up and Down California in 1860-64.* New Haven, 1930.
Raymond, Rossiter W. *Statistics of Mines and Mining in the United States and Territories West of the Rocky Mountains.* Washington, D.C., 1873 and 1875.

Secondary Sources

Jackson, W. Turrentine. *Report on the History of The Grover Hot Springs State Park Area and Surrounding Region of Alpine County.* California Division of Beaches and Parks, 1964. This is the most informative pu lication so far on the history of Alpine County.

Maule, William M. *A Contribution to the Geographic and Economic History of the Carson, Walker and Mono Basins in Nevada and California.* U.S. Forest Service, San Francisco, 1938. This is a thin ut very large volume well illustrated with maps and photos.

Alpine Heritage: One Hundred Years of History, Recreation, Lore in Alpine County, California, 1864 -1964. Centennial Book Committee, Markleeville. Revised edition, 1987. Highly illustrated with maps, sketches and photos.

Lewis Chalmers in Scotland:
Primary Sources
Aberdeen County Records: Old Parochial Register, Fraserburgh; Record of Marriages, 1712 to 1854; Valuation Toll, Fraserburgh,1859-60; Census Record, Fraserburgh, 1861.
Aberdeen Herald, July 6, 1850; Dec. 6 and 13, 1851; July 16, 1853 and July 29, 1863.
Aberdeen Journal, Oct. 9, 1850 and March 5, 1904.
Secondary Sources
Letters to author from George A. Dey, Fraserburgh, County of Aberdeen, Oct. 26 and Nov. 2, 1995.
Family Tree of James Kelman, great, great grandfather of Lewis Chalmers
Cranna, John: Harbour Treasurer. *Fraserburgh: Past and Present.* Aberdeen, 1914.
Brief biography of Lewis Chalmers, Alpine County Museum.

Chalmers in Alpine:
Primary Sources
Letter books of Lewis Chalmers, 1867 to 1882 (with gaps), Alpine County Museum, Markleeville.
Typewritten copies of Lewis Chalmers letters, Dec. 2, 1883 to Jan. 27, 1884, courtesy of Richard Race, Yerington, Nev.
Letters of Oscar F. Thornton, 1872 to 1874, Alpine County Museum, Markleeville.
Alpine Chronicle, Nov. 16, 1867 to July 27, 1878.
Alpine Miner, June 18, 1870 and Jan. 27, 1872.
Alpine Signal, Aug. 14, 1878 to June 27, 1879.
Bancroft Scraps, California Mining, v. 1, p. 189 (Chalmers letter, Jan. 15, 1869); v. 2, *Alpine Miner,* Jan. 18 and Oct. 24, 1868; *New York Tribune,* Jan. 12, 1869; *Mining and Scientific Press,* Sept. 2, 1882.

Frank A.S. Jones:
Primary Sources
F.A.S. Jones letters, Alpine County Museum, Dec. 20, 1863 to Feb. 3, 1869.
Capt. James Jones papers, 1863 to 1868, MS 1134, California Historical Society, San Francisco
Chalmers letters, Alpine County Museum, Feb. 26, 1868 to May 20, 1870.
Alpine Miner, Sept.14, 1867
Alpine Signal, Sept.11, 1878.
San Francisco Call, Dec. 21, 1895, p. 3, col. 6

Report of First Annual General Meeting of Shareholders, Imperial Silver Quarries Co., Ltd., Palmerston Bldg., London, April 20, 1868 (Bancroft Library).
Prospectus of Imperial Silver Quarries Co., Ltd., 1868, and Abridged Prospectus, 1869 (Bancroft Library).

Secondary Source

Memorial and Biographical History of Northern California. San Francisco, 1891, p. 607 (Biog. of F.A.S. Jones).

Chalmers and Antoinette Laughton:

Primary Sources

Typewritten copies of Chalmers letters, June 18 to Oct. 22, 1877 and Dec. 2, 1883 to Jan. 27, 1884, provided by Richard Race, Yerington, Nev
Letters of Lewis Chalmers, March 19 to April 24, 1878, Alpine County Museum, Markleeville.
Letter, Lewis Chalmers to Dominic Bari, Aug. 28, 1891, Alpine County Museum.
Monitor Argus, Dec. 13, 1880 and Sept. 26, 1881.
Alpine Argus, June 6, 1885
Oakland Tribune, Dec. 29, 1913
San Francisco Chronicle, Dec. 30, 1913
Gardnerville Record-Courier, May 20, 1904; Jan. 2 and 16, 1914
Coyan, Elizabeth Pearl Ellis. *An Oral History.* Historical Society of Alpine County, 1987
Hawkins, Harly. *Douglas-Alpine History* (a reminiscence), n.d.
County Archives: Assessor's Records—Assessment Roll, 1887, p. 29. Recorder's Office—Indentures and sheriff's sales, Deed, Lewis Chalmers to Dominic Bari, Aug. 31, 1885.
Census of the United States, County of Alpine: Tenth Census, 1880; Twelfth Census, 1900; Thirteenth Census, 1910.

4. THE TOWN WITH NO LAW

The principal sources for the stormy saga of Benton, Mono County, are the newspaper files of that and other towns, 1865 to 1885, and the records of two noteworthy cases:

Townsite of Benton vs. Eugene C. Kelty, Docket File 180, U.S. Land Office, Independence, California, 1877. In the National Archives, Washington, D.C.

Robert Ellon vs. William M. Lent et als, Ninth U.S. Circuit Court, File No. 2153, San Francisco, 1879. In the National Archives – Pacific Sierra Region, San Bruno, California.

Albert Mack and the Growth of Benton
Primary Sources

Benton vs. Kelty (cited above). Contains testimony by pioneers concerning the early settlement of Benton.

Bancroft Scraps, California Mining, vol. 2 (*Mining & Scientific Press*, Dec. 17, 1865; July 14, 1866); pp. 279 and 280, editorial correspondence, May 5, 26, 1866, no source.

Sacramento Union, May 23,1866; June 24, 1873

Inyo Independent, April 8 and 15, Nov. 7, 1874; April 22 and July 17, 1875; April 8 and 15, May 27, Aug. 5, Dec. 16, 1876; July 28, 1877; Jan. 19 and Sept. 21, 1878; April 12, October 4 and 18, 1879; Nov. 13, 1880; March 5, 1881; March 11, 1882.

Weekly, Semi-Weekly and *Tri-Weekly Bentonian*, Aug. 16, Oct. 18, 1979; Jan. 22, Feb. 14 and April 26, 1880; Jan. 19, 1881

Mono Weekly Messenger, Feb. 8 and 22, March 1, 1879

Mammoth City Herald, Oct. 22, 1879; March 20, July 3, 1880

Mammoth City Times, Oct. 22, 1879

Weekly Bodie Standard, June 19, 1878

Daily Bodie Standard, Aug. 15,1879 (source for naming Benton after Rev. J.E. Benton)

Mining & Scientific Press, San Francisco, Oct. 25, 1884; June 13 and July 11, 1885

Daily Stock Report, San Francisco, Dec. 22, 1879

Daggett Scrapbook, Vol. 2, p. 110. California History Room, State Library, Sacramento

Mack, Albert Christian. *Memoirs*. San Francisco, 1900. MS in Society of California Pioneers, San Francisco

Jenner, Demila. Notes on the History of Benton (principally from Mono and Inyo County newspapers, 1870-99) Kindly lent to author by Ms. Jenner

Meadows, Lorena Edwards. *A Sagebrush Heritage*. San Jose, 1973

Third Annual Report of the State Mineralogist. Sacramento, 1883

Wasson, Joseph. *Complete Guide to the Mono County Mines*. San Francisco, 1879

Raymond, Rossiter W. *Statistics of Mines and Mining in the United States and Territories West of the Rocky Mountains*. Wash., D.C., 1875

Secondary Sources
History of Mission Lodge No. 169. Free and Accepted Order of Masons, State of California, Diamond Jubilee, 1863-1938, p. 66, "Biography of Albert C. Mack".
Chalfant, W.A. *Outposts of Civilization.* Boston, 1928
Myrick, David F. *Railroads of Nevada and Eastern California.* Berkeley, 1963, vol. 1
Datin, Richard C., Jr. *Story of Benton's Mines.* MS transcribed from *Owens Valley Herald,* June 18, 1909

Attempt of E.C. Kelty to Preempt the Town of Benton
Primary Sources
Benton vs. Kelty (cited above). Contains 351 handwritten, legal-size pages of testimony, depositions, briefs, correspondence and judgment in hearings before the Register and Recorder of the U.S. Land Office, Independence and Benton, 1877
Inyo Independent, June 30 and July 28, 1877
Mono Weekly Messenger, April 19, 1879
Mammoth City Herald, Dec. 3, 1879
Weekly Bentonian, Apil 19, 1880
Jenner, Demila. Notes on the History of Benton (cited above)

Bill Lent and the Comanche:
Primary Sources
Ellon vs. Lent et als (cited above). 173 handwritten, legal-size pages of subpoenas, complaints, answers, replications and decree, 1879
Record of suits filed against Comanche Mill & Mining Company by J.B. Badger, Dec. 23, 1878, and twenty others, Jan. 24, 1879, in Superior Court, Mono County, Bridgeport, California
Mining Records, Blind Springs Hill Mining District, vol. A, 1865-1875. First meeting of miners, March 23, 1865. Recorded claims of Diana, Comanche and extensions. Mono County Recorder's Office, Bridgeport, California
Inyo Independent, April 22 and Aug.19, 1876; Feb. 17 and June 9, 1877; March 23, Dec. 14 and 21, 1878; Jan. 4 and 18, Feb. 1 and 15; April 19 and Nov. 15, 1879; Feb. 21, 1881
Mono Weekly Messenger, Feb. 1 and 15, March 1 and 8, 1879
Semi-Weekly Bentonian, Jan. 22, 1880
Mammoth City Herald, Sept. 13 and Nov. 5, 1879; May 19 and June 23, 1880
Goodwin, C.C. *As I Remember Them.* Salt Lake City, 1913 (includes reminiscences about William M. Lent)

Pacific Coast Annual Mining Review and Stock Ledger. San Francisco, 1878 (includes personal sketch of William M. Lent).

Secondary Sources

San Francisco Chronicle, October 18, 1904 (obituary of William M. Lent)
Alta California, Feb. 28, 1891 (obituary of William H. Sears)
The Bay of San Francisco, vol. 1, Chicago, 1892, p. 442 (biography of William H. Sears)

5. MINING THE STOCKHOLDERS

How mines were bought and sold, and how capital was raised to develop mines, were subjects of frequent newspaper comment during the silver era, especially if the editor "viewed with alarm".

Primary Sources

Gold Hill News, Oct. 29, 1863
Monitor Gazette, Alpine County, May 13, Nov. 4, Dec. 23, 1865; Jan. 6 and 13, 1865
Coso Mining News, Darwin, Nov. 6, 1875
Inyo Independent, Jan. 4 and 18, 1879; Feb. 21 and Dec. 25, 1880
Mammoth City Herald, July 10, Aug. 28, 1880

Two excellent contemporary descriptions of the San Francisco mining stock exchanges are in:

Pacific Coast Mining Review and Stock Ledger, San Francisco, 1878
San Francisco Daily Stock Report, Dec. 22, 1879
King, Joseph L. *History of the San Francisco Stock and Exchange Board,* San Francisco, 1910, and New York, 1975. Mr. King was associated with the Board from its beginnings, and became chairman.

Secondary Sources

Bancroft, Hubert Howe. *History of California,* Vol. 7, pp. 667-81. *Chronicles of the Builders,* vol..4, p. 17
Sears, Marian V. *Mining Stock Exchanges, 1860-1936.* Missoula, Montana, 1973

6. THE JINGLE OF NADEAU'S TEAMS

Much of this chapter is drawn from two previous writings by the author: 1. *City-Makers,* Doubleday, New York, 1948, and Trans-Anglo Books, Costa Mesa, Calif., 1965. 2. "King of the Desert Freighters," in *The Westerners Brand Book,* No. 11, 1964, Los Angeles Corral of the Westerners. Principal sources for these, as well as other sources in subsequent research, are:

Remi Nadeau and his freighting business
Primary Sources

Inyo Independent, July 9, 1870; Jan. 21, Nov. 11, 1871; July 20, Nov. 30, Dec. 7, 1872; April 19, May 10, June 14, 1873; Aug. 22, Oct. 31, 1874; Jan. 30, 1875

Los Angeles Star, Sept. 3 and 30, Nov. 24, 1870; July 21, Aug. 4, 1871; Nov. 20, Dec. 6, 1872; July 13, 1873

Los Angeles News, Sept. 9, 1870; June 9, 1871.

Los Angeles Express, Oct. 9, 1871; Nov. 4 and 19, 1872

Kern County Courier, June 8, July 13, 1872.

San Bernardino Weekly Times, Nov. 20, 1880.

Los Angeles Times, Jan.16, 18 and 20, 1887.

Los Angeles Herald, Jan. 16, 1887.

San Francisco Call, Jan. 20, 1887.

Alta California, Jan. 16, 1887

Indenture, July 25, 1868, Prudent Beaudry to Remi Nadeau, town property at corner, Fifth and Olive Sts., Los Angeles

Death Certificate, Remi Nadeau, Los Angeles. Date of death, Jan. 15, 1887. Certificate dated Jan. 17.

Newmark, Harris. *Sixty Years in Southern California, 1853-1913*, Third Edition, Boston and New York, 1930

Spalding, William Andrews. *Los Angeles Newspaperman*. Ed. by Robert V. Hine, San Marino, 1961.

Spalding, William Andrews. *History and Reminiscences of Los Angeles City and County*, Los Angeles, 1931

Burton, Mary June. "What Became of the 20-Mule Freighters?" in *Los Angeles Times Sunday Magazine*, Oct. 13, 1935, pp. 16-17. Includes reminiscences of Melvina Lapointe Lott, niece of Remi Nadeau and wife of his wagon boss, Austin Lott.

Nadeau, George A. (son of Remi Nadeau) *Dictation*, San Francisco, March 21, 1888. H.H. Bancroft Collection No. 55605.

Gorman, Mrs. J.S., Moorpark. Interview by author, April 1946

Richardson, Walter L., Porterville, Calif. Letter to author, Jan. 23, 1949

Delameter, John. Interview by author, 1941.

Remi Nadeau (grandson of Remi Nadeau the freighter). Interview by author's mother, Marguerite M. Nadeau, Sept. 1941.

Bell, Mary Rose (daughter of Remi Nadeau). Reminiscences given orally to family members, 1920s.

Mitchell, Frank, Soledad Canyon, Calif. Interview by Marguerite M. Nadeau, 1941, and by author, April 6, 1946

Chrisman, Marilyn Nadeau. Index of property transfers, Remi Nadeau and others in Nadeau family, 1868-1884. Compiled from Los Angeles County records, Recorder's Office

Secondary Sources

Chalfant, W.A. *The Story of Inyo*, revised edition. Bishop, Calif., 1933.

Wilson, John Albert. *History of Los Angeles County, California*. Thompson & West, Oakland, 1880.

Guinn, James M. *A History of California and an Extended History of its Southern Coast Counties*, two volumes. Los Angeles, 1907.

Guinn, James M. *A History of California and an Extended History of Los Angeles and Environs*, two volumes. Los Angeles, 1915.

Hill, Laurance L. *La Reina: Los Angeles in Three Centuries*. Security Trust & Savings Bank, Los Angeles, 1929

Loyer, Fernand and Beaudreau, Charles. *Le Guide Français de Los Angeles et du Sud de la Californie*, English Edition. Los Angeles, 1932.

Adventures in Business, Dec.21, 1945. Published weekly by Knott's Berry Place, Buena Park, Calif.

Nadeau, Marguerite M. Scrapbook of clippings, photos and memorabilia, compiled 1929-1941.

Nadeau, Remi. *Ghost Towns & Mining Camps of California*. Los Angeles, 1954. Fourth Edition, Santa Barbara, 1992

Salt Lake Trade
Primary Sources

Los Angeles Star, April 6 and 20, Nov. 1, 1861; Jan. 4 and 25, 1862; Jan. 3, Dec. 12, 1863; Fe b. 13, 1864.

Los Angeles News, Nov. 1, 1861; Feb. 28, Dec. 5, 1862; Jan. 23 and 30, Feb. 25 and 27, March 25 and 27, Dec. 16 and 23, 1863; Feb. 14, 1865; Jan. 9 and 23, March 2, 9 and 13, June 19, July 17, 1866; Jan. 29, Feb. 5 and 15, April 26, 1867.

Deseret News, Salt Lake City, Mar. 12, April 23 and 30, Sept. 3, 1862.

Newmark, Harris. *Sixty Years in Southern California, 1853-1913*, Third Edition. Boston and New York, 1930.

Cerro Gordo Freighting Company
Primary Sources

Inyo Independent, June 21, 1873; Mar. 7, May 23, June 6, Dec. 19, 1874; May 1 and 22, 1875; Oct. 28, 1876; April 21, May 12, 1877; Nov. 2, 1878; Dec. 31, 1881; Jan. 7, 1882.

Los Angeles Star, June 7 and 8, 1873; July 14 and 15, 1873; Jan. 6, 1875.

Los Angeles Herald, July 17, 1874

Los Angeles Express, June 5, 1873; Oct. 15, 1874; Jan. 10, 1876, Dec. 29, 1877
Sacramento Union, Oct. 4, 1873
Kern County Gazette, Dec. 25, 1875
Kern County Courier, June 21, 1873
Panamint News, Mar. 18, 1875
Accounts Ledger, Cerro Gordo Freighting Company, Dec. 1878 to May 1882. Kindly provided by Ken and Elaine Miller, Ridgecrest, Calif.
Morris, Oscar. *Recollections of my trip to California in 1875.* Typescript, San Bernardino Public Library
Leuba, Edmond. "Bandits, Borax and Bears" and "A Frenchman in the Panamints", *Calif. Hist. Society Quarterly*, June and Sept. 1938
Jenkins, W.J., Oakland. Interview by author, Feb. 1946

Nadeau in Arizona
Primary Sources
Inyo Independent, June 7, 1879, May 21, July 2, 1881
San Bernardino Weekly Times, Dec. 25, 1880
Bishop Creek Times, April 15, 1882
Tombstone Epitaph, Oct. 28, 1881
Ninety-nine year lease, Nov. 12, 1880, Wyatt S. Earp, James C. Earp, Virgil W. Earp and R.J. Winders to R. Nadeau. First North Extension Mountain Maid Mine, Tombstone District, Arizona Territory (surface lease, no mineral rights).
Ninety-nine year lease, April 21, 1881, same parties, town property on Safford St., Tombstone, A.T.
Accounts Ledger, Cerro Gordo Freighting Company, Dec. 1878 to June 1882

Nadeau Hotel
Primary Sources
Los Angeles Times, June 10, 1882; July 24, 1883; Mar. 4, 1886; July 6, 1886; Jan. 10, 1932, part 5.
Los Angeles Herald, May 9, July 6, 1886; July 2, 1929; Oct. 6, 1931
Willard, Charles Dwight. Letter to mother, Los Angeles, Oct. 31, 1886. Henry E. Huntington Library, San Marino
Advertising sheet for auction sale of furniture and fixtures, Nadeau Hotel, Oct. 28-29, 1931

Divorce, remarriage and the contested Nadeau estate
Primary Sources
Remi Nadeau, plaintiff, vs. Martha F. Nadeau, defendant, 7th Judicial District Court of California, Los Angeles County, June 30, 1879
Marriage certificate, Remi Nadeau and Laura M. Jones, Los Angeles, Oct. 11, 1879. J.F. Powell, Justice of the Peace
Death certificate, Martha F. Nadeau, Los Angeles, Jan. 18, 1904
Los Angeles Times, Jan. 17, 1886; April 14, Sept. 18, 1887; Jan. 3 and 26, 1890; July 16, 1894, p. 8; Dec. 10, 1909, p. 16 (obituary of Laura M. Nadeau)
Tombstone Epitaph, Nov. 5, 1881
Advertising sheet for auction sale of portion of Nadeau Vineyard Tract, Dec. 3, 1887
Conveyance of Nadeau Hotel property, R. Nadeau to Laura M. Nadeau, and waiver of Laura M. Nadeau to any other part of Remi Nadeau estate, Los Angeles, Dec. 30, 1886.
Last Will of Remi Nadeau, Jan.1, 1887
In the Matter of the Estate of Remi Nadeau, deceased, in the Superior Court of Los Angeles, (file no. 5733), beginning Jan. 20, 1887. Final report and accounting, Nov. 16, 1892.
Joseph F. Nadeau, George A. Nadeau, Mary R. Bell and Julia F. Tilton, plaintiffs, vs. Laura M. Nadeau, defendant Superior Court, Los Angeles County, March 2, 1889 (file no. 5877). Dismissal, June 26, 1894.
Appraisement of the Estate of Laura M. Nadeau, Incompetent. Richard Crews, guardian of the estate, July 9, 1909.

7. THE FIGHTING MCFARLANES
Johnny Moss and the birth of Ivanpah
Primary Sources
San Bernardino Guardian, June 18, Aug. 20, Sept. 10 and 17, Oct. 1 and 8, Nov. 5, 1870; Feb. 11, Sept. 30, 1871
Alta California, San Francisco, July 15, 1870; Sept. 19, 1871
Prospectus, Piute Mining Company, Inc., San Francisco, June 30, 1870
Secondary Source
Malach, Roman. *Adventurer John Moss.* Kingman, AZ, 1977

McFarlane brothers and the growth of Ivanpah
Primary Sources
San Bernardino Guardian, April 29, Aug. 5, 1870; Sept. 30, Dec. 9, 1871; April 13, 1872; April 26, May 31, Oct. 4, 1873; Feb. 21, 1879

Bancroft Scraps, California Mining, vol. 3. *San Diego Union*, Oct. 16, 1873

Inyo Independent, Oct. 21, Nov. 25, 1876 (Indian troubles)

Mining & Scientific Press, Feb. 2, Dec. 6, 1884

Calico Print, May 31, 1885

Dellenbaugh, F.S. "Record of a Sketching Tour to Northern Arizona & Southern Utah, 1875-76." MS 215, F.S. Dellenbaugh Collection, Box 4, Folder 17. Typescript in Southern Arizona Division, Arizona Historical Society, Tucson

Vale, William Adams. "Log of a Trip to Ivanpah and Resting Springs", March 31 to April 19, 1880. Typescript, San Bernardino Public Library

Raymond, Rossiter W. *Statistics of Mines and Mining in the United States and Territories West of the Rocky Mountains.* Washington, D.C., 1875

Secondary Sources

Belden, L. Burr. "Ivanpah Silver Sent to Wales But Pays Profit". In *San Bernardino Sun*, April 6, 1952

Holladay, Fred. "The Silver King". In *Odyssey*, pub. monthly by the City of San Bernardino Historical Society, March 1979

Holladay, Fred. "A Sterling Character Named McFarlane". In *San Bernardino Sun*, March 30, 1986

Illustrated History of Southern California. Lewis Publishing Co., Chicago, 1892, p. 504

Guinn, James M. A *History of California and an Extended History of Its Southern Coast Counties.* Los Angeles, 1907. vol. 2, p. 1531

Hensher, Alan. "Ivanpah—Pioneer Mojave Desert Town". *Heritage Tales*, City of San Bernardino Historical and Pioneer Society, Annual 7, 1984, p. 36

Myrick, David F. *Railroads of Nevada and Eastern California.* Berkeley, 1963, vol. 2

Ivanpah Consolidated and John McFarlane
Primary Sources

San Bernardino Weekly Times, March 20, May 8, 1880

Alta California, San Francisco, May 23, 1881, p. 4

Los Angeles Herald, May 24, pp. 1 and 2, May 25, pp. 2 and 7, 1881

San Bernardino Valley Index, May 27, 1881

W.S. McFarlane *et als*, plaintiffs, vs. Ivanpah Consolidated M & M Co., defendant, Superior Court, County of San Bernardino, Nov. 14, 1881. (File No. 253). Trial, Jan. 25 and Feb. 10, 1882; Judgment, Feb. 23, 1882

Secondary Sources
Lingenfelter, Richard. *The Hardrock Miners: A History of the Labor Movement in the American West, 1863-1893.* Berkeley, 1974

Vredenburgh, Larry M. and Shumway, Gary L. *Desert Fever: An Overview of Mining in the California Desert.* Canoga Park, CA, 1981

History of San Bernardino County (San Diego County). Wallace W. Elliott & Co., 1883

Resting Springs
Primary Sources
Inyo Independent, June 23, Nov. 11, 17 and 24, 1877; Nov. 2, 1878; July 5, Aug. 9, 1879; June 12, Oct. 30, 1880

San Bernardino Weekly Times, Aug. 24, 1878; Feb. 22, 1879; May 8, 1880

Secondary Source
Lingenfelter, Richard E. *Death Valley & the Armagosa: A Land of Illusion.* Berkeley, 1986

Providence
Primary Sources
Los Angeles News, June 1, 1863

Inyo Independent, April 24, 1880

San Bernardino Weekly Times, May 8, July 3, 1880; June 3 and 17, July 1, 1882; Jan. 20, Feb. 24, 1883; Jan. 16, 1886

Calico Print, Feb. 5, March 15, 22 and 29, April 19, May 24 and 31, June 7, July 19, 1885

Mining & Scientific Press, June 13 and 20, Sept. 26, 1885

Andrew McFarlane, Providence, to S.W. Holladay, San Francisco, March 19 and July 1, 1881. In Samuel Wood Papers, Huntington Library

Secondary Sources
Hensher, Alan. "Providence". *Heritage Tales,* City of San Bernardino Historical and Pioneer Society, Annual 7, 1984, p. 69

Vredenburgh, Larry M. and Shumway, Gary L. *Desert Fever: An Overview of Mining in the California Desert.* Canoga Park, CA, 1981

Myrick, David F. *Railroads of Nevada and Eastern California.* Berkeley, 1963, vol. 2

8. THE SILVER SENATORS

Richard C. Jacobs and the discovery at Panamint

Primary Sources

Records of Panamint Mining District, Book A, p. 4., Inyo County Recorder's Office. Feb. 1, 1873, first four claim notices. Feb. 10, Panamint Mining District organized. April 24, 1873, Stewart's Wonder and Wonder of the World mines located.

Inyo Independent, Dec. 28, 1872, July 12, Aug. 30, 1873

Alta California, San Francisco, April 3, 1873

Kern County Courier, Bakersfield, Nov. 25, Dec. 13, 1873

Los Angeles Herald, March 24 and 26, 1874

Page, Milo. "Old Panamint History," in *Inyo Register,* July 19, 1906

Sen. John P. Jones

Primary Sources

Goodwin, C.C. *As I Remember Them.* Salt Lake City, 1913

Panamint News, March 2, 1875

Alta California, June 4, 1875

Los Angeles Times, Sept. 15, 1894

Pacific Coast Mining Review and Stock Ledger, San Francisco, 1878

Secondary Sources

Angel, Myron. *History of Nevada.* Thompson & West, Oakland, 1881

Guinn, James M. *History of California and an Extended History of Los Angeles and Environs,* Vol. 2, 1915, p. 388

Nadeau, Remi. "History of Santa Monica", in *Santa Monica Evening Outlook,* 75th Anniversary Edition, 1950

Phelps, Alonzo. *Contemporary Biographies of California's Representative Men,* vol. 1, San Francisco, 1881-2, p. 223

Sen. William M. Stewart

Primary Sources

Stewart, William M. *Reminiscences.* New York, 1908

Goodwin, C.C. *As I Remember Them.* Salt Lake City, 1913

Secondary Sources

Mammoth City Herald, Oct. 11, 1879

San Francisco Chronicle, April 24, 1909 (obituary)

Phelps, Alonzo. *Contemporary Biographies of California's Representative Men,* Vol. 1, San Francisco, 1881-2, p. 165

Shuck, Oscar T., ed. *History of the Bench and Bar of California,* Los Angeles, 1901, p. 505

Elliott, Russell R. *Servant of ower: A olitical Biography of Senator William M. Stewart.* Reno, 1983

Lyman, George D. *The Saga of the Comstock Lode.* New York, 1934

Trenor W. Park
Primary Sources
Alta California, Nov. 6, 1863, p. 2
San Francisco Call, Dec. 21, 1882, p. 3 (obituary)
Sacramento Bee, Dec. 21, 1882, p. 2
Secondary Source
Bancroft, Hubert Howe. *History of California,* Vol. 7. San Francisco, 1890, p. 666

The Panamint Boom
Primary Sources
Inyo Independent. Most issues from July 4, 1874 to May 8, 1875, contain stories about Panamint. In addition: July 10, Oct. 30, Dec. 11, 1875; July 13, 1878; Sept. 13, Oct. 1, Dec. 6 and 13, 1879; Oct. 9, 1880; May 7, Sept. 17, 1881; January 21, 1882

anamint News. Every surviving issue between Nov. 26, 1874 and May 18, 1875 provides information on Panamint. In addition: Oct. 21, 1875

Alta California. Jan. 25, Feb. 1 and 18, May 21 and 25, 1875

Bishop Weekly Times, May 20, 1882

Eureka Daily Sentinel, Eureka, NV, Jan. 12, March 18, 1875

Los Angeles Express, Nov. 9, 1875

San Bernardino Guardian, Feb. 6 and 20, 1875

Bancroft Scraps, California Mining, vol. 3, Set W 5113. *San Diego Union,* Nov. 12, 1874; *Sacramento Union,* March 20, 1875; *San Francisco Bulletin,* Jan. 5, March 13, 1876

Leuba, Edmond. "A Frenchman in the Panamints", *California Historical Society Quarterly,* vol. 17, Sept. 1938

Secondary Sources
Vredenburgh, Larry M. and Shumway, Gary L. *Desert Fever: An Overview of Mining in the California Desert.* Canoga Park, Calif., 1981, p. 233

Lingenfelter, Richard E. *Death Valley and the Armagosa: A Land of Illusion.* Berkeley, 1986

Nadeau, Remi. *City-Makers.* New York, 1948; Second Edition, Corona del Mar, 1965

Nadeau, Remi. *Ghost Towns & Mining Camps of California.* Los Angeles, 1954; Fourth Edition, Santa Barbara, 1992

Wilson, Neill C. *Silver Stampede.* New York, 1937
Belden, L. Burr. *Mines of Death Valley.* Glendale, 1966

9. PAT REDDY FOR THE DEFENSE
Discovery at Darwin
Primary Sources
Records of the Coso Mining District, Book A, p. 13, Dec. 3, 1874, Formation of the district. Mines recorded, starting Dec. 4, p. 18.
Deeds, Inyo County, Book F, pages 434 to 445, Dec. 16 to 19, 1874. Sales of Defiance and other lodes.
Inyo Independent, Nov. 21, Dec. 5, 12, 19 and 26, 1874; Jan. 2, 23 and 30, 1875
Panamint News, Nov. 28, 1874
Kern County Courier, Bakersfield, Dec. 26, 1874, Feb. 6, 1875
Los Angeles Express, Dec. 18, 1874; Jan. 28, 1875
Los Angeles Star, Jan. 8, 1875

Darwin and the Defiance Mine
Primary Sources
Inyo Independent, Feb. 6, 13, 20 and 27, March 6 and 13, April 3, 10 and 17, May 1, July 10 and 17, Aug. 14 and 21, Sept. 4 and 18, Oct. 9, Nov. 6 and 27, Dec. 25, 1875; Jan. 22 and 29, March 11 and 18, May 6 and 20, June 3, 10 and 17, July 1, 8 and 29, Aug. 19, Nov. 11, 1876; Jan. 6, 13 and 20, Feb. 10 and 17, March 3, April 7 and 21, Sept. 22, Nov. 3, 1877; April 6, May 18, June 29, Oct. 12, 1878; May 3 and 10, 1879; Feb. 14, May 22, June 5, July 17 and 24, Aug. 7 and 28, Sept. 17, Oct. 9, 1880; Oct. 1, Nov. 26, Dec. 3, 1881; March 31, 1883.
San Bernardino Argus, Dec. 21 and 28, 1874
Panamint News, March 2, 9 and 18, 1875
Coso Mining News, Darwin, Nov. 6 and 13, 1875, March 24, June 16, Aug. 4, 1877
Kern County Courier, Bakersfield, June 12, 1875; Jan. 8 and 15, July 13, Oct. 5, 1876
Kern County Ga ette, Bakersfield, Nov. 27, Dec. 11, 1875; Feb. 12, June 17, 1876
Southern Californian, Bakersfield, Oct. 14, Dec. 9, 1875; Jan. 6, Feb. 3, 1876
Los Angeles Express, April 16 and 19, 1875; Aug. 23, 1876
Los Angeles Republican, April 29, 1876
Santa Monica Outlook, Dec. 22, 1875

Mammoth City Herald, July 12, Oct. 23, 1880

Bishop Creek Times, April 22, 1882

Mining & Scientific Press, San Francisco, Aug. 9 and 23, Oct. 4, Nov. 15, 1884

Myrick, David. Notes on New Coso Mining District from the *Mining & Scientific Press,* Dec. 5, 1874 to Dec. 29, 1877

Mecham, Elizabeth. "Darwin, Inyo County, California—no other place in the world like it". Thirty-six page handwritten manuscript, July 12, 1964. Recollections of Mrs. Mecham, who lived at Darwin. In Eastern California Museum

Wells, Fargo & Co. *Robber's Record.* San Francisco, 1885 (includes reports of detectives James B. Hume and John Thacker, 1870-1884)

Secondary Sources

Palazzo, Robert P. *Darwin, California.* Lake Grove, Oregon, 1996

Palazzo, Robert P. "The Fighting Reddy Brothers of the Eastern Sierra." In *The Album,* Bishop, 1996

Pat Reddy—early life

Primary Sources

Certificate of Incorporation, Garryowen Gold and Silver Mining Company, Esmeralda Mining District, Nevada Territory, Sept. 16, 1863, recorded October 5, 1863

Gold Hill News, Oct. 28, 1863. Shooting of Reddy (name given as McReady) by John Mannix in Virginia City, and amputation of right arm. Oct. 31, 1863: Explanation of the George Lloyd shooting in Aurora, which was the cause of the Reddy-Mannix difficulty. Nov. 30, 1863: Mannix charged and released on bond in attempt to kill Reddy (spelling given as Ruddy).

District Court Records, Virginia City, March term, March 16, 1864, p. 190. Jno. Mannix and Thos. McAlpin appear before Grand Jury on Assault with intent to commit murder. April 22, 1864, p. 241. People vs. John Mannix and McAlpin. Arraignment in District Court under Grand Jury indictment for assault with intent to commit murder. Plea of not guilty. December Term, Dec. 4, 1864. Mannix and McAlpin allowed two days to plead.

Virginia City Union, April 23, 1864: Arraignment of John Mannix "for assaulting Patrick Reddy with a pistol".

Index to Deeds, 1861 to 1878, Recorder's Office, Mono County Courthouse, Bridgeport, Calif. Book C, Oct. 10, 1864, P. Reddy buys town lot in Montgomery, Mono County.

Index to Mining Deeds, 1863 to 1878. Recorder's Office, Mono County Courthouse, Bridgeport, Calif. Books K and L. From Oct. 31, 1864, to Aug. 8, 1866. P. Reddy is grantor or grantee in nine mine sales. Book K, Feb. 7 and March 10, 1865. P. Reddy is elected Mining Recorder, Montgomery Mining District.

Wasson, Joseph. "Pioneer Days", Reminiscence in *Bodie Weekly Standard News*, April 20, 1881. Quotes from the *Montgomery Pioneer*, Nov. 26, 1867: "Pat Reddy, of the Pioneer Saloon, is the man to deal out spiritual comfort during the cold weather. Give him a call."

Secondary Sources

Inyo Independent, July 20, 1878. Extract from the obituary of Michael Reddy, father of Pat and Ned Reddy, in the *Woonsocket Patriot*, July 4, 1878, which states that Michael emigrated to Boston in October 1823.

San Francisco Examiner, June 27, 1900, p. 6 (obituary of Pat Reddy)

San Francisco Bulletin, Nov. 14, 1896, p. 16, (biography of Will Hicks Graham); June 26, 1900, p. 12 (obituary of Pat Reddy)

Pioneer Card File, California History Room, State Library, Sacramento. Biography of Will Hicks Graham

Earl, Phillip I. "Pat Reddy: Defender of the Downtrodden." *Proceedings of the Fourth Annual Death Valley History Conference*, 1995

Pat Reddy in Inyo
Primary Sources

Inyo Independent, March 13, 20 and 27, April 10, 17, Oct. 16 and 30, Nov. 13 and 20, Dec. 11, 1875; Jan. 1, 1876, May 26, June 2 and 30, Nov. 10, 1877; May 4, 1878

Panamint News, March 13, Oct. 21, 1875

Coso Mining News, Nov. 6, 1875

Los Angeles Express, March 16, 1875

Jenner, Demila. Eighteen pages of research notes on Pat Reddy, mostly from primary sources.

Secondary Sources

Chalfant, W.A. *Outposts of Civilization*. Boston, 1928.

Wilson, Neill C. *Silver Stampede*. New York, 1937

Shuck, Oscar T., ed. *History of the Bench and Bar of California*. Los Angeles, 1901

Ned Reddy
Primary Sources

Index to Deeds, Mono County Courthouse, Bridgeport, Book C. E.A. Reddy sells property on Oct. 20 and on Dec. 4, 1865

Index to Mining Deeds, Mono County Courthouse, Bridgeport, Books K and L, May 12, 1863 to Oct. 28, 1866. E. A. Reddy buys or sells six mining properties.
Sonora Union Democrat, Sept. 7, 1867
Sonora Herald, Sept. 14, 1867
Inyo Independent, Jan. 2 and 14, 1871; Oct. 18 and 25, Nov. 8, 1873; Feb. 20, 1875

Secondary Sources
California: Fifty Years of Progress. San Francisco, 1900, p. 297 (biography)
San Francisco Call, April 11, 1901, p. 7 (obituary)
Palazzo, Robert P. "The Fighting Reddy Brothers of the Eastern Sierra." In *The Album,* Bishop, 1996

L.L. Robinson and the New Coso Labor Strike
Primary Sources
"Report of L.L. Robinson, President, of the New Coso Mining Company, on the Properties of the Company at Darwin, Inyo Co., Cal., Dec. 1875". San Francisco, 1875
Inyo Independent, Feb. 23, May 25, June 1 and 8, 1878
Bodie Weekly Standard, May 29, 1878
San Francisco Call, Jan. 22, 1892, p. 7 (obituary of J.J. Williams, New Coso Mining Co. superintendent)
Roberts, Oliver. *The Great Understander.* Aurora, Ill., 1931, p.230 to 258.
In this book, Oliver Roberts gives his account of the strike, the shooting of Delehanty, and the aftermath. While he provides some useful details, his story has serious misstatements. In the encounter with Delehanty, Roberts claims he shot a second man, which he did not. Afterwards, at Lookout, he claims he shot and wounded John McGinnis. It was Frank Fitzgerald who shot and killed McGinnis. In other instances where Roberts' statements can be checked against known history, they prove wanting. For example, he claims to have been an eyewitness to the robbery of Coyote Holes by the bandits, Vasquez and Chavez. This event took place in February 1874, months before Roberts arrived in the Eastern Sierra.

Secondary Sources
Pacific Coast Mining Review and Stock Ledger, San Francisco, 1878, p. 49 (biography of Lester L. Robinson)
Phelps, Alonzo. *Contemporary Biographies of California's Representative Men.* San Francisco, 1881-2, p. 213
Biographical Information File, California Room, State Library, Sacramento. Two biographies of Lester Ludyah Robinson
Lingenfelter, Richard E. *The Hardrock Miners.* Berkeley, 1974, p.141

10. ENTER GEORGE HEARST

George Hearst

Primary Sources

Hearst, George. *The Way it Was: Recollections of U.S.Senator George Hearst, 1820-1891.* In George Hearst Letters and Autobiography, 1877-1890, Bancroft Library

Pacific Coast Mining Review and Stock Ledger, San Francisco, 1878. Includes information on the Modoc and Minnietta Belle mines, and on George W. Kidd, a principal Modoc investor.

Inyo Independent, Feb. 6, 1875

Secondary Sources

Sacramento Union, March 1, 1891, p. 1 (obituary)

Dictionary of American Biography. New York, 1928-36, p. 315

National Cyclopedia of Biography New York, 1891, p. 487

Older, Fremont and Cora. *George Hearst, California Pioneer.* Los Angeles, 1966

Lyman, George D. *The Saga of the Comstock Lode.* New York, 1934

Robinson, Judith. *The Hearsts: An American Dynasty.* Newark, London and Toronto, 1991

Lookout and the Modoc and Minnietta mines

Primary Sources

Inyo Independent, March 4, May 13 and 27, July 15, Aug. 26, Sept. 16 and 23, Oct. 21, Nov. 18, Dec. 9, 1876; Jan. 13, April 7, May 19 and 26, June 23, July 28, Aug. 18, Oct. 27, Nov. 3, Dec. 1, 1877; May 25, Dec. 2, 1878; March 29, April 5, 12 and 26, Oct. 18, 1879; Feb. 7, May 1, July 10, 1880, Sept. 17, 1881; Jan. 7, 1882

Coso Mining News, Darwin, Nov. 13, 1875; April 22, May 6, 1876; May 12 and 26, Sept. 8, 1877

Los Angeles Express, June 26, 1875

Kern County Gazette, Bsakersfield, Dec. 25, 1875

Mining & Scientific Press, Dec. 30, 1882

Judgment Book A, Inyo County Recorder's Office, p. 323. R.C. Jacobs, George Reed and R. Nadeau, plaintiffs, vs. Minnietta Belle Silver Mining Co. and John R. Hite, filed Nov. 7, 1877

Judgment Book A, Inyo County Recorder's Office, p. 344. M.W. Belshaw, Egbert Judson and R. Nadeau – Cerro Gordo Freighting Co., plaintiffs, vs. Minnietta Belle Silver Mining Company, defendant. Judgment, July 5, 1878.

Roberts, Oliver. *The Great Understander.* Aurora, Illinois, 1931

Secondary Sources
Vredenburgh, Larry M. and Shumway, Gary L. *Desert Fever: An Overview of Mining in the California Desert.* Canoga Park, California, 1981
Belden, L. Burr. "Minnietta Proves Steady Producer for 85 Years", in *San Bernardino Sun-Telegram,* n.d.
Wright, David A. "Looking in on Lookout", in *The Album,* Vol 3, No. 2, April 1990

11. THE CONSCIENCE OF GENERAL DODGE

Mammoth: Discovery and Beginnings
Primary Sources
Inyo Independent, March 23, May 4, June 1, July 6, 1878
Bodie Weekly Standard, June 19, July 10, 1878
Mammoth City Herald, Nov. 22, 1879
Twain, Mark. *Roughing It.* Hartford, Conn., 1872
Wasson, Joseph. *Complete Guide to the Mono County Mines.* San Francisco, 1879

Secondary Sources
Report of the State Mineralogist, vol. 8, Sacramento, 1888, p. 373
Chalfant, W.A. *Tales of the Pioneers,* Stanford, 1942
Caldwell, Gary. *Mammoth Gold.* Ed. by Genny Smith. Mammoth Lakes, 1990
Nadeau, Remi. *Ghost Towns & Mining Camps of California.* Los Angeles, 1954. Fourth Edition, Santa Barbara, 1992

General George S. Dodge
Primary Sources
Letters to Dodge from various correspondents, 1862-65, and letters from Dodge to various recipients, 1865-1867. Copies provided to author by The Center of Military History, Department of the Army, Washington, D.C.
Letter to author from Terence J. Gough, Chief, Staff Support Branch, The Center of Military History, Department of the Army, March 23, 1995. This summarizes Dodge's military record from the files of the U.S. War Department, 1862-1865.
Alta California, San Francisco. March 11, 1877, Aug. 25, 1881
San Francisco Chronicle, Aug. 25, 1881
Oakland Daily Times, Aug. 25 and 29, 1881
Biography Information File, California History Room, State Library, Sacramento

Pacific Coast Annual Mining Review and Stock Ledger. San Francisco, 1878, p. 32

Certificate of Death, George S. Dodge, Oakland, Aug. 24, 1881. Signed Aug. 27, 1881

Secondary Sources

Orcutt, Marjorie A. and Alexander, Edward S. *A History of Irasburg, Vermont.* Rutland, VT, n.d., p. 45

Woodard, Bruce Albert. *Diamonds in the Salt.* Boulder, 1967

Mammoth: Rise and Fall
Primary Sources

Inyo Independent, Sept. 28, Oct. 5, 1878; April 12 and 19, May 10, Aug. 9, Sept. 6 and 13; Oct. 4 and 18, Dec. 13 and 20, 1879; Jan. 10, June 5, July 17, Oct. 30, Nov. 20, 1880

Bancroft Scraps, California Mining, vol. 3. *San Francisco Bulletin,* May 9, Aug. 14, 1878; *San Francisco Call,* June 25, 1879; *San Francisco Post,* Aug.16, 1879; vol. 4. *San Francisco Bulletin,* Jan. 23, 1880

Mono Weekly Messenger, Benton, Feb. 1, April 5, 1879

Lake Mining Review, Mammoth City, May 31, Aug. 23 and 30, Sept. 13, 1879

Mammoth City Herald, July 9, 1879 through Jan. 15, 1881

Mammoth City Times, Oct. 18 through Dec. 24, 1879; Feb. 14, 1880

Alpine Signal, Silver Mountain, April 16, May 14, 1879

Daily Bodie Standard, Sept. 9, Oct. 11, 1879

Alta California, San Francisco, July 24, 1879

Los Angeles Express, Jan. 14 and 22, Feb. 9, 1880

Bishop Creek Times, Oct 31, 1881

Mining & Scientific Press, Mar. 28, 1885

Sacramento Record-Union, Aug. 13, 1885

Bank of California v. Mammoth Mining Co., Sheriff's Register of Actions, June-Aug. 1881 (Sheriff's Sale), Mono County Courthouse, Bridgeport

Wasson, Joseph. *Complete Guide to the Mono County Mines.* San Francisco, 1879

Buck, Franklin. *A Yankee Trader in the Gold Rush.* Boston, 1930

Report of the State Mineralogist, Vol 8. Sacramento, 1888

Secondary Sources

Chalfant, W.A. *Tales of the Pioneers,* Stanford, 1942

Caldwell, Gary. *Mammoth Gold.* Ed. by Genny Smith. Mammoth Lakes, 1990

Nadeau, Remi. *Ghost Towns & Mining Camps of California.* Los Angeles, 1954. Fourth Edition, Santa Barbara, 1992

Reed, Adele. *Old Mammoth*. Palo Alto, 1982
DeDecker, Mary. *Mines of the Eastern Sierra*. Glendale, 1966

12. BAD MEN OF BODIE
The birth of Mono County gold camps, including Bodie
Primary Sources
Bodie Free Press, April 13, 1881; also in:
Bodie Weekly Standard-News, April 13 and 20, 1881
Sacramento Record-Union, Aug. 27, 1880
Browne, J. Ross. "A Visit to Bodie Bluff", *Harpers New Monthly Magazine*, Aug. and Sept., 1865
Secondary Sources
Cain, Ella M. *The Story of Bodie*. San Francisco and Sonora, 1956
Chalfant, W.A. *Outposts of Civilization*. Boston, 1928
DeDecker, Mary. *Mines of the Eastern Sierra*. Glendale, 1966
Johnson, Russ and Anne. *The Ghost Town of Bodie, As Reported in the Newspapers of the Day*. Bishop, CA, 1967
Loose, Warren. *Bodie Bonanza*. Las Vegas, 1979
Wasson, Joseph. *Complete Guide to the Mono County Mines*. San Francisco, 1879
Smith, Grant H. "Bodie: The Last of the Old-time Mining Camps". *California Historical Society Quarterly*, March 1925

Life and times in Bodie
Primary Sources
Sacramento Daily Record-Union, April 2, Aug. 1, Oct. 15, Dec. 12, 1878; Sept. 30, 1879
Bodie Weekly Standard, May 1, 15 and 29, June 5, 1878
Daily Bodie Standard, Aug. 11, Sept. 8, 10 and 25, Oct. 6, 13 and 24, 1879
Bodie Weekly Standard-News, Sept. 11, 1880; June 15 and 22, 1881
Bodie Daily Free Press, Jan. 19, 21 and 22, May 29, June 7 and 9, Aug. 18 and 19, 1881
Alta California, Sept. 28, 1878; June 16, July 7, 11 and 21, Dec. 29, 1879
San Francisco Daily Stock Report, Dec. 22, 1879
Bridgeport Chronicle-Union, Aug. 27, 1881
Phoenix Herald, April 30, 1879
Parr, J.F. ""Reminiscences of the Bodie Strike", *Yosemite Nature Notes*, Yosemite National Park, May 1928
Buck, Franklin A. *A Yankee Trader in the Gold Rush*. Boston, 1930

Secondary Sources
Eaton, Henry G. "Bodie Was a Swell Town", *Los Angeles Times Sunday Magazine,* July 31, 1932
Loose, Warren. *Bodie Bonanza.* Las Vegas, 1979

Pat Reddy in Bodie and San Francisco
Primary Sources
Inyo Independent, March 1, 1879; Feb. 28, Sept. 11, 1881; May 13, 1904
Bishop Creek Times, May 6, 1882
Carson Daily Index, March 4 and 23, April 24, May 10, 1886
San Francisco Examiner, Aug. 5, 1896; June 27, 1900
San Francisco Call, June 10, 1899; June 27, July 6, Aug. 11, 1900; April 11, 1901
San Francisco Bulletin, June 26, 1900
San Francisco Chronicle, June 26, 1900
San Francisco Star, June 30, 1900
Sacramento Bee, June 26, 1900
Tonopah Bonanza, Feb. 6, 1904
Mono County Assessor's Book, 1881, vol. M-Z, Mono County Historical Museum, Bridgeport

Secondary Sources
Chalfant, W.A. *Tales of the Pioneers.* Stanford, CA, 1942
Shuck, Oscar T., ed. *History of the Bench and Bar in California.* Los Angeles, 1901, p. 520
Bay of San Francisco, vol. 2. Chicago, 1892, p. 140
Lingenfelter, Richard. *The Hardrock Miners: A History of the Labor Movement in the American West, 1863-1893.* Berkeley, 1974
California: Fifty Years of Progress. San Francisco, 1900, p. 297 (biog. of Ned Reddy)
Southworth, John. *Pat Reddy, Frontier Lawyer.* Typescript, dated Sept. 1989. In Eastern California Museum
Earl, Phillip I. "Pat Reddy: Defender of the Downtrodden". *Proceedings of the Fourth Annual Death Valley History Conference,* 1995
Palazzo, Robert P. "The Fighting Reddy Brothers of the Eastern Sierra". In *The Album,* Bishop, 1996

George Daly and the Jupiter-Owyhee Fight
Primary Sources
Daily Bodie Standard, Aug. 25, 26, 28, 29, Sept. 1, 12, 20 and 22, Oct. 10 (detailed testimony) and Oct. 22, 1879
Bodie Weekly Standard-News, April 13, 1881
Daily Bodie Free Press, Aug. 23, 1881 (Daly's death)

Alta California, Aug. 21 and 22, 1881
San Francisco Chronicle, Aug. 21, 1881
San Francisco Call, Jan. 13, 1882
Secondary Source
Griswold, Don L. and Jean Harvey. *History of Leadville and Lake County, Colorado,* 2 vols. Denver, 1996 (includes excerpts from Leadville newspapers and other primary sources)

Stage Holdups
Primary Sources
Bodie Weekly Standard-News, Sept. 5, 11, 1880
Report of Jas. B. Hume and Jno. N. Thacker, Special Officers, Wells, Fargo & Co.'s Express ("Robber's Record"), San Francisco, 1885
Wells Fargo collection of clippings and correspondence concerning the highwayman, Milton A. Sharp (provided by Robert J. Chandler, Historical Services, Wells Fargo Bank, San Francisco)
Secondary Source
Nadeau, Remi. *Ghost Towns & Mining Camps of California.* Los Angeles, 1954. Fourth Edition, Santa Barbara, 1992

Vigilance Committee
Primary Sources
Daily Bodie Free Press, Jan. 15, 16, 18, 20 and 23, 1881
Bridgeport Union, Jan. 15, 22, 1881
Secondary Source
McGrath, Roger D. *Gunfighters, Highwaymen & Vigilantes.* Berkeley, 1984

Bodie & Benton Railroad
Primary Sources
Daily Bodie Free Press, May 25, 26, 27 and 28, June 2, 1881
Bodie Weekly Standard-News, May 4, June 8, 15, 1881
Alta California, Aug. 2, 1877 (biog. on Col. Thomas Holt), May 25, 26, 28 and 30, 1881
Bridgeport Union, May 28, 1881
Tombstone Epitaph, May 28, 1881
San Francisco Call, June 21, 1885 (obituary of Col. Holt)
Secondary Source
Myrick, David F. *Railroads of Nevada and Eastern California,* vol. 1. Berkeley, 1963 (provides a complete history of the Bodie & Benton Railroad)

13. THE LAST BONANZA

Discovery of silver, Calico
Primary Sources
Mecham, Charles L. "Brief History of the Discovery of Calico." *San Bernardino Sun*, "Covered Wagon Days Edition," Nov. 13, 1938, pp. 7 and 27.
Belden, L. Burr. "Discoverers Tell How Calico Ore Veins Found." *San Bernardino Sun-Telegram*, Feb.26, 1960, p. D-2. Incorporates earlier accounts of the Calico discovery by G. Frank Mecham and C. Jeff Daley, and by Charles L. and Ed Mecham.
Secondary Sources
Ingersoll, L.A. *Century Annals of San Bernardino County, California.* San Bernardino, 1904. Biographies of Lafayette Mecham, p. 664, and of Sheriff John C. King, p. 776
San Bernardino County Sheriff's Office, 1853-1973. San Bernardino, 1973. Includes biography of Sheriff John C. King

Growth and character of Calico
Primary Sources
San Bernardino Weekly Index, Nov. 5, 1881, "The Town at Calico Mountain" (typed copy in San Bernardino Public Library)
Bishop Creek Times, May 13, 1882, quoting *Santa Ana Standard*
San Bernardino Weekly Times, June 3, 1882; Dec. 12, 1884
Los Angeles Times, June 9, 1882
Calico Print, Oct. 21, 1882; Feb. 22, Mar. 8 and 29, May 10 and 31, June 7 and 21, July 19, 1885 (San Bernardino Public Library) and October 10, 1886 (original copy in Library of the Natural History Museum of Los Angeles County)
Mining and Scientific Press, Aug. 12, Sept. 2, Dec. 30, 1882; May 12, Aug. 11 and 18, Oct. 27, Nov. 17, Dec. 1 and 15, 1883; Jan. 5, 12 and 19, Feb. 16, May 31, July 12 and 19, Aug. 2, Sept. 27, Oct. 11 and 18, Dec. 20 and 27, 1884; Jan. 3, 24 and 31, Feb. 14, March 14 and 21, April 18 and 25, July 11, Aug. 8, 1885
Daggett Scrapbook, California Mining. *San Francisco Bulletin*, Feb. 12, 1901, p. 3 (obituary of William H. Raymond)
San Francisco Chronicle, Jan. 6, 1904 (obituary of Capt. James Powning)
Henry Harrison Markham Papers, Box 21 "Oro Grande Mining Company", Huntington Library
Verdict of Coroner's Inquest on the body of J.O. Harris, by J. A. Johnson, Justice of the Peace, Daggett, Dec. 7, 1884

Delameter, John A. "My 40 Years Pulling Freight," As told to John Edward Hogg. *Touring Topics*, Aug. 1930.

Lane, Lucy Bell. "Calico Memories," ed. by Alan R. "Lefty" Baltazar. Calico Historical Society, 1993

Mellen, Herman F. "Reminiscences of Old Calico" Southern California Historical Quarterly, June, Sept. and Dec. 1952

Secondary Sources

Baltazar, Alan R.. "Lefty". *Calico and the Calico Mining District, 1881-1907*. Calico Historical Society, 1995

Belden, L. Burr. "Silver Spurs Growth of Barstow, Calico", San Bernardino Sun-Telegram, Sept. 1, 1963, p. A-7

Calico Ghost Town. 56-page booklet pub. by Knott's Berry Farm, 1959

Hensher, Alan and Vredenburgh, Larry M. *Ghost Towns of the Upper Mojave Desert*, Third Edition, 1987

Ingersoll, L.A. *An Illustrated History of Southern California*. Chicago, 1890, p. 156 (biog. of Joseph LeCyr)

Keagle, Cora L. "Calico's Canine Carrier", *Desert Magazine*, June 1943, p. 16

Myrick, David F. *Railroads of Nevada and Eastern California*, vol. 2. Berkeley, 1963

Nadeau, Remi. *Ghost Towns & Mining Camps of California*. Los Angeles, 1954. Fourth Edition, Santa Barbara, 1992

Swisher, John. "Calico". Mojave River Valley Historical Society, Barstow, 1991

Steeples, Douglas. "Calico Silver and the Fabric of Western Development" Undated monograph, provided by San Bernardino County Archives.

Weber, F. Harold Jr. "Silver Mining in Old Calico". Mineral Information Service, May 1966

James Patterson and the claim disputes
Primary Sources

Calico Print, Feb. 8, Mar. 1, May 3, 10, 17 and 24, 1885

J.S. Doe, plaintiff, vs. W.H. Foster and Ben Tiley, defendants, filed Feb. 16, 1885, Justice Court, Belleville Township, San Bernardino County. Case certified up to the Superior Court, Judge James Gibson, March 4, 1885. Complaint, responses, testimony, judgment for defendants, July 17, 1885. In San Bernardino County Archives.

J.S. Doe, plaintiff, vs. D. Edwards, John Doe and R. Roe, defendants, filed March 5, 1885. Complaint, briefs, testimony, and final judgment for plaintiff, Sept. 12, 1887. In San Bernardino County Archives.

Annie Kline Townsend Rikert
Primary Sources
Mining and Scientific Press, Dec. 30, 1882; May 12, Aug. 11 and Sept. 15, 1883; Aug. 30 and Oct. 4, 1884

San Bernardino Weekly Times, Nov. 18, 1886

Tuolumne Independent, Sonora, Dec. 22, 1906

The People of the state of California vs. Annie K. Rickert, charged with assault with a deadly weapon with intent to commit murder. Before H.R. Gregory, magistrate, Superior Court, San Bernardino County (file no. 1433). Testimony, Nov. 11 to 20, 1886; Verdict, not guilty, April 20, 1887. In San Bernardino County Archives.

Secondary Source
Deane, Dorothy Newell. *Sierra Railway.* Berkeley, 1960

Patterson and the Dodson Robbery
Primary Sources
The Daily Courier, San Bernardino, Sept 22, 1889

Coroner's inquest on the body of Harry Dodson before H.R. Gregory, Justice of the Peace and ex-officio coroner, Calico, Sept. 12, 1889. Testimony, depositions and finding by coroner's jury. In San Bernardino County Archives.

INDEX

Aberdeen Journal, 49, 61
Abraham, Marguerite, 96
Achison, Topeka & Santa Fe RR, 250
Ackerman, John, 265ff
Alexander, Don George, 98
Alhambra Mine, 248, 259, 261-2
Allen's Camp, 103
Almada, Joaquin, 13, 15
Alpha Mine, 186-7
Alpine Chronicle, 53ff
Alpine County, 46ff
Alta California, 11, 46, 119, 146, 188, 201, 208, 215-6, 234
Alverson, Ben, 83
American Hotel, Cerro Gordo, 33, 35
Anglo-Californian Bank of S.F., 78ff
Antioch, 39-40
Antioch Ledger, 40
Arnold, P.N., 103
Arnot, N.D., 57, 59-60
Ashim, Bark, 165-6
Ashley, Dan, 83
Atlantic & Pacific RR, 111, 250
Atwood, Melville, 3-4
Auburn, 224
Aurora, Nev., 18, 65ff, 71, 160-1, 185, 205-6, 208, 213-4, 224-5, 228-9, 231
Austin, Nev., 7, 140, 148

"Bad Man from Bodie", 209, 215, 230, 252
Bakersfield, 100-1, 103, 134-5, 140, 158, 168
Balch, William, 81
Baldwin, Elias J. "Lucky", 90
Ball, B.E., 177
Ballou, Lorenzo D., 116
Barber, C.J., 178, 180ff
Barber, George, 249ff
Barrel Springs, 98, 102

Barrett, Lawrence, 113
Barry, James H., 237-8
Barry, Judge Richard C., 68
Barstow, 243
Barstow, Edward, 143, 148
Bean, E.P., 124
Beatrice Mines, 121
Beaudry, Prudent, 12, 96
Beaudry, Victor, 12ff, 22ff, 95-6, 100ff, 155-6, 158-9
Bechtel Mine, 208
Bell, Mary Rose, 95, 108
Belleville, Nev., 55, 109
Belmont, 19
Belshaw, Charles Mortimer, 40
Belshaw, Jane Oxner, 14, 32, 40
Belshaw, John T.C.S., 32
Belshaw, Mortimer W., 14ff, 22ff, 95, 100ff, 110, 134, 164, 180, 270
Beltran, Bentura, 29, 155ff
Bend City, 11
Bennett, Paul, 72ff
Benton, 67ff, 89, 103, 109, 162, 170, 185ff, 213, 258, 270
Benton, Rev. J.E., 67
Bentonian, 76-7, 82ff
Benton Station, 85, 235
Bernhardt, Sarah, 113
Bessie Brady, SS, 31, 100, 102, 159, 167
Bicknell, Dr. F.T., 143
Big Bonanza, 89
Big Pine, 19, 109
Bishop Creek, 18-9, 6, 73, 77, 109, 110, 165, 167, 186, 192, 197
Bishop Creek Times, 200
Bismarck, 246
Bismarck Mine, 246, 259
Blind Springs Hill Mining District, 66, 162
Bodey, William, 66, 205-6, 216

Bodie, 66, 70, 93, 183, 186ff, 205ff, 223, 270
Bodie & Benton RR, 235-6
Bodie Bonanza, 207
Bodie Bluff Consolidated Mining Co., 206
Bodie Daily Free Press, 225ff
Bodie Fire Co., 226
Bodie Mine, 207, 210ff, 214, 217, 239, 239
Bodie Miners Union, 209, 215, 217ff, 220, 226, 233, 235
Bodie Mining Co., 210
Bodie Mining District, 205ff
Bodie Railway & Lumber Co., 232
Bodie Standard, 172,188, 208, 210ff 220, 222
Bodie Standard-News, 213, 225, 231-2
Bonanza Kine Consolidated Mining Co., 130-1
Bonanza King Mine, 111, 130-1
Booth, Edwin, 113
Boushey, Stephen, 32
Brady, James, 100
Braslin, John, 209
Bridgeport, 162, 200, 210, 216, 220-1, 235
Bridgeport Union, 234
Briggs, John, 186
Briggs, R.M., 72ff
Broderick, David C., 78, 135, 238
Brown, J. Ross, 206
Brown, Robert, 127, 156
Brown, Prof. William D., 127, 156
Bruce, Jim, 143, 148
Bryan, William Jennings, 267
Buena Ventura Mine, 155-6
Bullfrog, Nev., 153
Bunker Hill Mine, 206-7
Burcham, Dr. Rose, 239
Burke, Peter, 217, 219
Burning Moscow Mine, 245-6

Cage, John D., 99
Cain, Jim, 240

Calico, 93, 110-1, 181, 245ff, 270
Calico Print, 130ff, 252-3, 255ff
Caliente, 103, 159-60, 168
California, SS, 68
California Mining Association, 40
California Stock and Exchange Board, 90, 93
Callahan, Charley, 131
Callahan, H.C., 131
Cameron, Elizabeth Ann Gordon, 49
Camp, Martha, 141, 143, 147-8
Camp Cady, 120, 123
Camp Independence, 11-2, 165-6
Candelaria, 109, 198
Carson City, Nev., 188, 193, 199, 211,
   221, 224-5, 228, 230, 232
Carson City Appeal, 166
Carson & Colorado RR, 41, 85, 109, 235
Cartago, 100-1, 159
Casa Diablo Hot Springs, 186
Caswell, Samuel, 115ff
Central Pacific RR, 18, 67, 97, 100, 138
Cerro Gordo, 12ff, 95, 98ff, 134, 139,
   145, 155ff, 163-4, 169, 180, 213, 270
Cerro Gordo Freighting Co., 31, 33, 35-6, 101ff, 127-8, 140, 144-5, 159, 178ff
Cerro Gordo Historical Society, 41
Cerro Gordo Landing, 31, 102
Cerro Gordo Social Union, 21
Chalfant, Pleasant A., 24, 30, 34, 37, 134, 213
Chalfant, W.A., 134, 239
Chalmers, John, 50-1, 53
Chalmers, Laura, 60, 61
Chalmers, Lewis Sr., 49
Chalmers, Lewis, 48ff, 128
Chalmers, Lewis William, 49, 62
Chavez, Cleovaro, 33, 108
Chester, Julius, 100-1
Childs, Jerome S., 177

Chinese, 20, 38, 45, 69, 85, 109, 130, 135, 139, 168, 193, 232ff, 237, 252
Christmas Gift Mine, 158, 167
Chrysopolite Mine, Colo., 222
City of Para, SS, 152
Clark, Edward, 188
Clark Mining District, 120ff
Classen, J.C., 81
Clear Springs Water Co., 143
Coeur d'Alene mines, Ida., 237, 258
Coffey, Ed, 179
Columbus, Nev., 100-1, 140
Comanche Mining Co., 66, 69-70, 77ff
Comstock, Henry, 4, 14
Comstock Lode, 1, 3ff, 46ff, 52, 57, 65, 78, 88, 104, 135ff, 157, 177, 186-7, 208-9, 247, 257-8
Confidence Mine, 177
Conklin, A.N., 162, 164
Constitutional Convention (Calif.)–1878-9, 93, 175, 213
Con Virginia Mine, 89
Cook, Cinderella, 248
Cook, J.B., 125-6
Coso Mining Co., 10
Coso Mining District, 10
Coso Mining News, 151, 159, 167ff
Cottonwood Landing, 31
Courtney, Morgan, 247
Cow Holes, 102
Coyote Holes, Inyo Co., 33, 98, 102
Coyote Holes, San Bernardino Co., 265
Crabtree, Lotta, 113
Creaser, John A., 83
Crocker, Charles, 138, 150, 262
Crocker, William H., 262
Crossman, James H., 120
Crown Point Mine, 136
Crystal Springs, 156, 158
Cuervo, Rafael, 155-6
Cuervo Furnace, 159
Cuervo Mine, 156
Cuervo Springs, 156

Daggett, 110-1, 264-5
Daggett, Lt. Gov. John, 250
Daly, George, 208, 210-11, 214, 217ff, 234
Da Roche, Joseph, 225ff
Darwin, 89, 103, 109, 151, 159ff, 164-5, 167ff, 177-8, 180ff, 213, 236-7, 254, 270
Darwin Wash, 9, 155, 157, 178
Death Valley, 11, 103, 127, 133
Defiance Mine/Furnace, 89, 156ff, 167, 170, 236
Delano, 101
Delehanty, C.M., 172ff
Dexter Mills, 211
"Diamond Hoax"–1871-2, 187-8
Diana Mine/Furnace, 66ff, 77, 79ff, 85
Dinan, John, 220
Dodge, Gen. George S., 70, 77-8, 90, 187ff, 194, 200ff, 270
Dodson, Harry, 264ff
Doe, John S., 253-4, 258, 263-4, 267, 270
Dogtown, 265
Donaghue, Delia, 146-7
Don Quixote Mine, 194
Downieville, 6
Dumont, Eleanor (Madame Moustache), 209, 270
Dunn, Tom, 163
Dwyer, P., 130

Earp, James C., 110
Earp, Virgil W., 110
Earp, Wyatt S., 110, 113
East Calico, 246, 248, 253
Edwards, D.F., 255ff
Egleston, Frank, 113ff
Eight-mile House, 99
Elder, Abner B., 15, 17, 26, 156
Ellon, Robert, 80ff
Elmira Mine, 66,
Ely, John H., 247
Emigrant Silver Mining Co., 174

Emma Silver Mining Co., 139
Empire Coal Mine, 40
Essington, Peter, 207
Eureka, Nev., 7, 127, 140, 148, 211, 237
Eureka Sentinel, 151
Ewing, Thomas, 130-1
Exchequer Gold and Silver Mining Co., 53, 57, 62

Farnsworth, Joe, 227ff
Fenner Station, 131
Field, Judge Stephen, 135, 238
Fisher, Galen M., 27ff, 36-7, 40
Fish Ponds, 243-4
Fitzgerald, Frank, 122-3, 182-3
Flood, James, 89
Florence, 111
Flores, Pablo, 11ff
Forks-of-the-Road, 102
Foster, William H., 254ff
Freighting, 30, 31, 96ff, 123, 144-5, 159, 178, 180, 189-90, 197, 231, 249, 251
Fremont, Gen. John C., 137
French, Dr. Erasmus Darwin, 9, 40, 157
Fresno Flats (Oakhurst), 192-3

Garfield, Pres. James, 216
Garfield Mine, 247
Garryowen Gold and Silver Mining Co., 160-1
Genoa, 5, 24
Giant Powder Co., 16, 40, 52, 89, 157
Gibson, Judge James A., 254-5
Gillette, Fred, 168
Goff, John, 217ff
Golconda Mine, 248, 259ff
Golconda Mining Co., 259, 262
Gold Hill News, 160
Goodyear, W.A., 32
Gordon, Louis D., 41
Gorman, J.S., 175
Gorman, Richard, 130

Gould, Jay, 150
Grace Church Cathedral, S.F., 59, 223
Graham, Will Hicks, 162
Grass Valley, 3, 53, 258
Green-Eyed Monster, 123, 125
Gunn, Jack J., 183
Gunsight, Lost Mine, 9, 177
Gunsight Mine, 127-8
Guptil, A.N., 180

Hagan, Bill, 172
Haiwee Meadows, 10
Hammond, John Hays, 189
Hannah, Judge John A., 34
Hardy, Col. William H., 194ff
Harpold, Bill, 246
Harrell, Jasper, 115-6
Harris, T.S., 142, 146, 151, 159, 168ff, 222
Harrison, Pres. Benjamin, 113
Harrison, V., 43
Hassen, Charles, 130
Havilah, 134
Hawley Station, 265
Hawthorne, Nev., 109
Hayward, Alvinza, 136
Headlight Mine, 188, 194, 200
Hearst, Sen. George, 3, 4, 90, 119, 157, 177ff, 183, 263, 270
Hearst, Phoebe, 263
Hearst, William Randolph, 183
Held, Anna, 113
Hemlock Mine, 134-5, 139, 150ff
Hennington's Ranch Station, 144
Hermit Valley, 43
Highland Mining District, 45
Hightower, George W., 69, 72-3
Hillsboro, N.M., 223
Hisom, Fred, 125-6
Hitchell, Dave, 231
Holliday, Doc, 110
Holt, Col. Thomas H., 232ff
Homestake Mine, 177
Honan, James, 179
Hopkins, Mark, 150

321

Hughes, Judge John R., 36
Hume, James B., 168, 225
Huntington, Collis P., 138, 150
Hutchins, Joe, 73

Imperial Silver Quarries Co., Ltd., 48, 50ff
Independence, Inyo Co., 13, 23-4, 28-9, 71ff, 109, 141, 155, 162, 164ff, 178
Indian Wells, 33, 98, 135, 140, 145, 157, 160, 168, 163, 180
Indians, 11, 12, 38, 66, 84, 120, 123-4, 128-9, 136, 235, 243
Inglewood, 112
Inyo Independent, 20, 23-4, 30, 34, 38, 69, 73, 77ff, 104, 139, 155, 165ff, 172, 175, 179, 181, 190, 197, 213, 239
Irwin, Will, 207, 210-11, 221
Isabelle Mine, 57
Ivanpah, 103, 120ff, 243, 270
Ivanpah Consolidated Mill and Mining Co., 124
Ivanpah Mill and Mining Co., 121, 124
IXL Gold and Silver Mining Co., Ltd., 53
IXL Mine, 46, 47, 53, 57

"Jack", canine mail carrier, 259
Jacobs, Richard C., 133ff, 179-80, 243
Jacob's Wonder Mine, 134, 146
Jamestown, 262-3
Jenks, Albert, 114
Johnson, John, 43
Jones, Cora, 114
Jones, Frank A.S., 47-8, 51ff, 61
Jones, Gertrude, 114
Jones, Henry, 135-6
Jones, Capt. James, 47-8, 51ff
Jones, Sen. John P., 89, 135ff, 145, 150, 270
Jordan, William L., 236

Judson, Egbert, 16, 36-7, 40, 52, 89, 95, 101, 110, 157, 171, 180
Junction Ranch Station, 157
Jupiter Mine, 217ff

Kalakana, King, 113
Kearney, Denis, 171, 175
Kearsarge, Inyo Co., 12
Kearsarge Mine, Mono Co., 70, 84
Keeler, 41, 98
Kelley, Samuel, 131
Kelty, Eliza J. Parker, 65, 72ff
Kelty, Eugene Conway "Pap", 65ff, 71ff, 83, 170
Kennedy, W.L., 133-4
Kentuck Mine, 136
Kenyon, Frank, 208
Kern County Courier, 159, 168
Kern County Gazette, 104
Kernville, 133
Kerrick Mine, 79ff
Keystone Mine, Inyo Co., 179
Keysville, 10
Kidd, Capt. George W., 178
King, Sheriff John C., 243ff, 267
Kirgan, J.F., 221, 227, 229
Knott, Walter, 267
Kremkow, John, 83

Lady Blanch borax claim, 259
Lake Mining District, 186ff
Lake Mining Review, 191, 194, 196
Land Offices, U.S., 71ff
Lang Station, 103
Langtry, Lily, 113
Las Vegas, 97
Laswell, Andy, 122
Lathrop, 238
Laughton, Antoinette, 55ff
Laughton, Henry, 55, 60, 62
Leadville, Colo., 222-3
Leadville Chronicle, 223
LeBlanc, Angelina, 39
Le Cyr, Joseph, 181, 254

Lent, William Mandeville, 70, 77ff, 90, 119, 187, 207, 210, 217, 236, 239-40
Leota Gold and Silver Mining Co., 153
Liliokalani, Queen, 113
Little Chief Mine, 222-3
Little Lake, 10, 107, 134
Lloyd, George, 161
Lloyd, Jack, 169
Lockberg, Lewis, 207
Lone Pine, 12, 18ff, 23, 32, 98, 109, 141, 157
Long Tom Mine, 121
Lookout, 89, 103, 109, 169, 172ff, 243, 270
Lookout Mine, 177
Loose, Edwin, 207, 213-4
Loose, Warren, 207
Lopez Station, 99, 102
Los Angeles, 12 *et passim*
Los Angeles Express, 39, 100-1, 149, 194
Los Angeles Herald, 134, 140
Los Angeles & Independence RR, 150-1
Los Angeles News, 109
Los Angeles Star, 101
Los Angeles Times, 39, 113ff, 257
Lost Cement Mine, 185-6
Lott, Austin E., 107ff, 111, 197
Lott, Melvina, 106ff
Lucky Jack Mine, 188
Lucky Jim Mine, 158, 167
Lundy, 236
Lynchings, 54, 229-30, 251, 269
Lyon's Station, 99

McAlpin, Thomas, 161
McCallum, George H., 170
McClain, W.T., 179
McClinton Mine, 188
McDonald, James, 173
McDonald, John, 148, 150
McDonald, Joseph, 218-9, 221
McDonald, P.S., 179
MacEwen, Ellen Miller, 49
McFarlane, Andrew, 121, 127, 130
McFarlane, Louis, 121
McFarlane, John, 121-2, 124ff
McFarlane, Tom, 121, 124, 127, 270
McFarlane, William A., 124ff, 130, 270
McGee, Bart, 135
McGinness, John, 173
MacGowan, Capt. Alexander B., 165
McKinley, Pres. William, 267
McKinney, Robert, 148
McLaughlin, Pat, 3
Mack, Albert, 67ff, 74, 78, 80, 83-4, 270
Mack, August, 67-8, 80
Mack, Elizabeth, 68ff, 84
Mack, Theodore, 67-8
Mammoth City, 89, 103, 109, 186ff, 222, 270
Mammoth City Herald, 82, 84, 191, 193, 194ff, 197ff
Mammoth City Times, 194, 196-7
Mammoth Mine, 99, 186, 199ff
Mammoth Mining Co., 188, 192, 194
Mannix, Jack, 161
Markham, Gov. Henry Harrison, 248-9, 251, 270
Marklee, Jacob, 43
Markleeville, 43, 46, 54ff, 61-2
Marlow, Jim, 255-6
May, Thomas, 73-4, 170
Mecham, Charles, 244
Mecham, G. Frank, 244
Mecham, Lafayette, 243
Mechanics Union, Bodie, 214-5, 220, 235
Melba, Nellie, 113
Mellen, Henry, 249-50
Mendez, Blas, 13
Merced, 100
Mexican miners, 11ff, 38, 168
Meyerstein, Caesar, 144
Michigan Tunnel and Mining Co., 47ff

Mill City, 189
Mill Station, 102
Miller, Ida, 255
Miller Springs, 32
Millner, James, 83ff
Mills, David O., 109
Mineral Park, Ariz., 103
Mineral Park, Mono Co., 189
Mining & Scientific Press, 257
Minnietta Belle Mine, 179-80, 183, 243
Mitchell Caverns, 111
Modoc Consolidated Mining Co., 89, 178, 182
Modoc Mine, 177ff
Mogul, 45, 54, 62
Mojave, 103, 107, 109, 127, 130, 178ff, 189, 211, 250-1
Mollie Stevens, 167
Monahan, Jim, 227
Monitor, Alpine Co., 44ff, 51ff, 62
Monitor Gazette, 46-7
Monitor Mine, Ivanpah, 121
Monitor and Northwestern Mine, 54
Mono Messenger, 76, 82, 191
Mono Mills, 232ff
Mono Mining Co., 211, 214-5, 217, 220
Monte Cristo Mine, 194, 200
Montgomery, 66-7, 69, 85, 161, 213
Moore, Maggie, 19, 34, 38
Mormons, 6, 96-7, 243
Morning Star Mine, 54, 62
Morrison, Bob, 71
Morrison, O.B., 179
Moss, John Thomas, 119ff, 123, 127
Mound House, Nev., 85, 109
"Mountain Mary", 46
Mountain View Mine, 179
Mount Bullion, 43, 47
Mud Springs, 102

Nadeau, George, 100, 108, 114
Nadeau Hotel, 112ff
Nadeau, Joseph F., 100, 115
Nadeau, Laura Hatch, 113ff

Nadeau, Martha Flanders Frye, 96, 113ff
Nadeau, Ozani-Joseph, 96
Nadeau, Remi, 31, 35, 38, 70, 95ff, 123, 127-8, 130, 144-5, 149, 151, 159, 167, 170, 178ff, 189, 195, 197, 237, 249, 251
Neagle, Dave, 135, 142, 148, 237-8
Needles, 111, 250
Nevada City, 3, 4
New Coso, 155ff
New Coso furnace, 159, 170ff, 182
New Coso Mining Co., 89, 158, 160, 167, 171ff
New Coso Mining District, 156
Newmark, Harris, 111-2
New York Mine, 156
New York Tribune, 52
Nine-Mile Station, 102
Noonan, John, 162
Noonday Mine, 208, 236
Norton, Bill, 143
Norton, Col. J.R., 159, 169
Norton, Col. L.A., 16

Oakdale, 262
Occidental Mines, 246-7
Ochoa, José, 12-3
Old Coso, 11, 155-6
Omega Tunnel, 23, 25ff, 37
Oakland, 61, 188, 201
Oakland Tribune, 201
Oriental Mines, 246, 251, 255ff
O'Riley, Peter, 3
Orizaba, SS, 15, 17
Oro Grande Mining Co., 110, 248ff, 263
Orus, SS, 68
Osborne, Jonas, 127
Ott, James J., 3
Overshiner, John, 248, 257
Owens Lake Silver-Lead Co., 22, 27, 34-5. 100
Owens Valley, 11ff, *et passim*
Owensville, 11
Owl Holes, 103

324

Owyhee Mine, 217ff

Pachoca, Chief, 123
Pacific Stock Exchange, 90
Paiutes, 11-2, 66, 84, 97, 120, 123-4, 192, 235, 243
Palace Hotel, 56, 159
Panamint, 103, 134ff, 156, 168, 179-80, 237, 243, 270
Panamint Junction, 160
Panamint Mining Co., 134
Panamint Mining District, 134ff
Panamint News, 142, 146, 149, 157
Panamint Station, 180
Park, Trenor W., 137ff, 146ff, 152, 270
Parker, George, 65ff
Parker, James A., 185-6
Partz, Dr. A.F.W., 66ff
Partzwick, 67, 69
Patterson, James Lewis, 247, 253ff, 264ff
Patterson Mining Claim, 255
Patti, Adelina, 113
Paul, R.H., 169
Perasich, Nick, 165-6
Peterson, Johnny, 244-5
Petiss, B.F., 51
Pine City, 189, 191-2
Pino Blanco Mine, 262
Pioche, Nev., 7, 134, 140, 142, 148, 158, 170, 172, 237, 247
Piute Company, 119
Placerville, 5, 43, 247
Plane, Pauly, 214
Poole, A.H., 81-2
Poole, Wesley, 43
Porter, Capt. G.L., 231
Porter, John L., 243
Poulett, Lord Earl, 48, 53
Powning, Capt. James, 83-4, 247, 258
Prescott, Capt. S.D., 195
Promontorio Mine, 155-6
Prostitutes, 6, 19ff, 45, 169, 209, 258-9

Providence, 111, 127, 130-1, 243, 270
Providence Mine, 127

Railroads, 4, 18, 39, 41, 67, 85, 97, 102, 109, 111, 126, 144-5, 157, 170, 232ff, 250-1, 262-3
Rains, George "Bulger", 163
Raines, Eliphalet P., 134-5, 141
Ralston, William C., 151, 187
Randsburg, 239
Raymond & Ely Mine, Nev., 170, 247
Raymond Mining District, Alpine Co., 45
Raymond Mining Claim, Calico, 255
Raymond, William, 247, 253, 258, 270
Red Cloud Mine, 208, 236
Reddy, Edward Allen "Ned", 25, 67, 135, 142, 148, 157, 159ff, 168, 173, 175, 191, 213, 228, 239, 247
Reddy, Emily Cochran Page, 161ff, 166, 212, 236ff
Reddy, Michael, 160
Reddy, Patrick, 24, 26ff, 36, 67, 74-5, 135, 141-2, 155ff, 170, 173ff, 212ff, 221ff, 228ff, 236ff, 270
Red Rock Station, 98, 102
Reed, Judge Theron, 165
Reno, 18-9, 54, 69
Resting Springs, 103-4, 127ff, 243, 265, 270
Reusch, Ernst, 54
Reynolds, Patrick, 217ff
Rikert, A.M., 248, 259, 261
Rikert, Annie Kline Townsend, 247, 259ff, 270
Riverside, 157
Roberts, George D., 222-3
Roberts, Jack, 231
Roberts, Oliver, 168, 172-3, 180, 228
Robinson, E.P., 66
Robinson, Lester Ludyah, 157ff, 167, 170ff, 270

Rocket, SS, 231, 234
Roosevelt, Pres. Theodore, 113
Rosamond, 98
Rose Springs Station, 102
Rosslyn, 263
Rowan, George W., 69, 73ff, 84, 191, 199
Runover Mine, 247, 264
Ruperez, Pedro, 157
Russell, Jim, 158
Russell, Lillian, 113
Russian Steve, 163

Sacramento Union, 46
Sacramento Valley RR, 157
St. George, Utah, 17
Saltoun, Lord, 49-50,
Salt Lake City, 56, 243
Salt Lake-Los Angeles Trade, 96-7
Salt Springs Station, 145, 180
San Bernardino, 12, 97, 103, 119ff, 135, 140, 144, 239, 243ff, 252, 254 *et passim*
San Bernardino Argus, 139-40
San Benardino Guardian, 120, 144, 158
San Bernardino Times, 248
San Carlos, 11
San Felipe Mine, 12, 24ff, 35
San Fernando, 102, 145
San Francisco, 4 *et passim*
San Francisco Bulletin, 82, 139
San Francisco Call, 158, 164, 237
San Francisco Chronicle, 40
San Francisco Daily Exchange, 201
San Francisco Examiner, 183
San Francisco Star, 237-8
San Francisco Stock and Exchange Board, 88ff, 146, 211
San Francisco Stock Report, 188, 235
San Lucas Mine, 12
San Pedro, 15, 18, 97ff, 108, 117, 122
San Quentin Prison, 168
Santa Maria Mine, 27-8, 36

Santa Monica, 151, 153
Saratoga Springs, 104
Savage Mine, 217
Searles, John, 103
Sears, William H., 79ff
Selby, Thomas H., 18, 95
Selby Works, S.F., 67, 95, 121
Sharp, Milton A., 224-5
Shaw, Frank, 73
Shaw, Jim, 73
Shepherd, John., 140
Sheridan, Gen. Philip H., 188
Sherman Silver Purchase Act, 1890, 267
Sherman, Gen. William T., 188, 248
Sherwin, J.L.C., 186, 190
Shoshone Indians, 128-9
Sierra Blanca Silver Mining Co., 66
Sierra Railway, 262-3
Silver demonitization–1873, 59, 92, 138
Silver King Mine, 110, 244ff, 255ff, 262, 267
Silver, Lowry, 244
Silver Mountain, Alpine Co., 43ff, 51, 53, 59, 62
Silver Mountain Mining District, 43
Silver Mountain, Inyo Co., 10
Silver Odessa Mining Co., 262
Simpson, John, 23ff, 33
Small, John, 148-9
Smith, "Pegleg", 105
Smith, Judge W.C., 143-4
Sonora, 68, 162ff, 211, 262-3
Southern Pacific RR, 39, 100ff, 111, 126, 130, 144-5, 150-1, 156, 159, 170, 250-1
Spadra, 144, 147
Staging, 19, 35-6, 44-5, 78, 85, 128, 140, 147, 157, 160, 168, 183, 187, 192, 212
Stage robberies, 168-9, 224-5
Standard Mine, 70, 206ff, 216-7, 239
Stanford, Gov. Leland, 150, 206
Star, Tom, 158

Stevens, Col. Sherman, 31, 167
Stevens, T.A., 227
Stewart, Bob, 133-4
Stewart, Jodie, 41
Stewart, Sen. William Morris, 137ff, 144ff, 191, 270
Stewart's Wonder Mine, 134
Stockdale, Jim, 231
Stockton, 10, 65ff, 71, 75, 262-3
Stockton & Tuolumne RR, 262
Stoughton, W., 255-6
Stratton, James, 178
Strawberry Flat, 5
Sullivan, John L., 113
Surprise Valley Mining and Water Co., 138-9, 141, 147
Sutro, Adolph, 57
Swansea, Inyo Co., 22, 27, 35, 100-1, 156
Sweepstake Mine, 253ff
Syme, Henry, 48, 53, 57, 59-60
Syndicate Mining Co., 206, 208, 210-11

Tarshish Mine, 54
Taylor, E.S. "Black", 66, 206
Taylor, John, 245
Taylor, Peter, 174
Tecopa, 127-8
Tecopa John, 265-6
Terry, David S., 135, 238
Thomas, Edward Hughes, 244-5
Thornton, Oscar F., 47, 51
Tiley, Ben, 254-5
Timm's Landing, 98
Tipton, 100
Tobler, H.G., 259ff
Tombstone, Ariz., 110, 114, 237, 257
Tombstone Epitaph, 235
Tombstone Milling and Mining Co., 110
Tomlinson, John J., 98
Tovey, Mike, 224
Tower Mine, 77, 80
Travis, Lola, 19, 21, 34ff, 169

Treloar, Johanna Londrigan, 225ff, 230
Treloar, Thomas Henry, 225ff
Trona, 145
True Blue Mine, 194
Twain, Mark, 87, 113, 185

Union, 13, 15-6, 22ff, 164
Union Mining Co., 16
Union Consolidated Mining Co., 36-7
Union Pacific RR, 97, 150

Vanderbilt, 127
Vasquez, Tiburcio, 33, 107-8
Victorville, 144, 248
Virginia City, 6, 14, 45, 48, 53, 136, 138ff, 157, 161, 172, 186, 191, 212, 225, 229, 231m 270
Virginia Enterprise, 208, 211
Virginia & Truckee RR, 85, 109, 232
Visalia, 10-11, 18, 115
Vulcan Mill and Mining Co., 81-2
Vulture Mine, Ariz., 110

Waddington, Wilson, 131
Wadsworth, Nev., 100
Wales, Jim, 168
Walker, William, 162
Walsh, Judge James, 3, 4, 14, 247, 258
Wasson, Joe, 189
Waterman, Gov. Robert W. 243
Water Station, 145
Watterson, George, 83-4
Webster Mining District, 45
Weiss, John, 59
Welch, Billy, 168, 172ff
Wells Fargo Express, 14, 85, 148, 153, 168, 224-5, 236, 264, 267
West Calico, 246, 253
Wheeler, M. Allison, 26ff
Whiteman, Gid, 185
White Pine, 7, 69
Whitmore, Mrs. R.K., 61

Whitney, Mt., 32
Wickenburg, Ariz., 110
Wildrose Canyon, 148-9, 179, 183
Wild Horse Mesa, 169
Willard, Charles Dwight, 112
Williams, J.J., 171-2, 181
Williams, John, 252
Williams, Nancy, 169
Willow Springs, 98, 102
Wilmington, 98
Wilson, J. Albert, 111
Wilson, John, 156-7
Wilson, L.M., 123
Winders, R.J., 110
Wiss, Dr. Gustave, 25ff, 134
Women, 2, 6, 19, 38-9, 126-7, 141, 192, 198, 237, 255, 270
Wonder Consolidated Co., 146, 152
Workingmen's Club, Darwin, 171ff, 237
Workingmen's Party, 171, 175
Wren, A.J., 186
Wyoming Mine, 134-5, 139, 149ff, 153
Wyoming Consolidated Co., 146, 152

Yager, Fred, 142
Yellow Aster Mine, 239
Yellow Grade, 17, 19ff, 27, 31, 36, 38, 98, 100-1, 106, 155
Yellow Jacket Mine, 78, 136
Ygnacio Mine, 12ff